D1072249

Particle Size Analysis in Industrial Hygiene

AMERICAN INDUSTRIAL HYGIENE ASSOCIATION

and

U. S. ATOMIC ENERGY COMMISSION

(Division of Technical Information)

MONOGRAPH SERIES ON

INDUSTRIAL HYGIENE

GEORGE D. CLAYTON, *Managing Editor; Series I*

AMERICAN INDUSTRIAL HYGIENE ASSOCIATION

ADVISORY COMMITTEE

REVIEWERS OF THIS MONOGRAPH

MONOGRAPH TITLES

Beryllium—Its Industrial Hygiene Aspects

Pulmonary Deposition and Retention of Inhaled Aerosols

Thorium—Its Industrial Hygiene Aspects

Particle Size Analysis in Industrial Hygiene

Particle Size Analysis in Industrial Hygiene

LESLIE SILVERMAN, 1914–1966
Late Professor of Engineering in Environmental Hygiene
Harvard University School of Public Health
Boston, Massachusetts

CHARLES E. BILLINGS
President, Billings & Gussman, Inc.
Waltham, Massachusetts

MELVIN W. FIRST
Professor of Environmental Health Engineering
Harvard University School of Public Health
Boston, Massachusetts

Prepared under the direction of the American Industrial Hygiene Association for The Division of Technical Information, United States Atomic Energy Commission

ACADEMIC PRESS New York and London 1971

ACADEMIC PRESS, INC.
111 Fifth Avenue, New York, New York 10003

United Kingdom Edition published by
ACADEMIC PRESS, INC. (LONDON) LTD.
24/28 Oval Road, London NW1 7DD

LIBRARY OF CONGRESS CATALOG CARD NUMBER: 73-154376

PRINTED IN THE UNITED STATES OF AMERICA

This volume is dedicated to the memory of Professors Leslie Silverman and Charles R. Williams, our late friends and colleagues at the Harvard School of Public Health. Both contributed importantly to this volume, Professor Silverman as senior author and Professor Williams as an official American Industrial Hygiene Association reviewer, but neither saw the final manuscript. We acknowledge with gratitude the vital role each played in producing this monograph, and we accept full responsibility for all errors and inadequacies that may remain in this, the published version.

Contents

Foreword

This monograph was prepared by the American Industrial Hygiene Association for the Division of Technical Information, United States Atomic Energy Commission. The publication is a part of the continuing effort of both organizations to extend the field of technical knowledge so as to safeguard the health and well-being of persons exposed to toxic or deleterious material. Production of this work is also compatible with a basic objective of the American Industrial Hygiene Association, which is to increase the knowledge of industrial hygiene through interchange and dissemination of technical information.

The measurement of particle size distributions and the interpretation of the role of particle size in physiological effects has been an important part of industrial hygiene dating back to very early history. The hectic episodes of silicosis in the early 1930's provided considerable impetus to studies on and measurements of dust particle sizes and behavior. The technical information on particle properties, kinetic behavior, sampling instruments, and interpretation presented in this monograph will be most valuable not only to industrial hygienists but also to personnel working in the many cognate disciplines dealing with particle behavior—both physical and physiological.

A considerable amount of previously unreported work is contained in this monograph. Consolidation of information, supplemented by extensive references to the original reports appearing in the literature, makes this work an important contribution to the reference library of individuals interested in all phases of environmental health.

We are indebted to the Atomic Energy Commission for their support of this monograph program. We are also grateful to the editor and all contributors to this publication for the many hours spent in developing the information contained herein. Our thanks are given as well to the competent reviewers of this publication and members of the AIHA Monograph Committee who advised wisely and assisted immeasurably in the development of "Particle Size Analysis in Industrial Hygiene."

F. W. CHURCH, *President*
American Industrial Hygiene Association

Preface

From the day of his birth, man is confronted with various particles of many sizes and in all kinds of environments. They are present in the soil which provides much of his food, the earth below, from which he extracts minerals for industry, the water he uses in commerce or in his diet, and, most vitally, in the air which he breathes. All these sources contain matter which gets dispersed into his environment in particulate forms. How these particles behave and are measured in physical, chemical, and biological systems involves many basic concepts and techniques. The practice of industrial hygiene is devoted to the recognition, evaluation, and control of those environmental factors or stresses arising in or from the work place, which may cause sickness, impaired health and well-being, or significant discomfort and inefficiency among workers or among the citizens of the community. One of these factors is airborne suspensions of finely divided solid and liquid particles produced as a result of industrial processes.

It is the purpose of this monograph to discuss the philosophy, methodology, and basic techniques in the behavior, measurement, and interpretation of particle size. In particular, the concepts given are presented in relation to their applications in industrial hygiene, health physics, and air pollution control, and to the assessment of biological problems in which particle size is a dominant or significant factor. One objective of a particle size analysis of a toxic material is to estimate the amount of pulmonary deposition of particulate materials in relation to aerodynamic behavior.

This monograph is not intended to be a comprehensive compendium of all known aspects of particle behavior, or to contain all techniques for particle size measurement and interpretation. Techniques have been selected that may be applied to problems of health and safety arising from particulate use and exposure, and emphasis has been given to the range of application and utility of these methods. Sufficient details of practice are presented so that the reader may obtain a representative sample of airborne or deposited material and prepare a realistic, reliable, and reproducible particle size analysis with some confidence. Insofar as possible, details of the practices of various investigators have been included so that the industrial hygienist may evaluate for himself the merits and disadvantages of many techniques available for particle size analysis.

A companion monograph of this series is concerned primarily with the behavior and deposition of particles within the pulmonary system. Only the physical factors involved in particle measurement have been covered in this present volume. Because work exposures to particulate matter can be evaluated properly only when all sizes in the worker's breathing zone are represented in the sample in their proper proportions, parameters which influence sampling in industrial and other environments have been emphasized. In addition, aerosol sampling and size analysis techniques that are needed to determine the efficiency of respiratory protective devices and of dust and fume control equipment have been covered in some detail.

It has become evident in recent years that many of the physical and chemical properties of particles, such as specific surface, adhesion, and chemical reactivity, are important with regard to their influence on biological systems. For example, the chemical nature of a particle's surface is an important factor in adsorbing or chemisorbing gases, and when certain of these, such as SO_2, become adsorbed on respirable soot particles, they may induce enhanced or synergistic reactions on inhalation. Also, very small sulfuric acid droplets in the atmosphere act as nuclei for condensation of moisture, and the resulting particles attain sizes capable of causing intense respiratory irritation and corrosion or other damage to materials. Man-made pollution is an important source of particulate matter in the atmosphere, but many particles of diverse size and composition have their origin in natural causes; these include volcanic eruptions and dust storms.

Many early scientific pioneers of the last century, such as John Tyndall in the 1870's, observed and described characteristic particle phenomena, such as light-scattering behavior, but recognition of the importance of particle size measurement and interpretation for health and safety has been confined largely to the twentieth century. The first significant effort to relate dust size as well as composition to pneumoconiosis was the early work of Higgins, Lanza, Laney, and Rice (1917). Prior investigations were concerned only with gross aspects of dust and disease rather than the identification of size ranges of importance. Some of the sampling techniques with which we now work are considered "standard" because they have been correlated with incidence or prevalence of certain occupational diseases caused by dust, but many of the sampling instruments that were developed for these early investigations of dusty lung disease were defective in that they showed pronounced bias with regard to the largest or smallest particles. It has become evident that aerosol sampling and analysis techniques and methods for appraisal of the environment must be improved or altered significantly if the behavior and characteristics of suspended material in the environment are to be better understood with respect to disease. It is hoped that the methods presented here in regard to size factors will provide a clearer understanding of many of the important variables that must be considered.

The methods of particle size analysis used in industrial hygiene are related to those used in other branches of modern technology. Texts and symposia on particle sizing in other fields contain descriptions of methods which are useful in some instances. And yet, the range of sizes encountered in industrial hygiene practice is quite large, and the very low concentrations of material necessary to produce hazards to man present problems which are unique. For example, one phase of environmental air analysis may require measurement of the sizes of condensation nuclei less than 0.1 μm in diameter at concentrations of 10^{-8} g/m^3, while another may require measurement of mineral dusts approaching 100 μm in diameter entering a particle collector at concentrations in excess of 1 g/m^3. Therefore, many methods of size analysis applicable to powdered material or droplets are mentioned in the following chapters, but only those having special application to the practice of industrial hygiene are described in detail. A number of instruments for particle size determination have been developed for specific problems. Most of these have had too limited application in industrial hygiene air analysis for the presentation of any critical judgment as to their ultimate usefulness. These are mentioned, and references to the original articles are provided for further study.

Because particle behavior is largely influenced by dimensional characteristics, the first chapter of this monograph is devoted to this aspect as well as a discussion of the behavior of particles in the presence of all the physical forces to which they may be exposed. Succeeding chapters deal with sampling methods for airborne particles and bulk granular materials, direct methods of measuring particle size, such as screening, optical sizing, and electron microscopy, indirect methods which give average size and size distribution, and descriptions of many new automatic devices for particle counting and sizing which have come into wide use during the past few years. These latter instruments have reduced much of the human element responsible for variations in counting reproducibility, but, despite their automatic character, they must be calibrated and rechecked by well-established aerosol sampling and particle sizing techniques.

The material contained in Chapter 6 is statistical in nature and is concerned with the assessment of the significance of particle size information. Important statistical parameters are presented with a description of the techniques used to define size and size distribution in a wide variety of interpretations. This monograph does not attempt to provide a discussion of the fundamental mathematical statistics underlying the very useful relationships which are commonly reported to describe the average size and the distribution of sizes about this average. Adequate sources for this information are cited in an appropriate section of the text. The final chapter presents numerous examples of the ways particle size information is employed in practice.

We wish to thank the four American Industrial Hygiene Association reviewers and Jess W. Thomas and David Sinclair, who reviewed the manuscript on behalf of the Atomic Energy Commission, all of whom provided valuable assistance.

M.W.F.
C.E.B.

Particle Size Analysis
in Industrial Hygiene

1

Properties of Particles in Relation to Size

1-1 Introduction

The behavior of particles in the air is controlled by their size, shape, density, and concentration, and the associated air motion. Information about these characteristics is required to predict particle motion. This chapter does not attempt to discuss in detail the fundamental physics and chemistry of aerosol systems except in general terms as related to the task of particle sizing. Detailed considerations of these factors are contained in texts on aerosol science. [1–3]

1-2 Particle Properties

Properties which control the physical and chemical behavior of the individual particles in an aerosol system include (1) size or spatial extent, (2) size distribution, (3) shape or form, (4) specific gravity or density, and (5) surface characteristics such as vapor pressure or electrical charge.

Many of the physical and chemical properties of finely divided solid and liquid substances (e.g., melting point, hardness, latent heat, and viscosity) are independent of particle size, whereas others (e.g., shape, solubility, chemical reactivity, vapor pressure, and color) vary to some degree as functions of size. These variations may be used as an index of size for specific materials.

Definitions of several terms used in particle size technology are presented in the Glossary. Figure 1.1 illustrates certain of these definitions.[4,5] Others are discussed in more detail in Chapter 3.

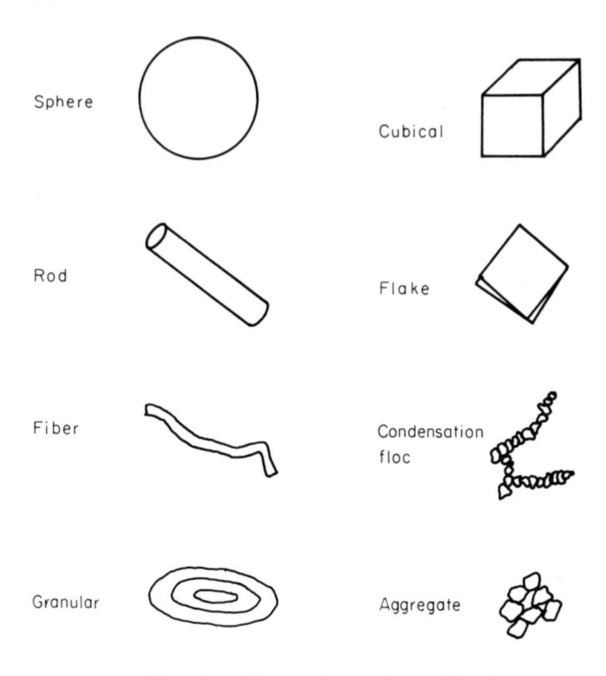

FIG. 1.1. Illustrations of particle form.

Figure 1.2 presents reported ranges of particle sizes for various aerosol materials and indicates the effective range of commonly used sizing methods. A linear property of the geometry of particles may be measured directly by a sieve analysis, or by comparison with a linear scale from optical or electron microscope preparations. Alternatively, it may be possible to measure a property of the particles which is related to particle size; this is commonly called particle size analysis by indirect measurement. For example, the aerodynamic diameter and the Stokes equivalent diameter (see Glossary, p. 303) are indirect measures of particle size inferred from measurements of particle terminal settling velocity. In many cases the use of an indirect measure of particle size is justifiable on the basis of simplicity for routine analysis, but, more important, an indirect measure, such as by terminal settling velocity analysis, will indicate the kinetic behavior of irregular particles suspended in a fluid. Indirect methods of size measurement utilize sedimentation and elutriation in air, water, or other fluids; impaction, electrostatic classification, surface adsorption, permeability to gas flow, and light scattering. These methods are described in detail in succeeding chapters.

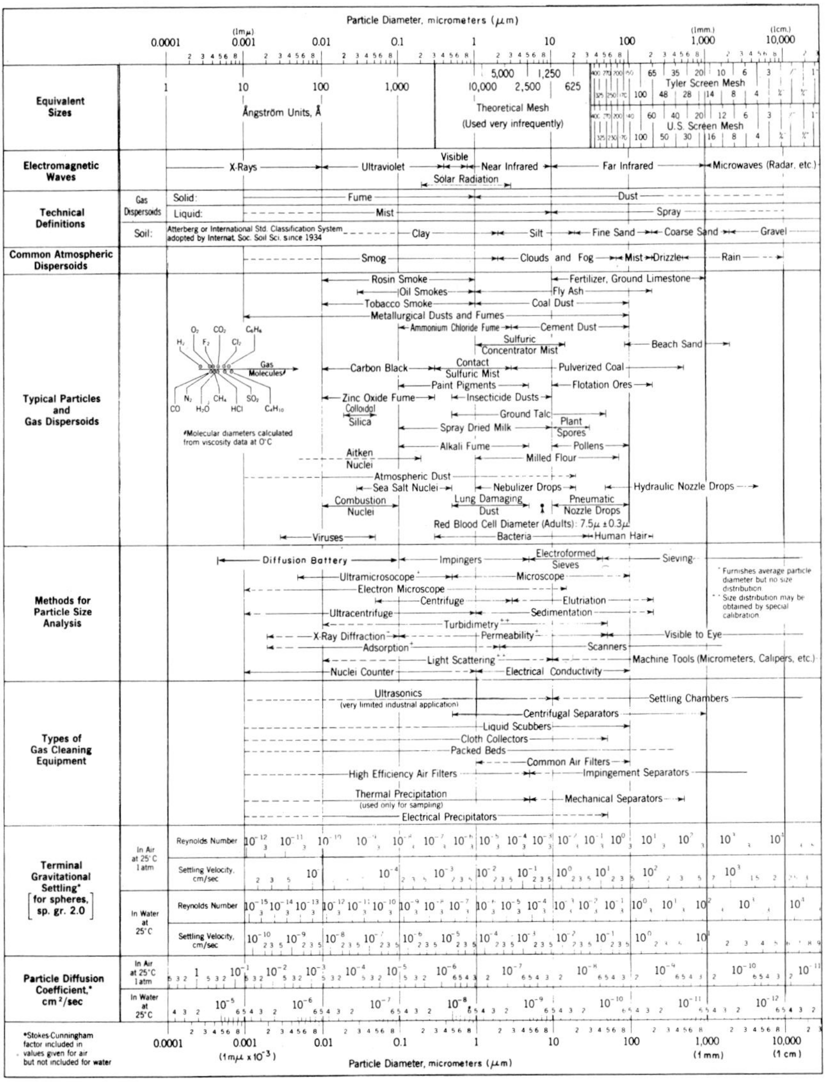

Particle Diameter, micrometers (μm)
0.0001 | 0.001 (1mμ) | 0.01 | 0.1 | 1 | 10 | 100 | 1,000 (1mm.) | 10,000 (1cm.)

Equivalent Sizes
1 | 10 | 100 | 1,000 Ångström Units, Å
5,000 | 1,250 | 10,000 | 2,500 | 625
Theoretical Mesh (Used very infrequently)
400 270 200 150 | 65 | 35 | 20 | 10 | 6 | 3
Tyler Screen Mesh
325 250 170 | 100 | 48 | 28 | 14 | 8 | 4
400 270 200 140 | 60 | 40 | 20 | 12 | 6 | 3
U.S. Screen Mesh
325 230 170 | 100 | 50 | 30 | 16 | 8 | 4

Electromagnetic Waves
X-Rays — Ultraviolet — Visible — Near Infrared — Far Infrared — Microwaves (Radar, etc.)
Solar Radiation

Technical Definitions
Gas Dispersoids: Solid: Fume — Dust
Liquid: Mist — Spray
Soil: Atterberg or International Std. Classification System adopted by Internat. Soc. Soil Sci. since 1934 — Clay — Silt — Fine Sand — Coarse Sand — Gravel

Common Atmospheric Dispersoids
Smog — Clouds and Fog — Mist — Drizzle — Rain

Typical Particles and Gas Dispersoids
O_2 CO_2 C_6H_6 H_2 F_2 Cl_2 Gas Molecules* N_2 CH_4 SO_2 CO H_2O HCl C_4H_{10}
*Molecular diameters calculated from viscosity data at 0°C
Rosin Smoke
Oil Smokes
Tobacco Smoke
Metallurgical Dusts and Fumes
Ammonium Chloride Fume
Carbon Black
Contact Sulfuric Mist
Paint Pigments
Zinc Oxide Fume
Colloidal Silica
Insecticide Dusts
Ground Talc
Spray Dried Milk
Alkali Fume
Aitken Nuclei
Atmospheric Dust
Sea Salt Nuclei
Combustion Nuclei
Lung Damaging Dust
Nebulizer Drops
Viruses
Bacteria
Red Blood Cell Diameter (Adults): 7.5μ ±0.3μ
Fertilizer, Ground Limestone
Fly Ash
Coal Dust
Cement Dust
Sulfuric Concentrator Mist
Beach Sand
Pulverized Coal
Flotation Ores
Plant Spores
Pollens
Milled Flour
Hydraulic Nozzle Drops
Pneumatic Nozzle Drops
Human Hair

Methods for Particle Size Analysis
Diffusion Battery — Impingers — Electroformed Sieves — Sieving
Ultramicroscope* — Microscope
Electron Microscope
Centrifuge — Elutriation
Ultracentrifuge — Sedimentation
Turbidimetry**
X-Ray Diffraction** — Permeability* — Visible to Eye
Adsorption* — Scanners
Light Scattering** — Machine Tools (Micrometers, Calipers, etc.)
Nuclei Counter — Electrical Conductivity
* Furnishes average particle diameter but no size distribution.
** Size distribution may be obtained by special calibration.

Types of Gas Cleaning Equipment
Ultrasonics (very limited industrial application) — Settling Chambers
Centrifugal Separators
Liquid Scubbers
Cloth Collectors
Packed Beds
Common Air Filters
High Efficiency Air Filters — Impingement Separators
Thermal Precipitation (used only for sampling) — Mechanical Separators
Electrical Precipitators

Terminal Gravitational Settling* [for spheres, sp. gr. 2.0]
In Air at 25°C 1 atm: Reynolds Number; Settling Velocity, cm/sec
In Water at 25°C: Reynolds Number; Settling Velocity, cm/sec

Particle Diffusion Coefficient,* cm²/sec
In Air at 25°C 1 atm
In Water at 25°C

*Stokes-Cunningham factor included in values given for air but not included for water

0.0001 | 0.001 (1mμ x 10^{-3}) | 0.01 | 0.1 | 1 | 10 | 100 | 1,000 (1 mm) | 10,000 (1 cm)
Particle Diameter, micrometers (μm)

FIG. 1.2. Common particle dispersions and methods of size measurement. (After Lapple, Ref. 6.)

1-2.1 SURFACE CHARACTERISTICS AND INTERFACIAL PHENOMENA

Properties of small particles related to their surface characteristics include specific area, rate of evaporation and condensation, electrostatic charge, adsorption, adhesion, and light scattering. In certain circumstances, changes in the environment of a particle during sampling and particle size analysis may change its size or state of aggregation or its surface characteristics. Such changes must be considered in the selection of a suitable sampling device or method for particle size analysis.

(a) *Surface Area.* One of the important characteristics of small particles is the rapid increase in surface area per unit mass as size decreases. For example, a drop of water with a diameter of 1 cm (0.52 g) has a surface area slightly over 3×10^{-4} m² (5.7×10^{-4} m²/g). When subdivided into 1- or 0.1-μm* droplets, this same volume of water has a total surface area of 3 m² (5.7 m²/g) or 30 m² (57 m²/g). The greatly increased area of exposed surface associated with fine particle suspensions leads to increased chemical reaction rate. Fine powders of organic and inorganic oxidizable materials (such as coal, iron, flour, sugar, and starch) burn vigorously or explode violently when aerosolized.[7] In addition to increasing oxidation and other chemical reactions, subdivision of some granular materials increases their toxicity.[8]

(b) *Evaporation and Condensation.* Evaporation or condensation is a diffusional mass transfer process which proceeds in proportion to the surface area exposed.[9,10] Equation 1.1 indicates the first-order approximation for the effects of temperature and partial pressure in the vicinity of the surface, which control the time required for a small particle (e.g., water) to evaporate into still air:

$$t = \frac{\rho_p D_p^2}{D_B (\Delta p)} \cdot \frac{RT}{8M} \tag{1.1}$$

where
t = time for a spherical droplet to evaporate, sec.
ρ_p = true density of particle, g/cm³.
D_p = particle diameter, cm.
D_B = diffusion coefficient of vapor from particle, cm²/sec.
Δp = difference between partial pressure at drop surface and in surrounding fluid, dynes/cm².
R = gas constant, 8.3×10^7 g·cm²/sec·g-mole °K.
T = absolute temperature, °K.
M = molecular weight of evaporating particulate material.

When there is a 1-mm Hg difference in partial pressures between the drop surface and air, the evaporation rate of water droplets at normal temperature and pressure is equal to about 10^{19} molecules per second per square centimeter for a 10-μm-diameter droplet.

*One micrometer (also micron) (μm) = 10^{-6} meter.

Finer particles can act as centers for condensation of moisture, leading to an increase in their size. This phenomenon has some bearing on the growth and deposition of fine particles in warm humid air during inhalation.

(c) *Electrostatic Charge.* Electrostatic charge represents an excess (−) or deficiency (+) of electrons on the particle. Most small particles have naturally acquired charges from electron transfer during contact or separation, or because of free ion diffusion. This charge may be assumed to reside on the particle surface in an adsorbed gas or moisture film.[1,11,12]

Mechanisms which produce a natural charge on aerosol particles are[11]:

1. *Electrolytic Mechanisms.* Electron exchange at a high-dielectric liquid–solid interface followed by separation, as from a jet, over a surface, or by impact of liquids on solids.
2. *Contact Potential.* Free electron transfer across a potential barrier because of a differential work function of two metals in contact.
3. *Spray Electrification.* Separation of liquids by atomization, which leads to formation of charged droplets due to ion concentration in the drops.
4. *Contact-Separation (Tribo-) Electrification.* Separation of dry non-metallic surfaces in contact (surface work function).
5. *Ion Diffusion in Gases.* Air ions created by electrical discharges in air, by natural radioactivity, or in a flame. These ions diffuse rapidly in air and become attached to particles.

Representative particle charges are given in Table 1.1. The number of charges acquired by particles is limited by the breakdown strength of the surrounding medium. In the case of dry air, this is about 8 esu/cm^2 or 1.6×10^{10} electrons/cm^2. It is possible to create charges which exceed

TABLE 1.1
NATURAL CHARGES ON SOME REPRESENTATIVE AEROSOLS[a]

	Charge distribution (%)			Specific charge[b] (esu/g)	
Material	Positive	Negative	Neutral	Positive	Negative
Fly ash	31	26	43	1.9×10^4	2.1×10^4
Gypsum dust	44	50	6	1.6	1.6
Copper smelter dust	40	50	10	0.2	0.4
Lead fume	25	25	50	0.003	0.003
Laboratory oil fume	0	0	100	0	0

[a]From White, Ref. 12
[b]1 esu $= 2 \times 10^9$ electron charges.

this value under certain conditions, but usual levels observed are considerably less. Figure 1.3 shows maximum likely particle charge and some experimentally determined values. Whitby and Liu, in reference 3, discuss the distribution of charge on particles between 0.01 and 1.0 μm in equilibrium with a bipolar ion atmosphere.

Charges of both signs may appear about equally after dispersion of small particles into a cloud, and the net charge of the aerosol as a whole may be quite small even when individual particles in the cloud are highly charged.[14–16] Collision and agglomeration of oppositely charged particles affects sedimentation rates of dust clouds and may lead to false size data when analyses are conducted by quiescent or stirred settling or by elutriation.

Acquisition of maximum theoretical charge gives rise to strong attractive coulombic and dipole forces between small particles. The presence of electrostatic charges is often a severe problem when handling particles less than 10 μm, since fine particles tend to clump. This makes it difficult to produce good air dispersions for subsequent sedimentation and restricts dry screening of powders with electroformed sieves. Undesirable charge effects may be counteracted by exposing the materials to emanations from an alpha-particle emitter which produces excess air ions for neutralization, or by wet sieving if a liquid is available in which the particles are not soluble. Powdered materials which are good electrical conductors (e.g., carbonyl iron spheres) can be grounded before and during sieving to reduce charge effects.

The presence of electrostatic charge on a particle gives rise to forces influencing its aerodynamic behavior in an electrical field. Classifiers have been designed to take advantage of the fact that a charge proportional to the size of a particle may be placed on it (see Chapter 5, p. 227). Theories of particle charging by diffusion and electrostatic field charging mechanisms have been reviewed by White[12] and by Whitby and Liu, in reference 3.

(d) *Light Scatter.* Scattering of a light beam arises from inhomogeneities, such as dispersed dust or water drops, in the fluid medium. Scattering is often accompanied by absorption, and both scattering and absorption remove energy from the beam of light. The quantitative response of the attenuated beam may serve to characterize the size of the particles. When viewed directly, the attenuation is called extinction. Whether scattering or absorption is mainly responsible for extinction can be judged by measuring light normal to the main beam[17] and at various angles. For homogeneous spherical particles,[18,19] the scattering coefficient is proportional to particle size, refractive index, and the wavelength of the incident light. Irregular

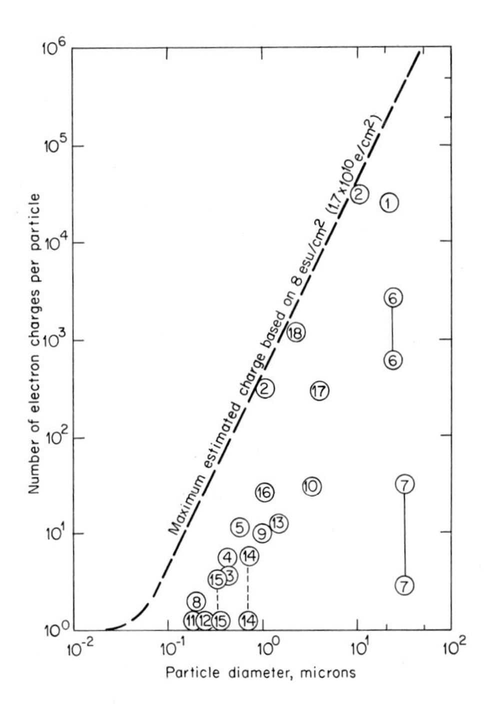

No.	*Ref.*	*Material*	*Remarks*
1	1	Theory	2 kV/cm, field
2	1	Theory	6 kV/cm, field
3	1	Ammonium chloride	Thermal
4	1	MgO	Combustion
5	1	Silica	25-psig air jet
6	1	H_2O	Fog, dry
7	1	H_2O	Fog, damp, unstable
8	15	Smoke	Burning tobacco
9	15	MgO	Combustion
10	15	Clay	Elutriation
11	15	Stearic acid	LaMer-Sinclair generator
12	15	Ammonium chloride	LaMer-Sinclair generator
13	15	Ammonium chloride	Sprayed alcohol solution
14	13	Dow polystyrene latex	Measured ion equilibrium
15	3	Theory	Calc'd ion equilibrium distribution
16	12	—	515 ions/cm³, $t \gg 1$, diffusion limited
17	12	—	6.7×10^4 ions/cm³, $t \gg 1$, diffusion limited
18	12	Oil	5.6 kV/cm, field limited

FIG. 1.3. Measured and theoretical single particle charge.

shapes and nonhomogeneous dispersions are more difficult to analyze theoretically, but empirical relationships may be developed for specific systems.

Light scattered from individual particles is approximately proportional to surface for particles larger than the incident wavelength; this relationship is used in airborne particle counters, described in Chapter 5, p. 206.

Light transmitted through a homogeneous dispersion of particles is attenuated according to the Lambert-Beer law (Bouguer law):

$$I = I_0 \exp(-abcd) \tag{1.2}$$

where I_0 = incident light intensity.
I = emergent light intensity (parallel).
a = particle extinction coefficient, dimensionless.
b = length of path through dispersoid, cm.
c = concentration of particles, number/cm^3.
d = particle projected area, cm^2.

Hodkinson, in reference 3, discusses the extinction and scattering of light by aerosols as applied in various instruments for optical measurement.

(e) *Adsorption.* Small solid and liquid particles are surrounded by a surface film of gas held by unbalanced electrical or chemical valence forces originating in the surface molecules. Adsorbed gas molecules held by surface-attractive forces may be present in proportion to their concentration in the surrounding gas phase up to the point of saturation for one complete monomolecular layer. Several additional layers have been reported on typical ambient aerosols[20]. The quantity of gas or vapor that may be adsorbed on particle surfaces is directly proportional to the exposed surface, and this fact may be used to measure particle size under certain conditions, as explained in Chapter 4, p. 183. Atomic or molecular configuration and polarity are additional factors affecting surface adsorption. Many surface characteristics of small particles, including electrical charge, adhesion, and evaporation, are modified by the presence of adsorbed gases.

(f) *Adhesion.* When a layer of liquid is spread between surfaces in contact, it produces adhering forces in proportion to the surface tension of the liquid and the radius of curvature of the liquid pool. Water vapor is adsorbed on many surfaces exposed to ambient atmospheres owing to hydrogen bonding or the presence of surface contaminants, so that the force of adhesion is directly related to humidity of the air in many instances. Therefore, reduction of the moisture content of dust may be required before dispersing it in air for sedimentation analysis. Since high electrostatic charges tend to accumulate at reduced humidity, a careful balance must be achieved for best results, this being a matter of experiment for a particular sample. Corn, in reference 3, and Zimon[21] discuss several practical aspects of other factors in the phenomenon of particle adhesion, such as van der Waals dispersion forces, electrostatic charge, external fields as in electrical precipitation, contact points, and roughness.

1-2.2 Specific Gravity

The specific gravity of a particle formed by attrition from a solid will be the same as that of the parent material. If it subsequently undergoes surface oxidation or hydration, or if it agglomerates into clusters, its density will change. When two or more solid nonporous particles cohere, the resulting particle has a different geometric form and includes void spaces. Therefore, density of the particle cluster will be less than that of the individual component particles. Data on density of metal oxide agglomerates are presented in Table 1.2. The density of an agglomerate may be less than one-tenth of the density of the parent material. Figure 1.4 shows agglomerates of NH_4Cl, illustrating inclusion of substantial void spaces.

TABLE 1.2
Particle Densities of Agglomerates[a]

Material	Floc density (g/cm³)	True density (g/cm³)
Silver	0.94	10.5
Mercury	1.70	13.6
Cadmium oxide	0.51	6.5
Magnesium oxide	0.35	3.6
Mercuric chloride	1.27	5.4
Arsenic trioxide	0.91	3.7
Lead monoxide	0.62	9.4
Antimony trioxide	0.63	5.6
Aluminum oxide	0.18	3.7
Stannic oxide	0.25	6.7

[a]From Whytlaw-Gray and Patterson, Ref. 22.

Difficulties are associated with the determination of individual aerosol particle densities, also, as illustrated by studies of the well-characterized research and test material, uranine dye. Stein *et al.*[23] reported a value for

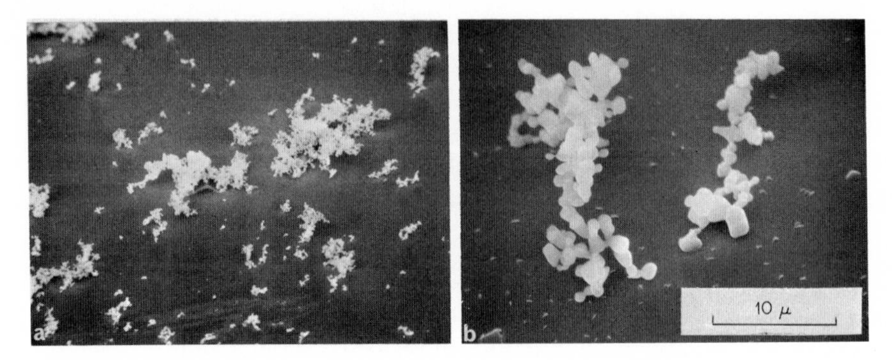

FIG. 1.4. Electron micrograph of ammonium chloride agglomerates. Courtesy of Dr. Parker Reist.

uranine density of 0.58 g/cm^3, compared to a pycnometer measurement of 1.53 g/cm^3. With a different method of generation, sampling, and analysis, for a one-to-two mixture of methylene blue and uranine, Sehmel[24] reported a value for the mixture density of 1.36 g/cm^3 as compared to a pycnometer measurement of 1.42 g/cm^3. McKnight and Tillery[25] report values for uranine density of 1.35 and 1.4 g/cm^3, depending on the analytical procedure employed. These reported differences apparently arise as a consequence of the method of generation and degree of drying of the liquid droplet containing the dissolved dye.

1-2.3 Shape Factors

Particle shape is influenced by the method of formation: if by comminution, attrition, or disintegration processes, the particle will resemble the parent material; if by condensation from a vapor (e.g., metal fume), the smallest unitary particle may be spherical or cubical. In many cases, condensation is followed immediately by solidification and formation of chain-like aggregates of particles (e.g., iron oxide fume, carbon black). After formation, changes in particle shape may occur because of crystallization, hydration, agglomeration, etc. Liquids form spheres which may coalesce on contact to form larger spheres. In most cases of industrial hygiene interest, particles are found to be irregular. Irregular particles may be assigned an arbitrary linear dimension in accordance with the definitions of size given in the Glossary. When irregular particle shapes are characterized by a single dimension which is measured in a consistent way, particle surface area and volume may be assumed to be functions of this same linear dimension; and area and volume can be estimated by applying appropriate shape factors. It is found empirically that the total surface area of a collection of particles, A_p, will be approximately equal to a constant multiplied by the square of some average diameter of the collection of particles, for the same material, over a reasonable range of sizes. Thus the proportionality factor is defined as the surface shape factor by

$$\alpha_s = A_p \left(\sum_i n_i D_{pi}^2 \right)^{-1} \tag{1.3}$$

It is possible to use any one of several different measures of particle diameter, D_p, and surface area, A_p, but the numerical value of the constant obtained from each combination of measures will relate to the specific method of measurement.

A similar empirical relationship is found to exist between the total volume of a collection of particles, V_p, and their average diameter, D_p, over a reasonable range of sizes. Thus, it is possible to define a volume shape factor as

$$\alpha_v = V_p \left(\sum_i n_i D_{pi}^3 \right)^{-1} \tag{1.4}$$

The values of V_p and D_p (and hence, α_v) again depend specifically on the method of measurement.

The derivation of shape factors has been described in detail.[26] Typical values of surface and volume shape factors obtained by various investigators using different methods are given in Table 1.3. For spherical particles $\alpha_v = \pi/6 = 0.52$ and $\alpha_s = \pi = 3.1$, as indicated in Table 1.3. The ratio α_s/α_v represents an index of particle shape, being 6 for spheres and cubes. According to Fair and Hatch,[29] the ratio for sands is about 6.1 for rounded particles, 6.4 for worn particles, 7.0 for sharp particles, and 7.4 for angular particles. In Table 1.3, the ratio of α_s/α_v for crushed quartz is about 8. Shape factors are necessary for converting size information derived from direct methods of analysis to aerodynamic diameter.

Particle shapes and physical data for hundreds of materials have recently been made available.[30] Many of these particle shapes may be encountered in industrial hygiene practice. The preferential orientation of suspended and deposited material may produce changes in their measurement as large as 50%.[31]

TABLE 1.3

VALUES OF SURFACE AND VOLUME SHAPE FACTORS FOR VARIOUS PARTICLES

Granular materials[a]	α_s	α_v	P.S.[b]
Filter sand	2.7	—	—
Crushed quartz	2.0 – 2.5	0.27 – 0.28	—
Quartz	—	0.140	2 – 64 μm
Calcite	—	0.135	1 – 69 μm
Granite	—	0.140	1 – 72 μm
White sand	2.1 – 2.7	—	—
Feldspar	—	0.26	—
Hornblende	—	0.02	—
Crushed quartz	—	0.17	0.19 cm
Crushed quartz	—	0.28	0.22 cm
Crushed quartz	—	0.27 – 0.28	0.26 – 0.36 cm[c]
Shapes[d]			
Spheres	π	$\pi/6$	—
Irregular-cubical	2 – 8	0.2 – 0.5	—
Flakes	1.5 – 2	0.02 – 0.1	—
Fibrous	π	$\pi/4$	—
Agglomerate flocs	0.2 – 2	0.01 – 0.1	—

[a]From Ref. 26 and 27.

[b]Geometric mean, as measured by various methods.

[c]Average volume diameter, $D_v = (\Sigma nD^3/\Sigma n)^{1/3}$.

[d]From Ref. 28.

1-3 Kinetic Behavior of Particles

The inertial motion of a particle relative to a defined fluid field can be used to determine the aerodynamic or kinetic particle size. In addition, the evaluation and control of airborne particulate hazards frequently depends on a knowledge of suspended particle motion with respect to the fluid motion. In particular, pulmonary deposition of some particles may be a function of their aerodynamic diameter. Design of industrial ventilation systems, such as hoods, ducts, or particle collectors, is governed by particle behavior in moving fluids. This section describes some of the simpler relationships which may be used to estimate aerodynamic particle size. Particle motion can result from one or a combination of many processes, as follows:

A. *Field forces*
   Gravitational
   Electrical
      Coulombic
      Image
      Dielectrophoretic
      Induced
      Space charge
   Magnetic

B. *Particle inertia and fluid mechanics*
   Drag
   Centrifugal or vortex flow
   Inertial (fluid flow around submerged object)
   Inertial (particle trajectory relative to fluid)
   Turbulent (convective transport of fluid)
   Shear gradient
   Coriolis

C. *Stochastic processes*
   Diffusion (due to concentration gradient of particle phase)
   Diffusiophoresis (due to concentration gradient of molecular species in gas)
   Thermal gradient

D. *Other factors*
   Photophoretic (photon pressure gradient)
   Sonic (alternating fluid pressure gradient)

1-3.1 Fluid Resistance

When a particle moves through a continuous real fluid, under the action of a force, it encounters a resistance to its motion due to stresses set up in the fluid. With small particles, at low velocities, it can be shown that the resistance is governed by the fluid viscosity, as given approximately by Stokes' relationship:

$$F_R = 3\pi\mu_f V D_p \tag{1.5}$$

where F_R = viscous resistance force of the fluid on the particle, dynes.
μ_f = viscosity of the fluid, g/cm-sec.
V = relative velocity of the particle with respect to the fluid, cm/sec.
D_p = spherical particle diameter, cm.

If the fluid is bounded by container walls, or is not continous, or if the particle is not spherical, the viscous resistance changes somewhat. As the relative velocity or particle size increases, inertial effects in the fluid become important. The relative magnitude of the inertial and viscous forces in the fluid can be shown to be related to the Reynolds number based on the particle diameter:

$$\mathrm{Re}_p = \frac{\rho_f V D_p}{\mu_f} \tag{1.6}$$

where ρ_f = density of the fluid, g/cm³.

Flash photographs of the flow around a submerged obstacle (revealed by means of suspended powder) indicate that the flow pattern changes as the Reynolds number is increased[32]. At very low particle Reynolds numbers (<1), flow is nearly symmetrical upstream and downstream of the particle, viscous forces in the fluid predominate, and streamline flow obtains. As the Reynolds number increases, fluid inertia produces eddies downstream of the body, and they eventually separate and appear in the wake.

The variation of fluid resistance (drag) with the Reynolds number for spheres, cylinders, and disks is shown in Fig. 1.5. These curves apply approximately to the steady uniform motion of a real continuous fluid of large extent relative to a characteristic length of the submerged body. The proximity of other bodies or containing walls, or variations in fluid motion in time or space, modify the resistance. The drag coefficient, C_D, is defined by

$$F_R = C_D A \rho_f V^2/2 \tag{1.7}$$

where A = projected area of particle normal to its motion. For spheres with $\mathrm{Re}_p < 1$, $C_D = 24/\mathrm{Re}_p$, and F_R is given as indicated in equation 1.5.

1-3.2 Particle Motion

(a) *Gravitational Sedimentation.* A practical method of size separation for small particulate materials is based on sedimentation in a gravitational field. Small particles released from rest and allowed to settle freely under the action of gravity in a still fluid will rapidly accelerate and attain an equilibrium, or terminal, velocity determined by the magnitude of the opposing forces of gravity and fluid drag. The time of fall of particles can be predicted or calculated by such methods as direct weighing or volume measurements, or by changes in the transmission of a light beam shining through the settling suspension.[19]

At equilibrium velocity, the downward gravitational force (F_g) on the particle equals the upward resistive force of the fluid on the particle, and

$$F_R = F_g \tag{1.8}$$

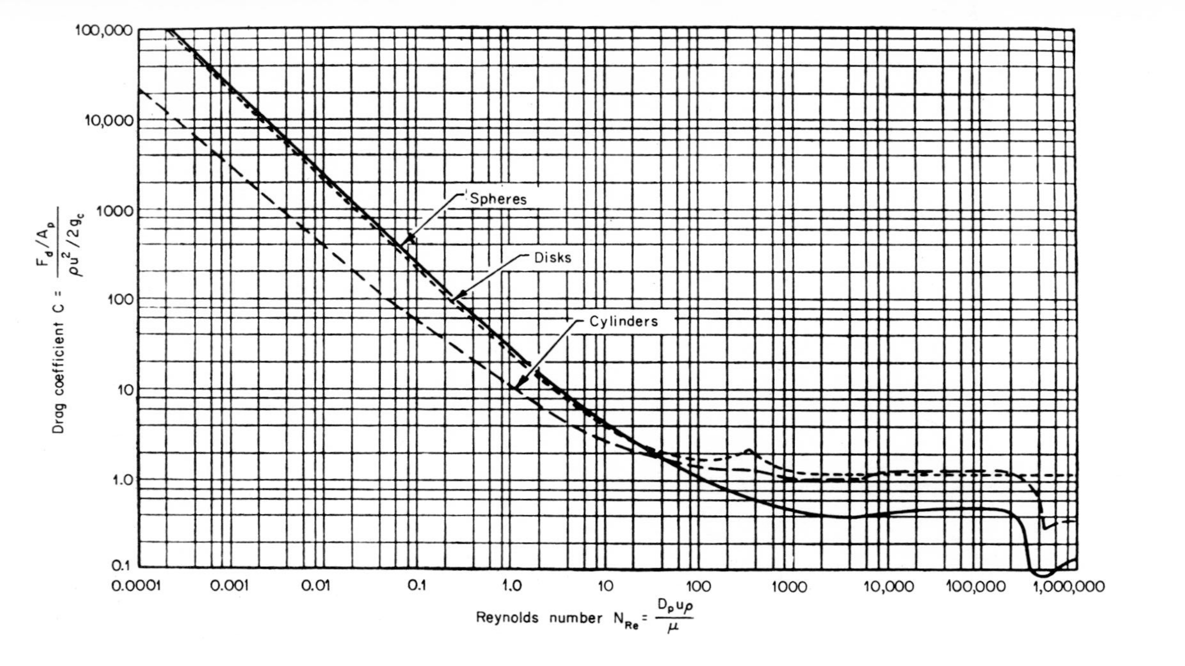

A_p = area of particle projected on plane normal to direction of motion, sq ft; C = overall drag coefficient, dimensionless; D_p = diameter of particle, ft; F_d = drag or resistance to motion of body in fluid, lb force; g_c = dimensional constant, 32.17 (lb) (ft)/(lb force) (sec^2); N_{Re} = Reynolds number, dimensionless; u = relative velocity between particle and main body of fluid, ft/sec; μ = fluid viscosity, lb/(ft)(sec); ρ = fluid density, lb/cu ft.

FIG. 1.5. Drag coefficients for spheres, disks, and cylinders. (From Lapple, Ref. 33.) Courtesy of McGraw-Hill Book Company, New York.

For particles below about 100 μm, which settle in streamline flow,

$$3\pi\mu_f D_p V = mg \tag{1.9}$$

where m = the buoyant weight of the particle (g), and g = acceleration due to gravity (cm/sec^2).

The buoyant weight of a spherical particle is given by

$$m = (\pi/6)\ D_p^3\ (\rho_p - \rho_f) \tag{1.10}$$

where ρ_p = the particle density, g/cm^3.

The terminal settling velocity becomes

$$V = D_p^2\ (\rho_p - \rho_f)\ g/18\mu_f \tag{1.11}$$

For particles settling in air, it is usual to neglect the buoyancy correction, since ρ_p is of order unity ($0.1 < \rho_p < 20$) and the air density is of order 10^{-3}. The buoyancy correction must be included in liquid sedimentation studies.

With appropriate values for the viscosity of ambient air (1.8×10^{-4} poise, g/cm-sec) and the gravitational acceleration, equation 1.11 can be reduced to

$$V = 0.003\ \rho_p D_p^2 \tag{1.12}$$

In this case, the terminal settling velocity in air is given in centimeters per second when the particle density is in grams per cubic centimeter and the particle diameter is in microns. For example, a 100-μm water droplet falls at 30 cm/sec (about 1 ft/sec), according to equation 1.12. If one calculates the droplet Reynolds number, it is found to be about 2, near the upper limit of the validity of assumptions in the Stokes relationship. By setting Re_p equal to 2, and solving equation 1.11 for V, it is evident that spherical particles can be assumed to fall in the streamline range when their diameter in microns, D_p, is less than $100/\rho_p^{1/3}$. Therefore, the largest spherical quartz particle ($\rho_p = 2.65$) that settles approximately in the Stokes range is about 70 μm.

Particles that are not spherical fall at slower rates than predicted by the Stokes relationship because of their larger projected surface area (A) per unit mass.[34] When sedimentation data are used to determine the size of irregularly shaped particles, the diameters obtained by using equation 1.11 are Stokes equivalent diameters. They represent the diameters of equivalent spherical particles of the same density that settle at the same rate. Estimates of the size of irregular particles from their Stokes settling rate give equivalent particle sizes smaller than those measured directly under the microscope[34]; i.e., irregular particles fall more slowly than spheres containing the same mass.

Wall effects are usually negligible for small particles, but a correction factor is available[35] which relates particle diameter, D_p, to settling tube diameter, D_t:

$$K_w = 1 + (9D_p/4D_t) + (9D_p/4D_t)^2 \tag{1.13}$$

and the resulting fluid resistance will ge given by K_w multiplied by the value for the unbounded fluid from equation 1.7. For small single particles in sedimentation tubes of usual dimensions (5 to 10 cm), values of D_p/D_t are less than 0.01 and the second-order term is negligible.

Proximity of particles to each other has also been shown to introduce significant differences in observed sedimentation rates.[35,36] Therefore, volume concentrations less than 0.05% are recommended.

For small particles, orientation during sedimentation in still fluid is of little concern, but for flow systems having Reynolds numbers greater than 1, orientation effects may occur. Attempts have been made to classify settling orientation.[37] In practice, many small particles will have inhomogeneities of composition and shape sufficient to produce a "heavy end" which influences orientation during sedimentation. Kunkel[34] observed that 10% of particles settled at an angle to the vertical.

Particles with diameters close to the mean free path of the molecules in the fluid are suspended in a noncontinuous medium. They begin to slip between the molecules and settle at a higher velocity than that predicted by equation 1.11. The Cunningham-Millikan correction for molecular slip is given in Fig. 1.6. Note that this correction is of major importance only for submicron particles at 1 atmosphere pressure when Brownian motion

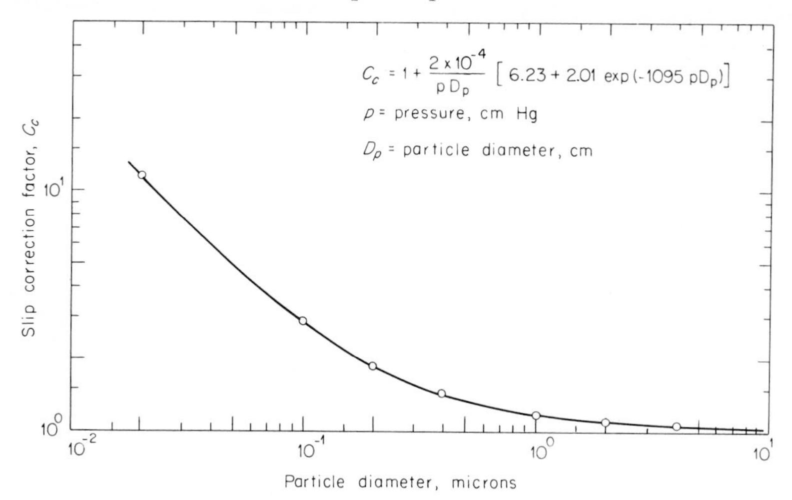

FIG. 1.6. Slip correction factor, C_c.

becomes important. Fig. 1.2 presents settling rates, diffusion coefficients, and other physical characteristics of small particles.

(b) *Motion in a Centrifugal Field.* When a small suspended particle travels in a circular path, it is acted upon by a centrifugal force and moves radially outward relative to the path. Centrifugal force, F_c, is proportional to the square of the tangential velocity ($V_{\tan}$) and inversely proportional to the radius of turn, R, as follows:

$$F_c \simeq \frac{V_{\tan}^2}{R} \tag{1.14}$$

Although the particle may be suspended in a rotating fluid which is in turbulent flow, it is assumed that particle movement outward is resisted by viscous drag and that Stokes's relation provides a reasonable approximation for the drag. Therefore, the equation of equilibrium motion for a small spherical particle in a centrifugal force field assumes the following form, whereby centrifugal force ($V_{\tan}^2/R$) replaces gravitational force (g) in Stokes' equation.

$$V_{\mathrm{rad}} = \frac{D_p^2 \rho_p}{18 \mu_f} \cdot \frac{V_{\tan}^2}{R} \tag{1.15}$$

In equation 1.15 it is assumed that $F_c = F_R$ (i.e., particle moves at constant terminal velocity) and $\rho_f \ll \rho_p$.

The ratio of centrifugal to gravitational force (i.e., $V_{\tan}^2/Rg$) is defined as the separation factor. Separation factors in dust-collecting cyclones which retain particles larger than 10 μm are usually less than 200, whereas conventional centrifuges, which can be used to precipitate submicron particles and large molecules in liquid suspension, have separation factors in excess of 5000.

(c) *Motion in an Electrostatic Field.* When particles greater than about 1 μm in diameter are passed through a corona discharge, they acquire charges from electrons and adsorbed gas ions in proportion to the square of the particle diameter, D_p, and the strength of the charging field, E_o.[12] For conducting particles, the saturation charge is

$$ne = \frac{3E_0 D_p^2}{4R} \tag{1.16}$$

where n = number of electron charges at saturation, and
e = charge of one electron (4.8×10^{-10} statcoulomb).

Particles of insulating materials acquire charges of 50 to 60% of this value. For particles less than about 0.2 μm, diffusion charging predominates,[3,12] and charges acquired by a particle are given approximately by

$$n = \frac{D_p kT}{2e^2} \ln \left(1 + \frac{\pi D_p c N_0 e^2 t}{2kT}\right) \tag{1.17}$$

where n = charges on an initially neutral particle after time t.
k = Boltzmann constant, 1.38×10^{-16} erg/molecule °K.
N_0 = ion density, ions/cm³.
c = ion velocity (root mean square), cm/sec.
t = time, sec.

TABLE 1.4
NUMBER OF CHARGES ACQUIRED BY PARTICLES[a]

Particle diameter (μm)	Field charging				Diffusion charging			
	Exposure time (sec)				Exposure time (sec)			
	0.01	0.1	1	∞[b]	0.01	0.1	1	10
0.2	0.7	2	2.4	2.5	3	7	11	15
2.0	72	200	244	250	70	110	150	190
20.0	7200	20 000	24 400	25 000	1100	1500	1900	2300

[a]From Lowe and Lucas, Ref. 38.
[b]Limiting charge.
Note: Calculated from equations 1.16 and 1.17 under the following conditions typical of a wire-in-tube assembly; $T = 300$°K, $N_0 = 5 \times 10^7$ ions/cm³, $E_0 = 2$ kV/cm, in air at atmospheric conditions at 40 kV with a discharge current of 40 μA/ft.

Typical charges acquired by particles of various sizes are shown in Table 1.4. These agree approximately with values reported by White.[12] Particles in a unipolar ion flux acquire sufficient charge in less than 0.1 sec to precipitate rapidly in a parallel-field condenser, and this method has formed the basis for the design of several electrostatic classifiers.[3,39] A commercial device using this mechanism for particle size analysis is discussed in Chapter 5, p. 227. The migration velocity of a spherical charged particle, V_e, in the direction of a collecting electrode can be obtained from the following expression, assuming that air resistance is given by Stokes' law:

$$V_e = (Ene/3\pi\mu_f D_p)C_c \tag{1.18}$$

where E = field strength in the collecting space (esu/cm), and
C_c = slip correction factor.

(d) *Impaction.* Inertial impaction devices cause an air sample to be drawn into a round or rectangular nozzle where the gas velocity is substantially increased. The jet from the nozzle is discharged against an adjacent flat surface, causing the air to diverge sharply as shown in Fig. 1.7; the numbers indicate the velocity pattern in the jet. Particles in the air stream have more inertia than the air, and tend to continue forward as the air passes off to the sides, causing some of the particles to impact on the surface. Since the air velocity over the collecting plate is high, the plate may be coated with a viscous material such as a silicone fluid to prevent particle re-entrainment. The distance from the jet outlet to the stationary plate governs the sharpness of curvature of the fluid stream and, with jet velocity, controls collection efficiency. This distance is usually maintained about equal to the characteristic dimension of the nozzle (e.g., jet diameter) for effective capture of small particles (i.e., those less than 1 μm).

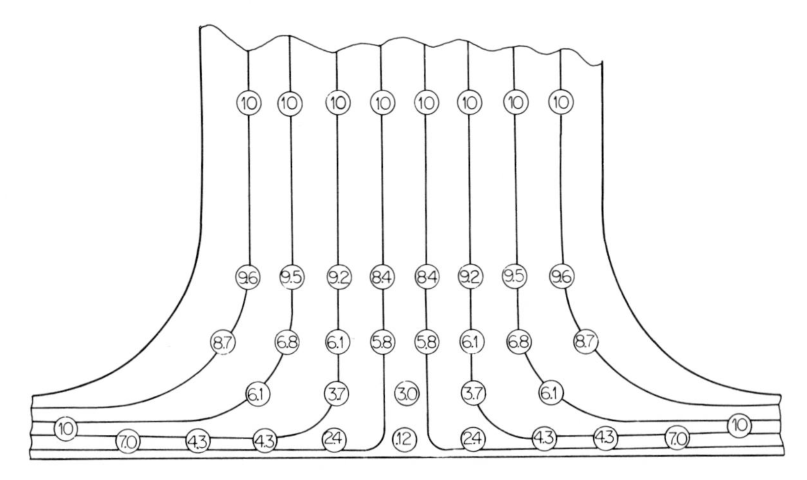

FIG. 1.7. Impingement of a free jet on a flat plate. (From Jorgensen, Ref. 40.)

Since the range of particle sizes removed by a single impaction stage is limited, size classification of a polydisperse cloud may be accomplished by arranging several stages with decreasing nozzle sizes in series. This arrangement is known as a cascade impactor.

The efficiency of impaction is usually presented as a function of the square root of the dimensionless impaction parameter (I):

$$I^{1/2} = \left(\frac{C_c \rho_p D_p{}^2 V_j}{18 \mu_f D_j} \right)^{1/2} \tag{1.19}$$

where D_j = jet diameter for round jet, or jet width for rectangular jet (cm), and V_j = the average air velocity at the jet outlet (cm/sec). The effects of jet design on efficiency and calibration are discussed by Mercer[41] for several single-jet cascade impactor configurations. A multihole commercial cascade impactor has been described[42] and used at high velocity for collection of ambient atmospheric aerosol particles.[43] The exact form of the efficiency curve as a function of $I^{1/2}$ for individual cascade impactors is provided in technical literature accompanying each device. Experimental efficiencies for round and rectangular jets were determined by Ranz and Wong,[44] as shown in Table 1.5. The ratio of plate spacing to jet dimension in experiments reported in this table was approximately 3. Decreasing this ratio to 1 or less steepens the slope of the $I^{1/2}$ – efficiency relationship, and shifts the characteristic impaction parameter (i.e., the value of the parameter at which efficiency is 0.5) to a smaller value.[41]

TABLE 1.5
IMPACTION EFFICIENCIES FOR ROUND AND RECTANGULAR JETS[a]

Impaction efficiency (%)	Parameter $I^{1/2}$ [b]	
	Round jet	Rectangular jet
0	0.20	0.31
10	0.28	0.40
20	0.32	0.46
30	0.35	0.51
40	0.37	0.53
50	0.38	0.57
60	0.40	0.60
70	0.42	0.63
80	0.43	0.67
90	0.50	0.71
100	0.56	0.83

[a]From Ranz and Wong, Ref. 44.
[b]Experimental values above were determined for a ratio of plate spacing to characteristic jet dimension of 3.

Classification of a particle cloud into discrete parts by use of the cascade impactor may be interpreted as measuring aerodynamic diameter. For fixed jet size, jet-to-plate spacing, and air velocity, the calibration data given in Table 1.5 may be expressed in terms of the aerodynamic size.

(e) *Motion in a Thermal Gradient.* An aerosol particle subjected to a temperature gradient between a heated and a cold surface will tend to move toward the colder surface.[1,2] The motion is caused by a thermal force arising from differential interaction of the particle with the gas molecules. Those approaching from the hot side have a higher average velocity (momentum) than those approaching from the cold side, producing a net force in the direction of the flux of thermal energy. Thermal precipitation of dust particles is frequently used to obtain a sample for subsequent analysis by light optical or electron microscopy.

At atmospheric pressure, thermal force is given by

$$F_t = (-9\pi)(D_p/2)(\mu_f^2/\rho_f T)(2 + k_i/k_f)^{-1}(\Delta T/\Delta x) \qquad (1.20)$$

where k_i = thermal conductivity of the particle material, cal/sec cm °K.
k_f = thermal conductivity of the air, cal/sec cm °K.
$\Delta T/\Delta x$ = temperature gradient in the air, °C/cm.

The other symbols are as previously defined.

Typical ranges of thermal conductivity are: aluminum, 0.5; cadmium, 0.2; steel, 0.1; stone and sand, 10^{-3}; water, 10^{-3}; glycerol and stearic acid, 10^{-4}; air, 10^{-4}.

By setting the thermal force equal to the resistive force of the fluid (equation 1.5), the velocity of a given particle may be determined as

$$V = (-3/2)(C_c\mu_f/\rho_f T)(2 + k_i/k_f)^{-1}(\Delta T/\Delta x) \qquad (1.21)$$

The particle velocity can be used in conjunction with the flow velocity and configuration to establish the necessary geometry of a thermal precipitator. Equation 1.21 indicates that the thermal motion is dependent on particle size when the slip correction, C_c, is greater than 1, which occurs for particles less than a micron in diameter in air at atmospheric pressure (see Fig. 1.6). Deposition of submicron particles (or deposition at reduced pressure) can thus occur as a function of size in a thermal field, although this phenomenon has not been widely applied as a means of sample stratification.

References

1. H. L. Green and W. R. Lane, *Particulate Clouds — Dusts, Smoke and Mists*, 2nd ed., D. Van Nostrand Company, Inc., Princeton, N.J., 1964.
2. N. A. Fuchs, *The Mechanics of Aerosols*, The Macmillian Company, New York, 1964.
3. C. N. Davies (Ed.), *Aerosol Science*, Academic Press, New York, 1966.
4. American Society for Testing and Materials, Tentative Standard D 1356-58T, Philadelphia, Pa., 1958.
5. H. Green, The Effect of Non-Uniformity and Particle Shape on Average Particle Size, *J. Franklin Inst., 204*:713 (1927).
6. C. E. Lapple, Characteristics of Particles and Particle Dispersoids, *J. Stanford Res. Inst., 5*:95 (1961).

7. Factory Mutual Insurance Co., *Handbook of Industrial Loss Prevention,* Chapter 42, McGraw-Hill Book Co., Inc., New York, 1968.
8. T. F. Hatch and P. Gross, *Pulmonary Deposition and Retention of Inhaled Aerosols,* Academic Press, New York, 1964.
9. N. A. Fuchs, *Evaporation and Droplet Growth in Gaseous Media,* Pergamon Press, London, 1959.
10. B. J. Mason, *The Physics of Clouds,* Clarendon Press, Oxford, 1959.
11. L. B. Loeb, The Basic Mechanisms of Static Electrification, *Science, 102*:573 (1945); see also L. B. Loeb, Static Electrification, in J. B. Birks (Ed.), *Progress in Dielectrics,* Vol. 4 *et seq.,* Heywood & Company, Ltd., London, 1962.
12. H. J. White, *Industrial Electrostatic Precipitation,* Addison-Wesley Publishing Company, Inc., Reading, Mass., 1963.
13. W. J. Megaw and A. C. Wells, Production of Monodisperse Submicron Aerosols of Which Each Particle Carries a Specified Number of Electron Charges, *Nature, 224*:689 (1969).
14. W. B. Kunkel, The Static Electrification of Dust Particles on Dispersion into a Cloud, *J. Appl. Phys., 21*:820 (1950).
15. J. M. Dallavalle, C. Orr, Jr., and B. L. Hinkle, The Aggregation of Aerosols, *Brit. J. Appl. Phys.,* Suppl. 3, p. S198 (1954).
16. J. H. Daniel and F. S. Brackett, Investigation of Small Airborne Charged Particles by an Electrical Method, *A.M.A. Arch. Ind. Hyg. Occupat. Med., 3*:505 (1951).
17. T. Sicotte and M. Rinfrette, Scattering of Light by Binary Liquid Mixtures, *Trans. Faraday Soc., 58* (474, No. 6):1090 (June 1962).
18. H. C. van de Hulst, *Light Scattering by Small Particles,* John Wiley & Sons, New York, 1957.
19. D. Sinclair, Optical Properties of Aerosols, *Handbook on Aerosols,* USAEC, United States Government Printing Office, Washington, D.C., 1950; reissued 1963, Report TID-4500.
20. B. M. Smith, J. Wagman, and B. R. Fish, Interaction of Airborne Particles with Gases, *Environ. Sci. Technol., 3*:558 (1969).
21. A. D. Zimon, *Adhesion of Dusts and Powders,* Plenum Press, Inc., New York, 1969.
22. R. Whytlaw-Gray and H. S. Patterson, *Smoke,* Edward Arnold and Company, London, 1932.
23. F. Stein, N. Esmen, and M. Corn, The Density of Uranine Aerosol Particles, *Am. Ind. Hyg. Assoc. J., 27*:428 (1966).
24. G. A. Sehmel, The Density of Uranine Particles Produced by a Spinning Disc Aerosol Generator, *Am. Ind. Hyg. Assoc. J., 28*:491 (1967).
25. M. E. McKnight and M. I. Tillery, On the Density of Uranine, *Am. Ind. Hyg. Assoc. J., 28*:498 (1967).
26. G. Herdan, *Small Particle Statistics,* Academic Press, New York, 1960.
27. J. M. Dallavalle, *Micromeritics,* 2nd ed., pp. 41-95, Pitman Publishing Company, New York, 1948.
28. K. T. Whitby and B. Y. H. Liu, Dust, Engineering Aspects, in *Encyclopedia of Chemical Technology,* Vol. 7, John Wiley & Sons, New York, 1965.
29. G. M. Fair and L. P. Hatch, Fundamental Factors Governing the Streamline Flow of Water through Sand, *J. Am. Water Works Assoc., 25*:1551 (1933).

30. W. C. McCrone, R. G. Draftz, and J. G. Delly, *The Particle Atlas,* Ann Arbor Science Publishers, Inc., Ann Arbor, Mich., 1967.
31. M. Corn, The Effect of Dust Particle Orientation on Particle Size Determined by Microscopic Techniques, *Am. Ind. Hyg. Assoc. J., 25*:1 (1964).
32. L. Prandtl and O. G. Tietjens, *Applied Hydro- and Aeromechanics,* McGraw-Hill Book Co., Inc., New York, 1934.
33. R. H. Perry, C. H. Chilton, and S. D. Kirkpatrick (Eds.), *Chemical Engineers' Handbook,* 4th ed., pp. 5-60, McGraw-Hill Book Company, Inc., New York, 1963.
34. W. B. Kunkel, The Magnitude and Character of Errors Produced by Shape Factors in Stokes' Law Estimates of Particle Radius, *J. Appl. Phy. 19*:1056 (1948).
35. J. S. McNown, H. M. Lee, M. B. McPherson, and S. M. Engez, Influence of Boundary Proximity on the Drag of Spheres, *Proc. 7th Inter. Congr. Appl. Mech.,* Vol. 2, Part 1, p. 17, (1948).
36. V. H. Kay and R. P. Boardman, Cluster Formation in Dilute Suspensions, in *Symposium on Interaction between Fluids and Particles,* Institution of Chemical Engineers, London, June 1962.
37. J. F. Heiss and J. Coull, The Effect of Orientation and Shape on the Settling Velocity of Non-Isometric Particles in a Viscous Medium, *Chem. Eng. Progr., 48*:133 (1952).
38. H. J. Lowe and D. H. Lucas, The Physics of Electrostatic Precipitation, *Brit. J. Appl. Phys.,* Suppl. 2 (1952).
39. G. Langer and J. L. Radnick, Development and Preliminary Testing of a Device for Electrostatic Classification of Submicron Airborne Particles, *J. Appl. Phys., 32*:955 (1961).
40. R. Jorgensen, *Fan Engineering,* 7th ed., Buffalo Forge Company, Buffalo, N. Y., 1970.
41. T. T. Mercer, On the Calibration of Cascade Impactors, *Ann. Occupat. Hyg., 6*:1 (1963).
42. A. A. Andersen, A Sampler for Respiratory Health Hazard Assessment, *Am. Ind. Hyg. Assoc. J., 27*:160 (1966).
43. R. E. Lee, Jr., and J. P. Flesch, A Gravimetric Method for Determining the Size Distribution of Particulates Suspended in Air, Paper No. 69-125 presented at the annual meeting of the Air Pollution Control Association, New York, N. Y., June 8-12, 1969.
44. W. E. Ranz and J. B. Wong, Impaction of Dust and Smoke Particles, *Ind. Eng. Chem., 44*:1371 (1952).

2

Sampling Methods for Particles

2-1 Introduction

Sampling instruments for particle sizing may be divided into two functional groups: (1) those which capture a bulk sample of dust, mist, smoke, or fume in a convenient form for transportation to the analytical laboratory, and (2) instruments which, while collecting a sample, effect in some fashion a size separation or classification of particles which becomes an integral part of the subsequent analytical procedure.

Some collection instruments are suitable only for sampling ambient air in workplaces or outdoors; others are suitable for removing a fraction of dust flowing in a pipeline; and some may be used for both purposes. When a size analysis of a bulk powder is required, it is only necessary to select a representative portion of the well-mixed mass.

A further differentiation, or classification, of sampling instruments for size analysis may be made with respect to whether the collected material is in a form suitable for *in situ* examination, or whether subsidiary manipulations and treatment (which may alter the size or state of agglomeration) are required. The advantages inherent in methods which permit collection of airborne particles or droplets for examination or measurement in their unaltered state are obvious.

Finally, it is important that the portion selected for examination be truly representative of the bulk of the material, since the quantity analyzed represents, in most cases, but an infinitesimal fraction of the parent material. Not only must the collection method give a representative sample of the material, but it must not alter the basic particle size of the collected dust; i.e., there must be no agglomeration or fracturing of particles and naturally existing clumps.

Particle concentration and the time available for sampling exert an important influence on the choice of sampling instrument. A number of instruments, such as the Konimeter, are suitable for obtaining instantaneous or "grab" samples of an aerosol containing light dust loadings, whereas others, such as a paper filter, may be used to collect large amounts of dust over periods as short as 5 to 10 min or as long as 30 days. Thus, it may be seen that the duration of a long-period sample, also termed an "average" or "integrated" sample, may be highly variable. Instantaneous sampling instruments provide an excellent means for following the variations which occur during a cyclic or irregular process, as well as for capturing a dust sample from an operation of very brief duration. Integrated samples, on the other hand, provide information on average conditions over some appreciable time interval.

As a general rule, particulate material flowing in a pipepline or stack will be present in high concentration (grams per cubic meter), whereas in the wake of stacks or close to industrial plants, ambient dust and mist concentrations are likely to be moderate (milligrams per cubic meter). Light dust concentrations (micrograms per cubic meter) are found in suburban and rural atmospheres.[1]

2-1.1 Isokinetic Sampling

Studies indicate that representative samples of particulate materials greater than a few microns in diameter may be obtained from rapidly moving air streams only when sampling is conducted so that the flow rate at entry to the sampling probe is at the same velocity as the main gas stream (i.e., isokinetic). When the velocity into the sampling probe is different from that in the main stream (i.e., anisokinetic), the size distribution and the total weight concentration of the particulate matter collected will be different from those in the main stream. The explanation lies in the behavior of particles having sufficient mass and velocity to possess appreciable momentum. At velocities less than main stream velocity, all large particles in the path of the probe (i.e., in the stream tube intercepted by the probe cross section) will be projected into the sampler opening, while a portion of the air or gas in this stream tube will be deflected around the probe, as shown in Fig. 2.1.

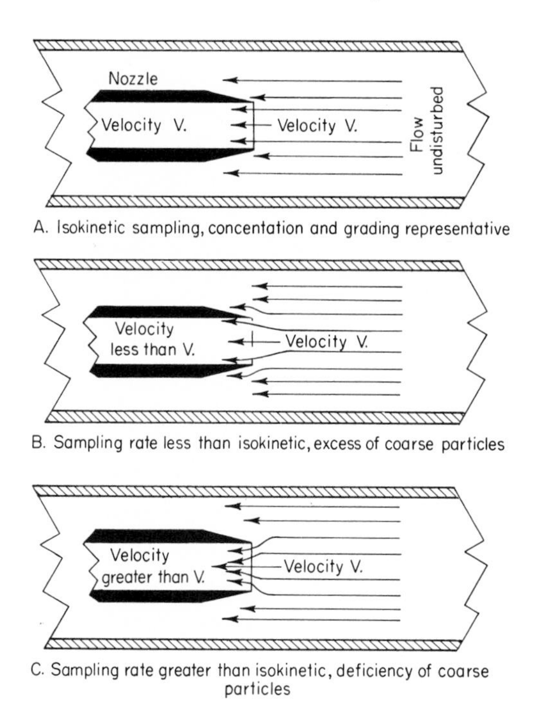

FIG. 2.1. Streamlines around a sampling probe. (From Stairmand, Ref. 6.)

This results in a size mixture entering the sampling tube that has more large particles than are in the sampled aerosol. In addition, the total weight of particulate matter in each cubic meter of sample will, for this reason, be greater than in the aerosol from which it was collected. When sampling velocity is greater than main stream velocity, all the air contained in the stream tube that is intercepted by the inlet to the probe will be sucked into the sampler, plus some air from the main stream outside this stream tube, as shown in Fig. 2.1. When air is drawn into the sampling probe from outside the intercepted stream tube, the smallest particles will follow the curved streamlines into the entry of the probe, but larger particles, because of their inertia, will fail to be drawn in. The result will be that the collected sample will have fewer large particles than the aerosol from which it was drawn and the total weight concentration of particulate matter will be less, as well.

Samples collected at other than isokinetic velocity are likely to be in error by an amount that is related to the terminal settling velocity of the particles, the magnitude of the deviation from isokinetic sampling velocity, and the gas flow pattern at the sampling nozzle. The latter refers principally to sampling tube yaw relative to the direction of the main stream flow, a zero angle of inclination being ideal. Various investigators differ as to the magnitude of the error produced by sampling at other than isokinetic velocity[2-4] but agree that it is significant for particles greater than 5 μm. Deviations from isokinetic sampling rates probably do not significantly affect results for particles below 1 μm in size.[5]

Stairmand[6] is the source of the data shown in Table 2.1, which illustrate the effect of incorrect sampling velocity for particles 1, 10, and 100 μm in diameter.

TABLE 2.1
EFFECT OF INCORRECT SAMPLING VELOCITY[a, b]

A. Sampling conditions		
Condition	Experiment I	Experiment II
Duct speed (ft/sec)	30	30
Sampling speed (ft/sec)	15	45
Sampling speed/duct speed	0.5	1.5
B. Sampling results	Sampled concentration/duct concentration	
Particle diameter (μm)	Experiment I	Experiment II
100	1.99	0.67
10	1.54	0.82
1	1.013	0.96

[a]For particles of density 1 g/cm³ suspended in air at 20°C and 1 atm.
[b]From Stairmand, Ref. 6.

Diamond *et al.*[7] tested the effects of anisokinetic sampling rates in a 17-in.-diameter duct with foundry dust containing 7.4% by weight of particles 0.20 μm and 92.6% of particles above 0.20 μm. Their findings, summarized in Table 2.2, show that in sample pairs 2 and 3, a variation of 26% below isokinetic sampling velocity influences the particle size distribution. There is a decrease in the 0- to 4-μm range and a corresponding increase in the 4- to 20-μm range. There is an increase in total weight of 30% over a paired isokinetic sample. In the sample pairs 4 and 5, collected 85% above isokinetic sampling velocity, there is an increase in the collection of fine particles, and the weight of total samples collected is 39% below the paired isokinetic sample.

TABLE 2.2
WEIGHT AND PARTICLE SIZE DISTRIBUTION OF DUCT SAMPLES[a, b]

Sample pair	Sampling rate	Sampling velocity	Stack velocity	Total weight (g)	Size distribution by weight (%)		
					0–4 μm	4–20 μm	20+ μm
1	1 cfm	1292	1297	0.242	28.5	6.5	65.0
	1 cfm	1292	1297	0.198	32.7	5.0	62.3
2	1 cfm	1292	1297	0.536			
	1 cfm	957	1297	0.684	22.8	11.2	66.0
3	1 cfm	1292	1297	0.112			
	1 cfm	957	1297	0.148	24.6	10.2	65.2
4	1 cfm	1292	1297	0.675			
	1 cfm	2390	1297	0.414	34.4	12.1	53.5
5	1 cfm	1292	1297	0.758			
	1 cfm	2390	1297	0.460	32.0	17.5	50.5

[a]Collected at isokinetic and anisokinetic velocities.
[b]From Diamond *et al.*, Ref. 7.

Usually, it is not sufficient to utilize an average velocity when withdrawing a sample isokinetically from a duct or stack, since velocity is far from uniform in a duct cross section downstream of an elbow or other disturbance to flow. Extreme variations in velocity and dust-loading gradients occur in the discharge of cyclone dust collectors and other centrifugal air-handling machines. As an example, Figs. 2.2a and 2.2b show velocity and dust profiles in the 36-in.-diameter exit duct of a cyclone dust collector. It may be seen that the vortical flow pattern has tended to concentrate most of the dust and air flow into a layer only a few inches in thickness along the duct wall. Obviously, a displacement of only a fraction of an inch will, in this case, produce profound changes in air velocity and dust loading entering a sampling probe and will greatly affect the quantity and quality of the collected dust sample.

Whereas there can be little question concerning the importance of using isokinetic sampling rates when extracting a representative sample for determining both total weight and particle size distribution from a flowing air or gas stream inside a duct or stack, less is known about the effects of sampling rate on size distribution when collecting dust from the atmosphere, indoors and out. Conditions are somewhat more uniform indoors, as ambient air velocities seldom exceed 100 to 200 ft/min except in the immediate vicinity of air supply grills, floor fans, or other ventilation sources. Outdoor air movements may differ widely in speed as well as direction, and this often produces marked variations with time in the character and maximum size of airborne particulate matter.

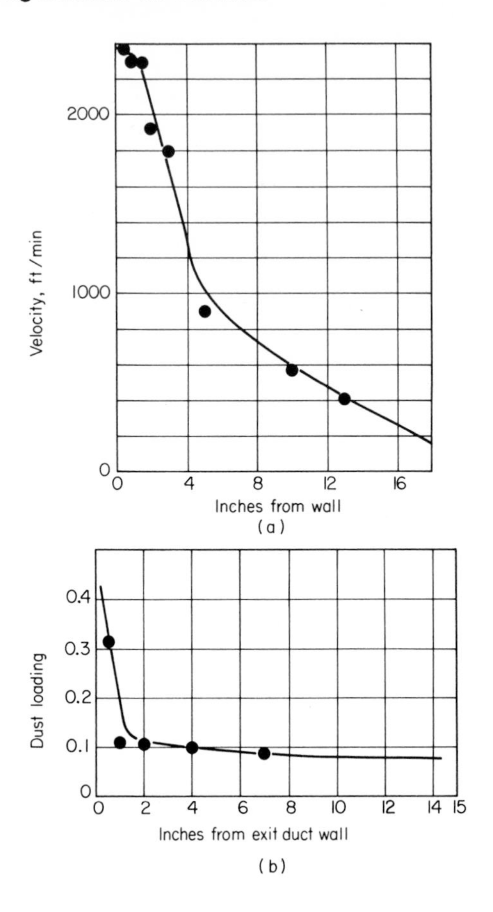

FIG. 2.2. Cyclone exit-duct velocity distribution *(a)* and dust-loading contour *(b)*. (From Silverman *et al.*, Ref. 8.)

May and Druett[9] studied the effect of wind speed on the collection characteristics of their dust sampling trap (called a pre-impinger) which operates at a constant rate of 11 lpm through a fixed-entry orifice. At zero wind velocity, intake efficiency was high for all particle sizes. It was also high for a wind speed of 12 mph which was the sampler inlet velocity. At wind speeds one-half isokinetic, intake efficiency declined for particles greater than 10 μm. Efficiency became greater than 100% for particles larger than 10 μm when wind speed was substantially greater than intake speed. It is of great importance to note that May and Druett found intake efficiency to be close to 100% for particles smaller than 10 μm under all conditions tested, and it may be concluded that anisokinetic sampling from open spaces has little effect on intake efficiency for particles in the size range of hygienic importance.

The effect of the angle of yaw of the sampling orifice to the wind direction has been found by Watson[10] to be important at high wind speeds but to be insignificant at wind speeds close to zero—speeds which are more characteristic of unconfined spaces, indoors and out, than of aerosols flowing in ducts. The effects of sampling probe yaw are more noticeable for large particles than for small, as shown in Fig. 2.3.

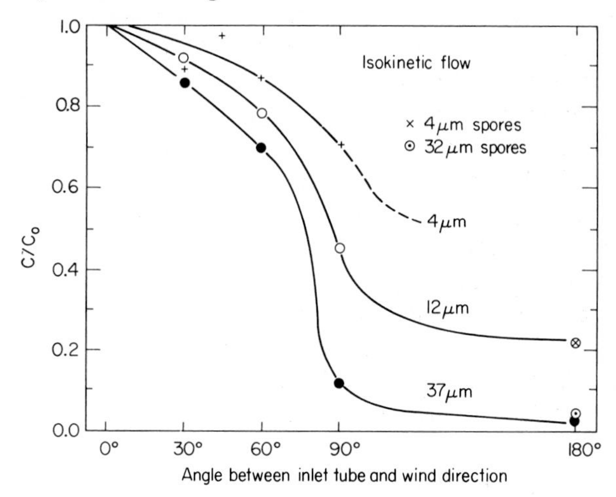

FIG. 2.3. Effect of probe angle of yaw on isokinetically sampled concentration for particles 4 to 37 μm. (From Watson, Ref. 10.)

A sampling device which is substantially less than 100% efficient for all sizes of dust, fumes, or mists is of limited value for size analysis, since the fraction that escapes usually contains the finest particles, and size measurements on the fraction retained may be greatly in error.[5]

2-2 Particle Collection Methods Which Do Not Classify or Size-Segregate the Sample

Samples of dust for sizing may be obtained in ambient air and from ducts and stacks with a number of instruments routinely used by industrial hygienists for collecting air samples for count and weight analyses. Among these are standard and midget impingers, high-velocity jet impactors, midget cyclones, thermal and electrostatic precipitators, and fibrous and granular filters of many types.

When using cyclone-collecting devices, as well as most of the common filter samplers, the dust must be redispersed for size analysis. Complete redispersal is virtually impossible, since naturally occurring agglomerates (as well as those formed by the collection method) are broken apart and particle attrition occurs in the process. Nevertheless, bulk samples collected on filters or by mechanical devices are generally adaptable to almost all methods of size analysis, and for this reason dusts are often collected in this manner.

Dispersed samples suitable for size analysis in the state in which they are collected may be obtained with a number of instruments, such as the thermal precipitator, the electrostatic precipitator, and certain filters frequently used by industrial hygienists. The principal advantage of this group of sampling instruments is that they make it possible to examine particles in more or less the same status as that in which they existed in the aerosol prior to entry into the sample device. The total number of particles that may be collected in these instruments without serious interferences or overlapping of granules or droplets is, unfortunately, quite small, and the situations in which these methods may be used are limited to very light particulate loadings and brief sampling periods. These limitations will be discussed in greater detail in the sections that follow which describe the instruments and how they are used for collecting samples for particle size analysis.

2-2.1 Settling Chambers

Sample collection by settling may be conducted dynamically or statically. In the dynamic method, an aerosol flows continuously, at low speeds, through a horizontal chamber, and particles settle by gravity onto quiescent collecting trays, or bins, in accordance with equation 1.11. The distance a particle must settle to reach the collecting tray, combined with the time the aerosol is retained within the chamber (the ratio of chamber length to horizontal speed), determines the smallest particle that may be collected at close to 100% efficiency. As a general rule, dynamic settling chambers are unsuitable for collecting particles below 5 μm, although somewhat smaller particles can be retained when particle specific gravity is unusually great, as with lead and uranium metals and oxides. Collecting devices of this type find their greatest use as presamplers to eliminate from the final air sample particles too large to be of hygienic significance (i.e., those >5 to 10 μm). Although not of special interest as a means of collecting particles for size analysis, precollectors that effect a size separation simulating that of the respiratory tract have assumed increasing importance in evaluating the toxicity of the atmospheric environment.

A typical instrument using the dynamic settling chamber principle is the Casella Hexhlet Sampler*, shown in Fig. 2.4. The instrument comprises two

FIG. 2.4. Hexhlet settling chamber dust collector. Courtesy of C. F. Casella & Co., Ltd.

main compartments: first, an elutriator, in which particles larger than 7 μm are collected, and second, a paper filter, on which particles 7 μm and below are retained. Although in many cases a complete size analysis will be made on only one of the fractions, it is usually desirable to know the weight fraction discarded.

The elutriator is a rectangular open-ended box containing a large number of thin aluminum plates slotted into the side walls. The length, breadth, and height of each shallow air passage thus formed is so proportioned that with a flow rate of 50 lpm particles above 7 μm will settle onto a plate before reaching the end. The large number of independent air passages prevents the buildup of particles, and the elutriating properties of the entire unit remain unaltered during the sampling period.[11]

Dust fall jars, sticky papers, and sticky slides are often used to collect settled dust, pollens, and other particulate matter sampled by methods of dynamic gravitational settling for weight and count determination. The volume of air from which the collected material comes is variable and uncertain with these sampling methods, but, more important, only particles greater than 10 to 20 μm in diameter are captured with any degree of certainty. Therefore, they are not recommended for collecting particles for sizing.

Static settling chambers are exceedingly simple in design. They consist of

---

*C. F. Casella & Co., Ltd., London, N.1, England.

a sedimentation column in which a known volume of dust-laden air may be trapped by immersing the clean, open column in the aerosol to be sampled and then closing the two ends when the atmosphere inside the column has reached equilibrium with its surroundings. The instrument is then set aside for quiescent gravitational settling of the particles onto a collection slide at the base of the chamber. A sedimentation cell described by Green[12] is shown in Fig. 2.5. The chamber *(A)* is 3.6 cm in diameter and 5 cm high and is equipped with means *(B* and *C)* for closing the ends rapidly after moving the open instrument through the air to trap a sample. Green recommended a settling period of 3 hr to allow for the complete settlement of particles visible under the light microscope onto microscope coverslips *(E* and *E')* that are held in place by swivel springs *(D* and *D')*.

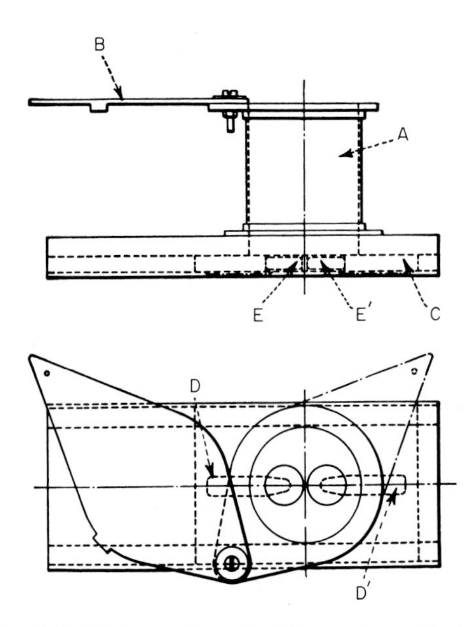

FIG. 2.5. Sedimentation cell. (From Green, Ref. 12.)

Sample collection in air is simple and rapid; collection from pipelines is possible but less simple, since the open sampler must be introduced into the flowing stream and be closed by remote control to trap a representative sample.

Particles settling slowly in a gravitational field may agglomerate, be attracted to the chamber walls by electrostatic forces, or remain suspended for unusually long periods because of convection currents generated in the settling column by temperature differences. In addition, gravity settling yields a particle distribution which differs from that in the aerosol because it is generally impractical to wait for all the smallest particles to settle out. However, correction factors based on conditions of tranquil or stirred settling and an assumed size distribution pattern may be determined, but since settling, in practice, is neither completely tranquil nor stirred, only an approximate correlation can be calculated readily. Particles which settle onto the bottom glass slide in a Green (or similar) instrument may be examined *in situ* with a microscope. If the dust concentration in the aerosol is sufficiently high, the rate of change of scattered light at a fixed height in the chamber may be utilized during settling to measure the size parameters (see p. 163). In general, particles are not altered during collection, and this is an advantage for many investigations.

The settling process may be speeded up and small particles precipitated onto the collecting slide of the settling chamber by using centrifugal force generated in a high-speed centrifuge after the largest particles have been precipitated by gravitational settling.

2-2.2 Impingement and Impaction Samplers

High-velocity jet impactors such as the Owens jet dust counter, Konimeter, and Bausch and Lomb dust counter are familiar to most industrial hygienists. They have been described and illustrated by Drinker and Hatch,[13] who also include proper procedures for their use. These devices are suitable for collecting particles of hygienic signficance (i.e., those less than 5 to 10 μm in diameter), and dust deposits from these instruments may be used for size analysis as well as for dust counting. Dust deposited on the collecting slides is suitable for direct microscopic sizing, and the collection efficiency of both the Owens and the Bausch and Lomb instruments is satisfactory down to the limit of resolution of the light microscope. However, careful analytical techniques are required because the dust ribbon deposited by the jet is nonuniform with respect to size and number, and it is necessary to examine many representative microscope fields in all sections of the dust ribbon. Drinker and Hatch[13] have pointed out that, in practice, the high jet velocities required to collect submicron-sized particles produce "disaggregation of . . . flocculated particles," and this characteristic of high-speed jet impactors must be kept in mind when they are used for collecting particles for size analysis.

The Bausch and Lomb dust sampler and counter, which contains a built-in microscope, is typical of this class of instrument. It is shown in Fig. 2.6. The physical laws governing the performance of jet impactors have been reviewed in Chapter 1.

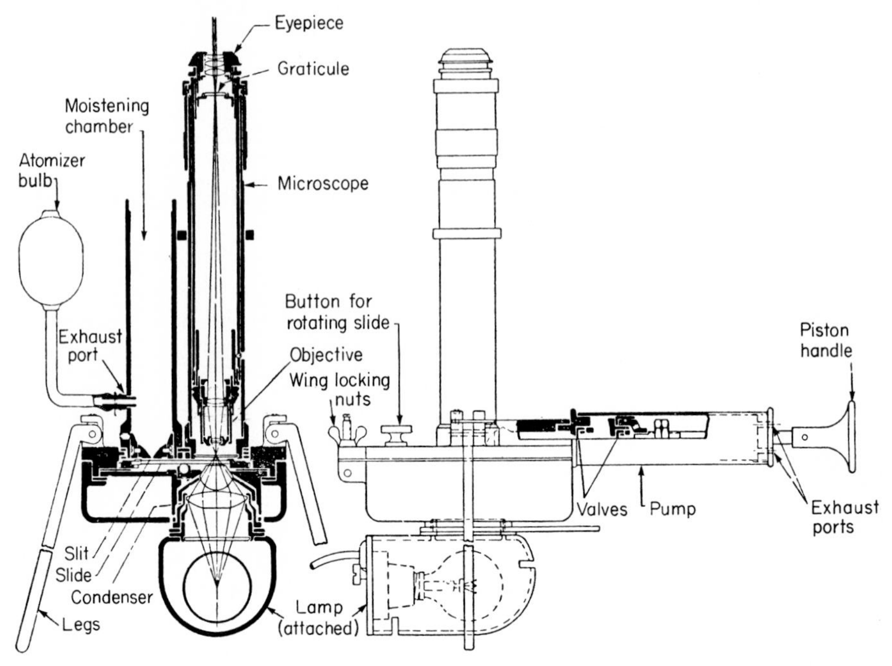

FIG. 2.6. Bausch and Lomb dust sampler and counter. Courtesy of Bausch and Lomb Co.

The Greenburg-Smith impinger dust sampling instrument[14] and the smaller midget impinger are used routinely by industrial hygienists. Impingers combine the particle arresting principle of the high-speed jet impactor with dust collection and retention in a liquid medium, such as water or alcohol.

A modified version of the standard instrument is shown in Fig. 2.7. It operates at 1 cfm and develops an impingement velocity of close to 100 m/sec through the 2.3-mm air inlet nozzle. The flat bottom of the flask, located 5 mm below the nozzle, is the impinging surface. It is filled with 75 ml of liquid.

Standard and midget impingers have poor efficiency for particles less than a half-micron in size. In addition, the dust concentration may be too low for efficient size analysis, or the impingement liquid may be an incompatible medium for the method of size analysis. As is true for the high-velocity, dry, jet-type instruments already described, disaggregation of flocculated particles occurs during impingement. Because of this characteristic, when large numbers of flocculated particles are present in the aerosol, the average size determined after impinger sampling is likely to be far smaller than the cloud present in the sampled air. At usual sampling rates, however, shattering of

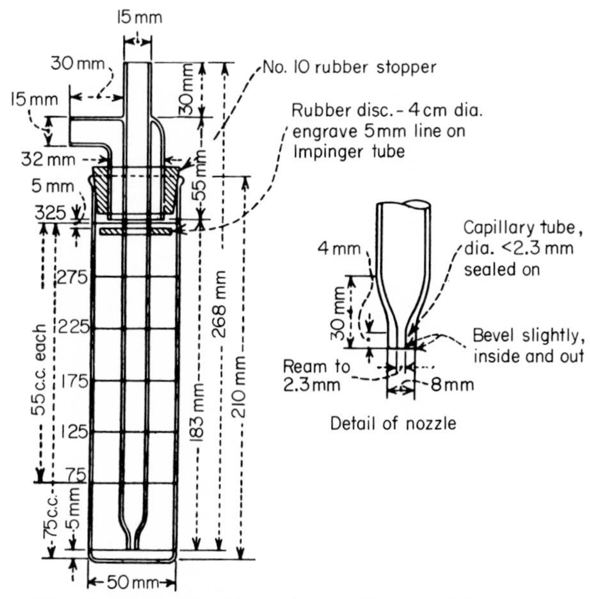

FIG. 2.7. Modified Greenburg-Smith impinger. (From Drinker and Hatch, Ref. 13.)

discrete particles in the impinger was found by Silverman and Franklin[15] to be negligible.

2-2.3 THERMAL PRECIPITATOR

The thermal precipitator was an outgrowth of Aitken's[16] observations that a dust-free area exists around a hot wire and that small particles can be removed from air quantitatively by precipitating them onto a cold surface adjacent to a hot surface. Green and Watson[17] used this principle to develop the dust sampling instrument illustrated in Fig. 2.8a. It consists of two brass blocks and insulating shims which form a narrow channel (0.95×0.05 cm) through which the aerosol sample is drawn at a velocity of 140 cm/min. A heated wire (0.025-cm diameter), suitably insulated, is centrally located between the cool adjacent walls. When the central wire is maintained at 100°C, the thermal gradient in the air passage is of the order of 4000 degrees per centimeter, due to the close spacing. The sharp thermal gradient produces differential molecular bombardment and migration of aerosol particles to the adjacent cool surfaces (thermophoresis), where they are collected on glass microscope coverslips, to which they adhere by molecular attraction. Deposition effectiveness is inversely proportional to particle size, thermal conductivity of the particle, and aerosol flow rate, as noted in Chapter 1.

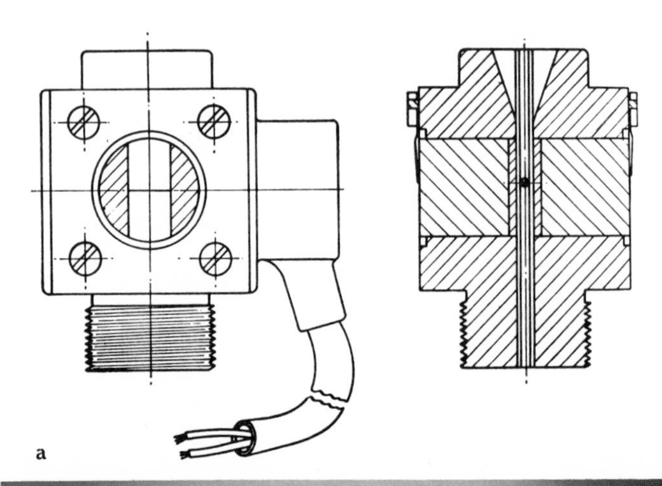
a

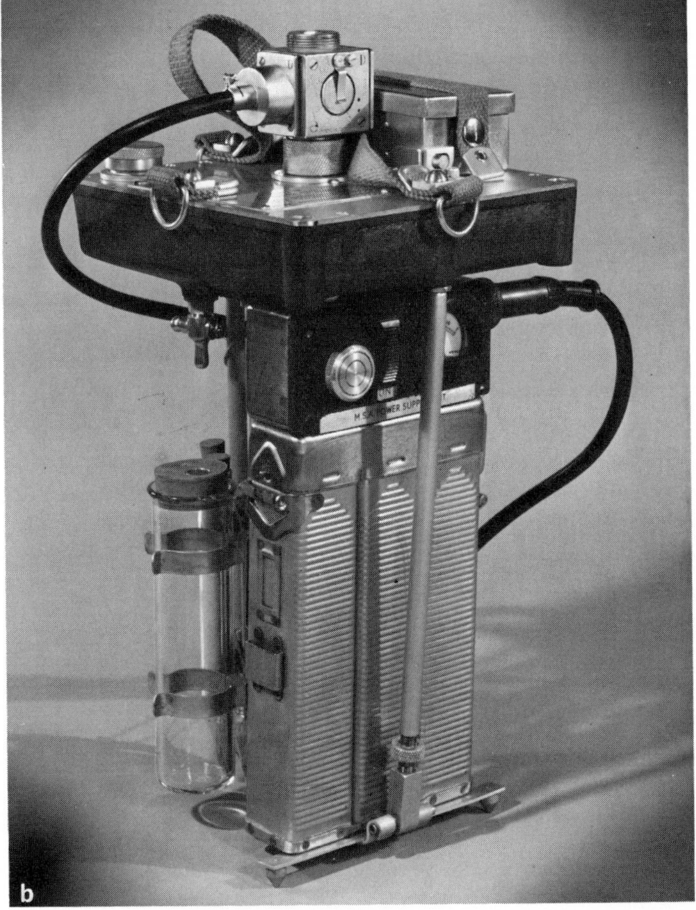
b

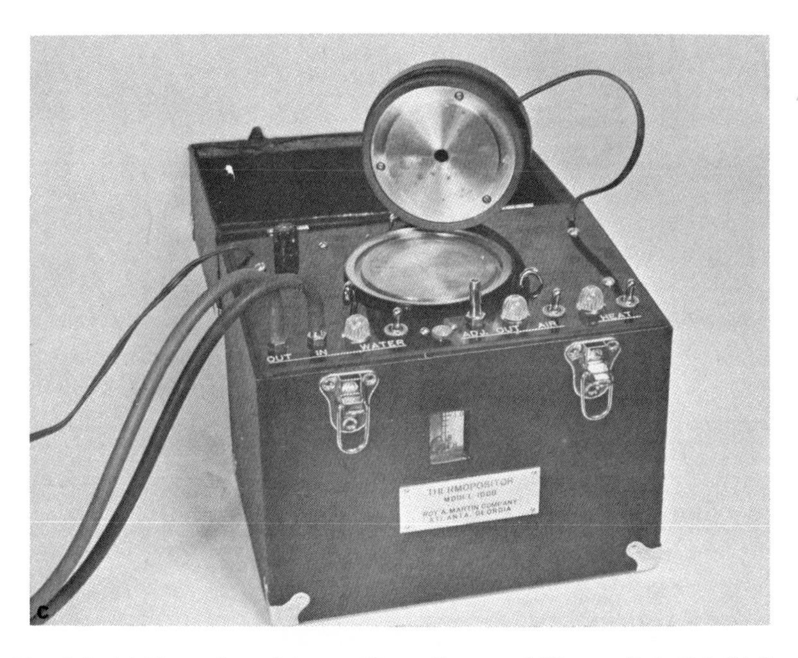

FIG. 2.8. *(a)* Thermal precipitator. (From Green and Watson, Ref. 17.) *(b)* Casella thermal precipitator. Courtesy of C. F. Casella & Co., Ltd. *(c)* Thermopositor. Courtesy of Micromeritics Instrument Corp.

The thermal precipitator collects essentially all submicron particles. Particles larger than 20 μm are more difficult to collect because of their greater inertia and consequent tendency to resist crossing flow streamlines in response to a weak thermal force, but parallel-plate configurations (i.e., one heated, one cooled) can be employed for more effective precipitation. Gas flow rate in the thermal precipitator illustrated in Fig. 2.8a is unusually low (i.e., 7 ml/min). Limitations of the thermal precipitator, in addition to low flow rate, include (1) the unknown effects of heat on deposited materials which may be altered in some way by long exposure to a thermal gradient (i.e., evaporation of liquids and oxidation of organic materials) and (2) deposition of particles in a nonuniform manner with respect to particle size, requiring examination of representative areas of the entire dust deposit.

The oscillating thermal precipitator was developed to provide a uniform deposit by slowly moving the hot wire over the slide[18] or by moving the sample collecting slide to and fro under the hot wire,[19] but our experience with currently available commercial models has been disappointing. To produce the desired uniform dust deposit, there is a need for considerable redesign to maintain continuity of sample air flow and smooth oscillatory motion. A rotary thermal precipitator appears to have merit, but no commercial models are available. It has been found that some improvement in uniformity of deposition can be achieved by substituting a broader wire.

When using the thermal precipitator for sampling from ducts, it is usually advisable, because of the low flow rate, to sample an aliquot of the fraction extracted from the main stream flow inside the duct (i.e., a sample of a sample). Hodkinson *et al.*[20] have shown little effect on particle collection from air flowing past the inlet during sampling of mine air for particle sizes of hygienic concern. Procedures for obtaining an aerosol sample for electron microscopy with the thermal precipitator are described in Chapter 3, p.

The most valuable features of the thermal precipitator are as follows: it is able to collect particles as small as 0.1 μm and below at close to 100% efficiency; collection efficiency is unaffected by high or low dust concentrations; and dust is collected in the same form as it existed in the air—i.e., because of the gentle precipitating action, aggregates are not broken up, nor are fragile granules shattered. The collected dust sample is suitable for direct viewing under the light or electron microscope.

The Casella model* of the thermal precipitator (Fig. 2.8b) utilizes a built-in water reservoir and calibrated jet aspirator to pull 7 ml/min of air through the collection slot. The wire may be heated with a small dry cell, making the instrument completely self-contained, easily portable, and safe for use in explosive atmospheres.

Kethley *et al.*[21] developed a thermal precipitation instrument which uses a pair of parallel hot and cold disks 7.5 cm in diameter. Air is introduced at the rate of 1 lpm through the center of the hot disk and flows radially outward, depositing particles on the cool surface, which may be covered with glass, paper, metal foil, or transparent film. When one disk is cooled and the other is heated, temperature extremes may be avoided and the instrument may be used for sampling viable bacteria and liquid droplets. However, care must be taken to prevent water vapor condensation on the cooled disk. A commercial version of this instrument, called the Thermopositor,† is available (Fig. 2.8c).

2-2.4 Electrostatic Precipitator

The electrostatic precipitator, like the thermal precipitator, collects almost 100% of all particles less than 5 μm in size. One of the original precipitators described by Drinker[22] is illustrated in Fig. 2.9. The precipitation tubes were

---

*C. F. Casella & Co., Ltd., London, N.1, England.

†Micromeritics Instrument Corp., Norcross, Georgia.

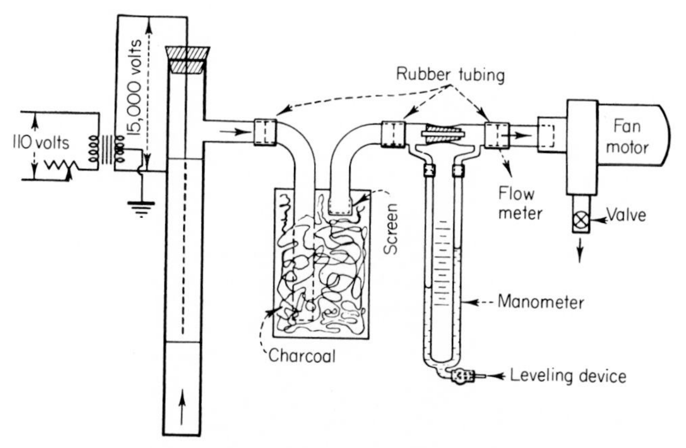

FIG. 2.9. Electrostatic precipitator. (From Drinker, Ref. 22.) Courtesy of *J. Ind. Hyg.*

made of Pyrex glass and operated at 15,000 volts obtained from a luminous tube transformer. The central electrode of gold-plated drill steel was held in alignment by means of a rubber stopper. The outer electrode was made of 12-mesh copper screen, wrapped tightly around the tube and secured in place with adhesive tape. A hand vacuum-cleaner unit was used to draw 50 lpm of air through the 3-cm-diameter precipitation tube, and a charcoal trap was included to remove ozone formed in the corona.

A commercially available electrostatic precipitator manufactured by the Mine Safety Appliances Company* is illustrated in Fig. 2.10. The tubular aluminum collecting electrode surrounds an ionizing wire electrode, and both are contained in a portable sampling head with a motor-blower assembly which draws 3 cfm through the collecting tube. The instrument operates on 110- to 115-volt, 60-cycle current and develops about 10,000 volts at the electrodes with the aid of a self-contained transformer-rectifier assembly.

Samples from the electrostatic precipitator may be examined directly under the microscope if some form of transparent liner has been used in the tubular electrode during collection. However, this is not entirely satisfactory, as the lining material usually will not lie on a flat plane even when cut into small strips. Another method is to place microscope slides between the ionizing and collecting electrodes. Bourne and Fosdick[23] have described just such an application using a hemacytometer cell instead of a plain glass slide.

The electrostatic precipitator has a tendency to agglomerate particles, but of greater importance is the tendency of the tubular form of the instrument to make a size separation of the particles along the axis of the tube such that a

*Mine Safety Appliances Co., Pittsburg, Pennsylvania.

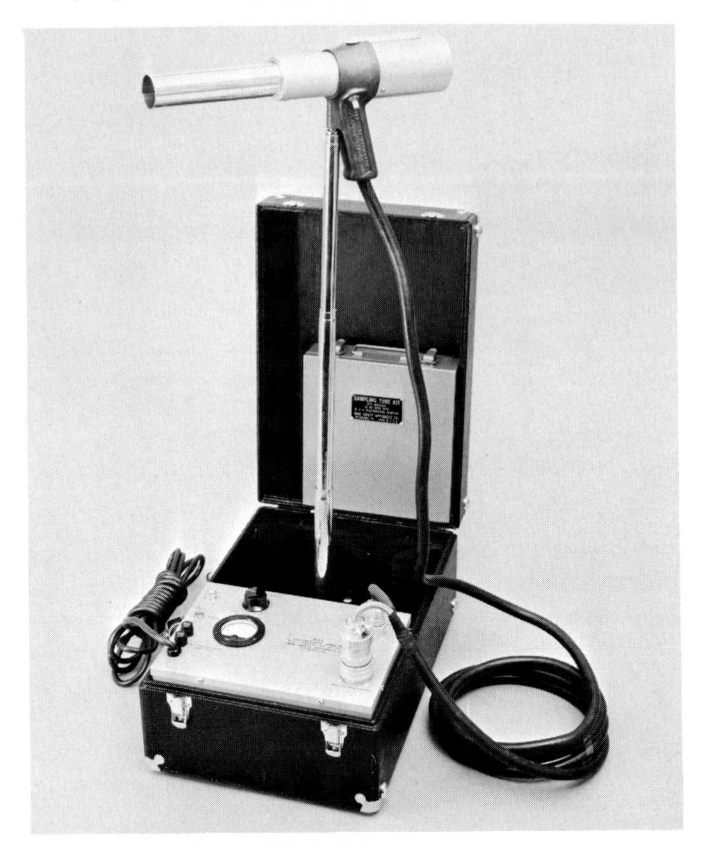

FIG. 2.10. Electrostatic precipitator. Courtesy of Mine Safety Appliances Co.

larger proportion of the finer particles are found in the front part of the tube, with a larger proportion of the larger sizes graduating toward the discharge end, although very large particles (>20 μm) will be found concentrated at the upstream end also. For a representative sample, therefore, entire strips along the longitudinal axis of the tube must be examined thoroughly. Electrostatic precipitation with a point ionizing source greatly reduces, but does not entirely eliminate, size gradations in the dust deposit. Bouton *et al.*[24] developed the point source electrostatic precipitator illustrated in Fig. 2.11. They indicated that with this apparatus an entire dust sample could be deposited on an electron microscope grid without size selectivity, thereby greatly simplifying the selection of representative fields for size analysis.

Other designs of point-to-plane electrostatic precipitators have been described by Baum,[25] Arnold *et al.*,[26] Morrow and Mercer,[27] and Billings and Silverman.[28] Reist[29] has confirmed that the size distribution collected in a

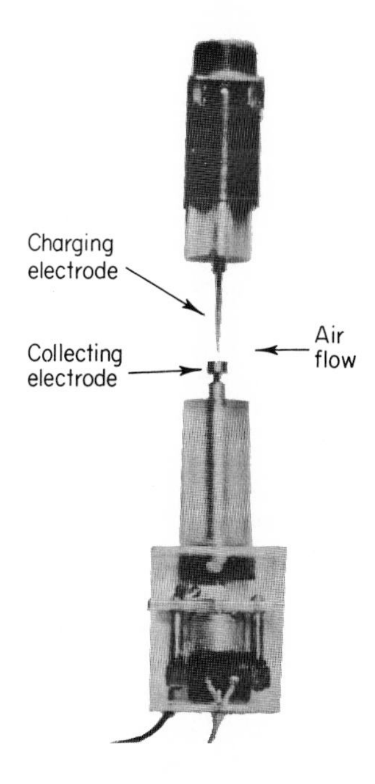

FIG. 2.11. Modified electrostatic precipitator. (From Bouton *et al.*, Ref. 24.)

point-to-plane electrostatic precipitator "represents reasonably well the actual particle size distribution of the test aerosol."

Use of electrostatic collection of particles is limited to nonexplosive atmospheres. Sampling in moisture-saturated atmospheres may require reduced voltages because of excessive arcing. Procedures that may be used to obtain an aerosol sample for electron microscopy with the electrostatic precipitator are described in Chapter 3.

2-2.5 FILTER SAMPLERS

(a) *Soluble Filters.* Porous beds of soluble or volatile materials, such as salt, sugar, resorcinol, and anthracene, have been used for many years to collect airborne dust by filtration. After sampling, the filter bed is dissolved or volatilized and the dust is left behind.

Commercial sugar contains large numbers of dust particles, and it has been found that dust-free crystals may be prepared only at the expense of great care and faultless technique. A mat-like filter made from filaments of polystyrene as small as 0.3 μm in diameter is commercially available under the trade name Microsorban.* This filter has high efficiency for submicron particles. After dust sampling, the filter fibers may be completely dissolved in organic solvents such as benzol and toluol, or, if high heat will not affect the collected sample, the filter may be evaporated in a muffle furnace.

Viles[30] was able to prepare nearly dust-free fiber beds from pure salicylic acid crystals which were capable of retaining virtually 100% of tobacco smoke. When placed in an oven at 120°C, the filters disappeared completely in a few hours, leaving nonvolatile particles deposited on the surface of a supporting microscope slide. When the size of the sample was controlled so that there was little overlap of particles, the remaining dust deposit closely approximated the condition of the particles in their original airborne state with respect to agglomeration. Volatile filters have special application for collecting low loadings of small, nonvolatile particles for microscopic sizing. Soluble filter materials have less value for size analysis, since solution of the filter and concentration and recovery of the collected dust may drastically change the size (through solubility) and state of agglomeration of the residual particles.

(b) *Membrane Filters.* These are manufactured from cellulose ester gels that are dried in the form of thin (150-μm) porous films of controlled pore size. They retain airborne dusts as small as 0.001 μm. Dust deposited on the membrane filter surface penetrates only to a depth of approximately 10 μm, and collected dusts may be examined microscopically *in situ.* Membrane filter material has a property of great value for light microscopy in that its index of refraction (1.51) is close to that of microscope immersion oil. When impregnated by a drop of immersion oil, the filter disappears optically, leaving deposited dust in sharp relief. If the size of the sample is controlled so that there is little or no piling up of collected particles, dust will deposit on the surface of the filter in the state in which it existed in air; i.e., shattering is eliminated and agglomerates are not broken.[31] Membrane filters may be used for specimens to be examined by light or electron microscopy. Methods of preparing membrane filters for electron and light microscopy are described in Chapter 3.

Membrane filters are available in a number of pore sizes ranging from 8 μm to 0.01 μm. Pore volume occupies approximately 80% of the total filter volume, and for the smaller pore sizes (i.e., those less than 1 μm in diameter) pore diameters vary only slightly from the average size. Air flow resistance of the small-pore membranes is high, and the filters must be supported in special holders to prevent rupture of the delicate, thin film.

---

*Gelman Instrument Co., Ann Arbor, Michigan.

Smooth disks of sintered glass, metal, or carbon granules have been used as filter supports for air sampling, but tightly stretched, fine-mesh, stainless-steel support screens (Fig. 2.12) serve adequately. Because of the high negative pressure required to draw 1 cfm of air through a 47-mm-diameter aerosol filter (i.e., 40 mm Hg), special care must be taken to seal the fragile membrane into the filter holder. Flat Teflon gaskets and rubber O-rings have proved best for this service.

For air containing 10 million particles per cubic foot, a sample volume of 1 cu ft will deposit approximately 100 particles per oil immersion microscope field on a 47-mm-diameter filter, a convenient number for microscopic particle sizing. If too many dust particles have been collected to be accurately sized because of interference from particle overlapping, the membrane may be dissolved in dust-free acetone and a drop of the diluted suspension mounted on a glass slide for microscopic examination.

(c) *Paper and Deep-Bed Fibrous Filters.* Many types of filters in addition to membrane filters, are used to collect large samples of airborne dusts and other particles for size analysis. In all cases, sufficient dust must be accumulated on the surface of these filters so that the removable dust cake, the part that will be used for sizing, contains a correct distribution of sizes. In other words, the nonrecoverable dust in the interstices of the filter, as well as the dust lost through the filter in the early stages of sampling before the high retention surface dust cake is formed, must represent only a negligible fraction of the total dust cake recovered for analysis. Obviously, it will be impossible to distinguish naturally occurring agglomerates from those formed after deposition of particles in the dust layer. Therefore, prior to sizing, the dust cake must be thoroughly dispersed into unitary particles without using

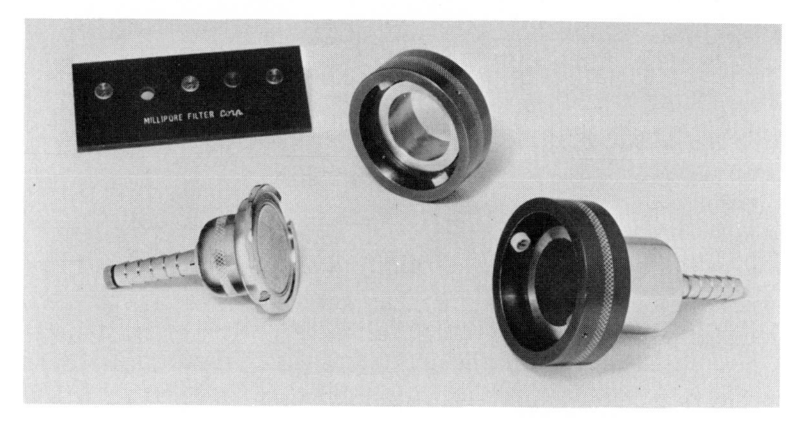

FIG. 2.12. Membrane filter holder. Courtesy of Millipore Filter Corporation.

sufficient force to fracture single grains. Many of the sizing methods described in Chapters 3 and 4 (pp. 78 and 137) may be used with bulk samples; suitable methods of dispersing bulk dust are given there in considerable detail, since this is an important step in the analysis.

Collection efficiency is an important consideration in selecting a filter for particle size sampling, since loss of the finest dust fractions may influence the validity of the analytical results adversely. Dust collection efficiencies of a number of commonly used sampling filters before formation of a highly efficient filter deposit on the surface are shown in Table 2.3. Somewhat different results will be obtained with other aerosols and at different filtration velocities. Hard-surfaced filter papers such as Whatman No. 44 resist dust penetration and make it relatively easy to strip off the collected dust layer, but air flow resistance is high initially and increases rapidly with dust deposition. Soft, porous papers such as CWS-6 and HV-70 have exceptionally good particle collection characteristics and only modest rates of air flow resistance increase, but it is difficult to separate the dust for analysis without considerable contamination by filter fibers. Binderless, all-glass sampling papers made from superfine fibers (e.g., MSA 1106B and AEC all-glass) have the same high efficiency and low-pressure loss characteristics as CWS-6 and HV-70, but the harder, smoother surface makes it much easier to recover the accumulated dust cake.

Papers manufactured from superfine glass fibers are preferred for sampling in high humidity conditions, for temperatures up to 900°F, and for sampling corrosive atmospheres. At temperatures above 900°F, Fiberfrax* papers are available that will withstand temperatures as high as 3200°F for short periods.

Thimble-shaped filters are convenient for use with sampling probes for extracting dust samples from ducts and stacks. A typical sampling device for this purpose is shown in Fig. 2.13. It may be noted that the filter thimble is located in the nose of the sampler rather than outside the duct. This arrangement avoids loss of particles of unknown size in the bends and walls of a long probe. Whatman† and S and S‡ cellulose filter thimbles are suitable for sampling at temperatures as high as 350°F for brief intervals (5 to 10 min) but should not be used at temperatures above 250°F for long periods. Glass fiber thimbles are available from S and S‡ which are resistant to most corrosive chemicals except hydrogen fluoride and strong alkalies and can tolerate temperatures up to 900°F for long periods. Ceramic thimbles,

---

*Carborundum Corp., Niagara Falls, New York.

†W. & R. Balston, Ltd., Maidstone, Kent, England.

‡Schleicher and Schuell Co., Keene, New Hampshire.

TABLE 2.3
COLLECTION EFFICIENCY OF COMMONLY USED SAMPLING FILTERS

Filter medium	Flow rate 20 linear ft/min		Flow rate 50 linear ft/min	
	Atmospheric dust count efficiency (%)[a]	DOP optical efficiency (%)[a]	DOP optical efficiency (%)[a]	Uranine dye weight efficiency (%)[b]
Whatman No. 1	50.0	57.0	89.0	—
4	15.0	23.0	38.0	93.4
32	99.1	99.5	99.7	—
40	85.1	84.0	97.2	—
41	26.5	23.0	33.0	97.2
41H	24.0	19.0	35.0	—
42	98.8	99.2	>99.9	99.6
44	97.0	98.6	99.8	—
50	92.0	97.0	99.8	—
540	67.0	65.0	—	—
S and S No. 640	13.0	15.0	33.0	—
HV 70 9 mil	96.5	96.5	98.3	98.2
18 mil	99.5	99.3	99.5	—
MSA Type S	46.0	41.0	49.0	—
1106 B	—	—	—	>99.9
Millipore HA	—[c]	>99.9	99.8	—
AA	—[c]	>99.9	>99.9	>99.9
RA	—	—	—	>99.9
Gelman AM-4	—	—	—	>99.9
AM-1	—	—	—	96.2
S and S Ultrafilter	—[c]	—	—	—
Hurlbut glass paper	—[c]	>99.99	>99.99	—
CWS No. 6	—[c]	>99.99	>99.99	—
AEC No. 1	—[c]	>99.99	>99.99	—
AEC glass–asbestos	—[c]	>99.99	>99.99	—
AEC all–glass	—[c]	>99.99	>99.99	—

[a]Data from Ref. 32. DOP test aerosol, 0.3–μm monodisperse droplets.

[b]Data from Ref. 33. Uranine dye aerosol, 0.27–μm mass median diameter. Standard deviation 1.8.

[c]No particles found after 6 hr of running.

available from Norton Co.,* are suitable for temperatures in excess of 2000°F. The hard, smooth inner surfaces of ceramic thimbles make quantitative recovery of the dust relatively simple. Thimbles formed from cellulose or glass fibers have poor efficiency for submicron-sized particles when clean,

*Worcester, Massachusetts.

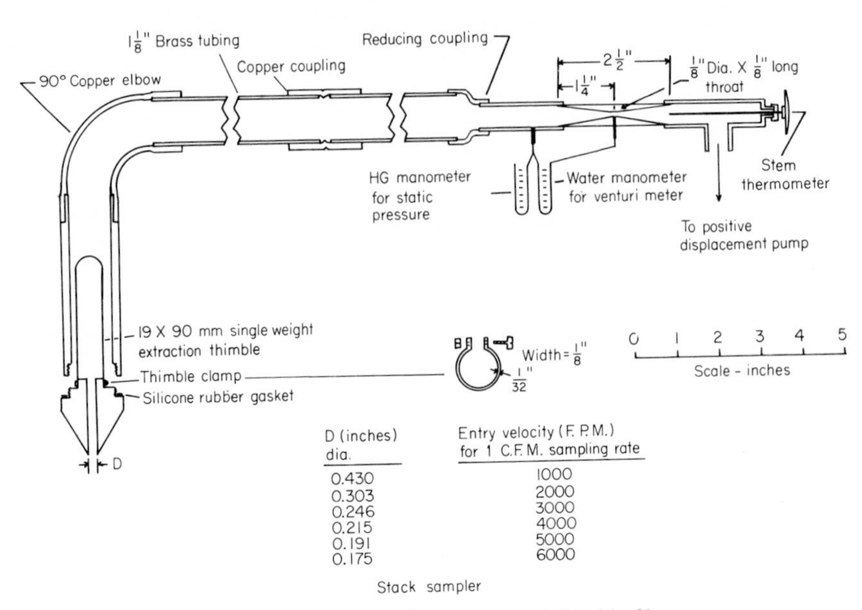

FIG. 2.13. Stack sampling probe and thimble filter.

but they rapidly accumulate a surface dust layer which retains small particles quantitatively.

Filter holders similar to those described for use with membrane filters (Fig. 2.12) are suitable for filter paper disks of all types. For larger air volumes, the high-volume sampler shown in Fig. 2.14, with an adapter to accommodate an 8 × 10-in. glass filter, is used widely inside industrial plants and for outdoor sampling for air pollution evaluations.[35]

FIG. 2.14. High-volume air sampler. (From Silverman and Viles, Ref. 34.)

Deep beds of fibrous filters composed of natural (cotton), synthetic (Dynel, Saran), or mineral fibers (glass, silica, Fiberfrax) of small diameter make excellent low-resistance, high-efficiency dust filters. Nevertheless, they are of little value for dust sampling for particle size analysis, since the collected dust tends to become size-segregated as it deposits throughout the depth of the very porous filter bed (with the smallest particles depositing near the discharge end and the coarsest near the entry end). This makes it virtually impossible to recover a representative sample of the dust for size analysis. In all respects, except high air flow resistance, many of the high- and ultra-high- efficiency filter papers (cellulose, glass, and ceramic) described earlier are greatly superior for securing dust samples for size analysis and should be selected over deep porous beds of filter fibers for this purpose.

2-3 Collection Methods Which Classify Particles into Size Groups

Instruments which effect a size separation of the particles during sampling include cascade impactors, winnowing devices such as the Conifuge and Aerosol Spectrometer, and samplers utilizing spiral and helical air passages. None of these sampling instruments make a sharp size separation (as might be expected from a sieve), but rather particles are deposited on a target in a continuous or discontinuous size array, or size spectrum. The advantages associated with size-selective samplers are related to simplifications in the analytical procedures that they make possible. These are reviewed in the sections which follow.

2-3.1 Cascade Impactors

May's cascade impactor[36] is a jet impactor developed for the specific purpose of effecting a size separation of airborne particulate materials. The instrument (Fig. 2.15) employs a series of jets of decreasing width. As the jet area decreases progressively, whereas air is drawn through the instrument at a constant volume rate, the velocity (and therefore the efficiency of collection for smaller and smaller particles) increases at each successive stage. In this fashion, the total aerosol sample is divided into a number of size-segregated but overlapping distributions, each appearing on a separate stage. After calibration, size–number distributions can be determined by microscopic counting of the particles deposited on each stage, and this instrument may be used advantageously for aerosols with particle size distributions covering a wide range.

The cascade impactor has an important advantage for materials that can be analyzed easily by chemical or physical methods, because once each stage

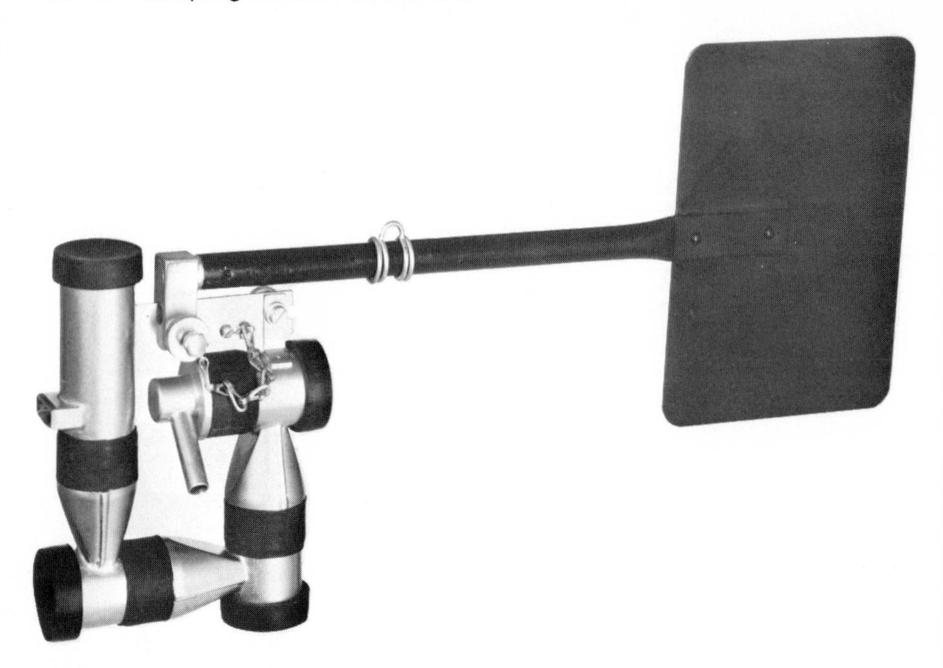

FIG. 2.15. May's cascade impactor, original Casella commercial model. (From May, Ref. 36.)

has been calibrated for a given material (i.e., after the size median of each stage has been determined by microscopic examination), it is necessary only to determine the total weight of material collected on each stage in order to translate this information, in conjunction with the calibration values, into a size distribution by weight for the original aerosol. This device is suitable for use with dust or liquid droplets. Collection characteristics of each of the four jets of May's cascade impactor when sampling liquid droplets of unit density at an aerosol flow rate of 17.5 lpm are given in Fig. 2.16. May[36] states that this graph "shows the percentages of particles of each size which pass through the four jets. When the four jets are operated in series, the maximum drop sizes found on the second, third, and fourth slides are 20 μm, 7 μm, and 3 μm, respectively, these being the minimum sizes impacted with 100% efficiency by the previous slide." An advantage of using graded impaction stages is that liquid droplets, naturally occurring agglomerates, and fragile particles are not likely to be shattered by exposure to excessive impaction velocities, since the largest particles are deposited in the initial low-velocity stages.

The penetration curves for May's cascade impactor when operated at normal temperature and pressure, shown in Fig. 2.16, may be corrected for

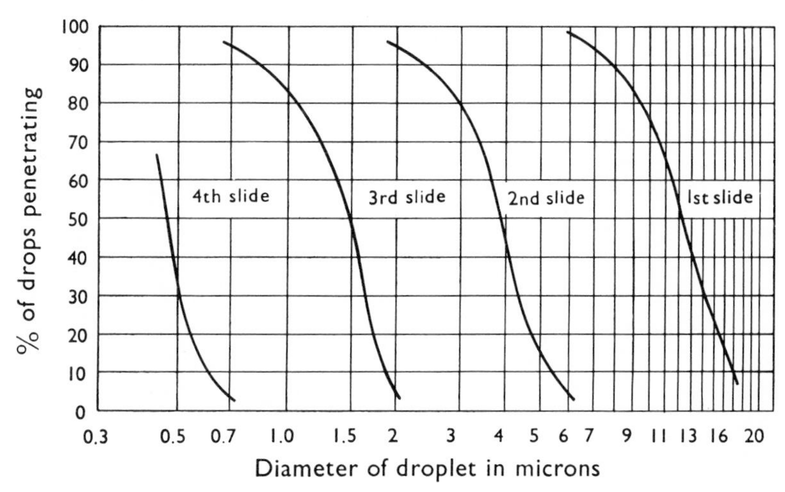

FIG. 2.16. May's cascade impactor: collection characteristics. (From May, Ref. 36.)

other sampling flow rates and for particles of different specific gravity with the following equation, which relates impaction parameter *(I)* for equal impaction efficiency to particle density *(ρ)*, particle diameter *(D)*, and jet velocity *(V)* in the following manner:

$$I \simeq \rho D^2 V \qquad (2.1)$$

Thus, from Fig. 2.16, the percentage of 3-μm drops of unit density penetrating stage 2 is 80% (i.e., 20% collection). Therefore, the size of silica particles (specific gravity 2.5), 20% of which will be retained on this stage, is given from equation 2.1 as $I = (1)(3)^2V = 2.5(D^2)V$. As velocity through the second jet is unchanged, the diameter of silica particles will be

$$[1(3)^2/2.5]^{1/2} = 1.9\ \mu\text{m}$$

The penetration curves shown in Fig. 2.16 apply only to May's model of the cascade impactor. Similar instruments, described by Mercer,[37] Cohen and Montan,[38] Couchman,[39] and others, have a different number of stages and modified jet geometry.

Cascade impactors are available from commercial sources. The Casella* model of May's instrument is shown in Fig. 2.17 along with the Bendix-Unico† and Scientific Advances, Inc. (Battelle)‡ versions. The last is a six-stage instrument.

---

*C. F. Casella & Co., Ltd., London, N.1, England.
†National Environmental Instruments, Inc., Fall River, Massachusetts.
‡Scientific Advances, Inc., Columbus, Ohio.

The fourth and most efficient stage of May's cascade impactor removes solid and liquid particles having a density of 1 g/cm^3 down to approximately 1 μm in size. To collect finer materials and extend the range of the instrument, it is common practice to add a fifth stage consisting of an impaction jet operating at sonic velocity, a very efficient filter, or an electrostatic precipitator. All the commercial impactors shown in Fig. 2.17 are equipped with a final-stage filter holder. This is an aid in constructing size distribution curves, as it gives an additional point. It will be obvious that for any given aerosol the size mean of the first and last stages is dependent on the extreme ranges of the size distribution originally present, whereas material caught on the intermediate stages will have known size limits but the mean particle size on each stage will not be known with certainty except when the particle sizes follow an assumed statistical relationship, such as the logarithmic probability distribution, which occurs frequently in nature. For this reason the cascade impactor must be calibrated microscopically for each aerosol. It has become customary to calibrate impactors with monodispersed spherical particles to obtain a calibration in terms of aerodynamic diameters. In cases where a given aerosol need be sized only once, it is obviously more economical to do so by direct visual methods than to attempt to calibrate the stages of the

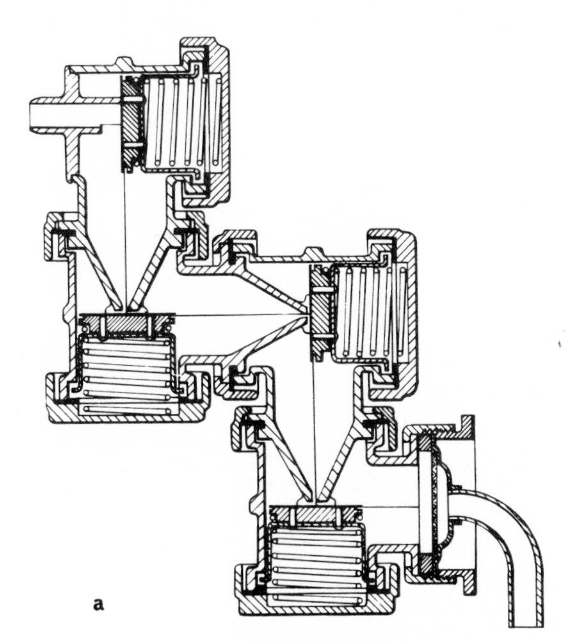

FIG. 2.17. Commercially available cascade impactors: *(a)* Casella cascade impactor. Courtesy of C. F. Casella & Co., Ltd. *(b)* Bendix-Unico cascade impactor. Courtesy of

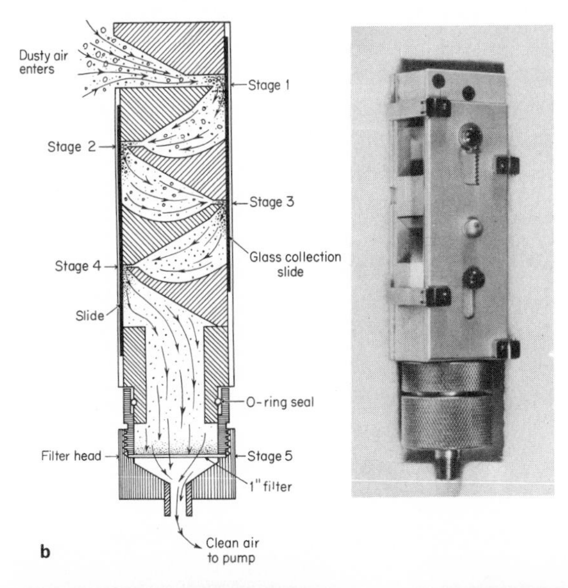

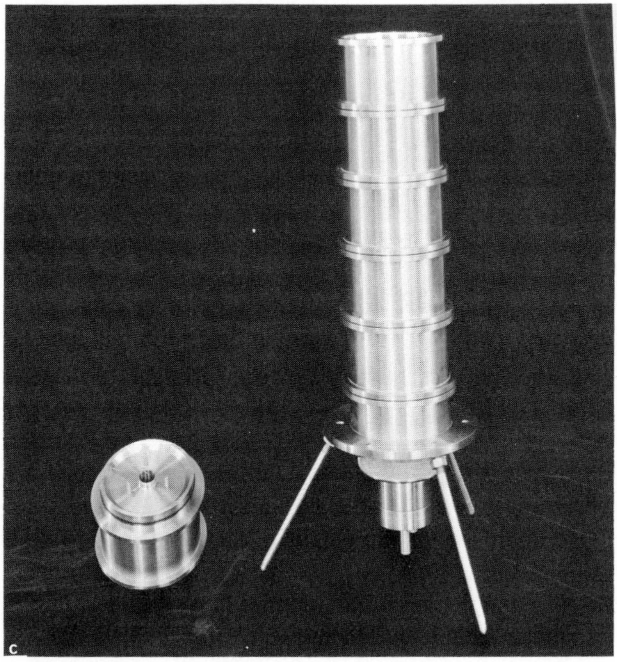

National Environmental Instruments, Inc. *(c)* Battelle cascade impactor. Courtesy of Scientific Advances, Inc.

impactor. On the other hand, for control work in connection with toxicological studies or for repeated size analyses in routine plant sampling, it is advantageous, especially when the nature of the particulate matter makes it possible, to calibrate the cascade impactor with the aerosol of interest and then proceed with size distribution analyses by simple chemical or physical techniques, such as radioactivity.

To apply the cascade impactor to isokinetic sampling from stacks and ducts, the first stage may be constructed in the nose of a stack sampling probe as shown in Fig. 2.18. The inlet end of the probe is removable, and openings of different sizes can be substituted so that isokinetic samples can be taken without altering the total air flow rate associated with the instrument calibration. Stages 2, 3, and 4 may be located outside the duct and connected to stage 1 by means of an extension tube of convenient length for the duct size.

It has been noted[40] that a sizable fraction of the test suspension deposits on the walls of all stages of May's cascade impactor. This is not confined to the first stage, as was stated by May,[36] but occurs throughout the instrument, even in the last stage when all the larger particles have been removed. Lippmann[41] also found losses up to 30% by number on the walls of all stages of the Bendix-Unico impactor.

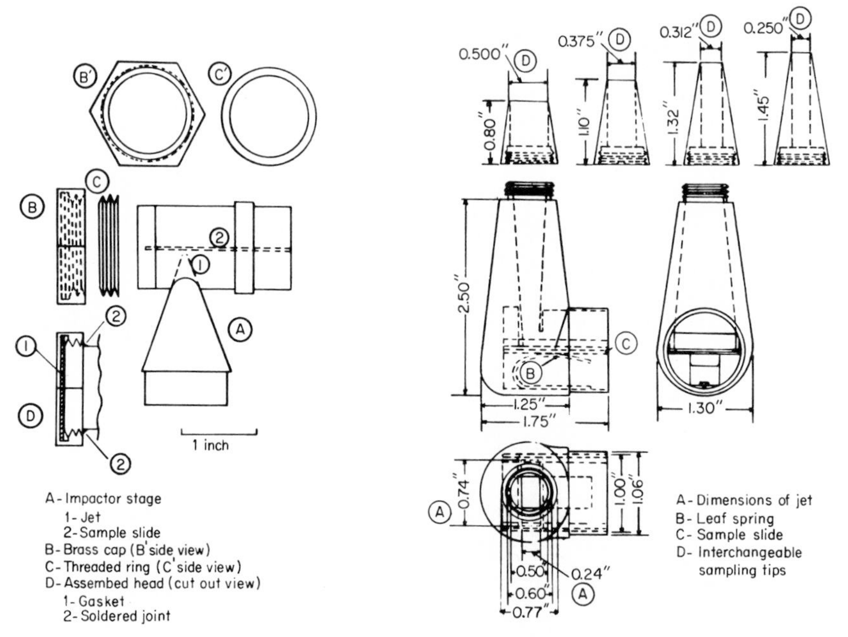

FIG. 2.18. Modified first stage of May's cascade impactor. (From First *et al.*, Ref. 40.)

Deposition is probably caused by turbulence induced after each jet stage, but in any case it should be recognized that the sum found on each of the slides does not represent the total sample. May[36] found that 20% or more of the total sample collected by four stages could be recovered from the walls of the initial stage when sampling clouds containing droplets >50 μm. This situation is accentuated when an inlet probe must be attached to the cascade impactor to withdraw a sample from a duct, and considerable deposition takes place before the aerosol reaches the first stage. Some typical measured results for the unit shown in Fig. 2.15 are shown in Table 2.4. A question naturally arises as to the size of the particles deposited in the probe and on the walls of each of the stages. Lippman[41] assumed that the material deposited on the walls is representative of the original size distribution and that it need not be included in calculating the percent of deposited material in each size range, but it has not been proved that this assumption is correct.

TABLE 2.4
DEPOSITION OF DUST ON CASCADE IMPACTOR INLET PROBE AND BODY WALLS

	Percent of total weight on walls and probe	
Sample No.	Commercial instrument[a]	Modified instrument
1[b]	83	49
2[b]	84	37
3[c]	20	14
4[c]	19	14
5[c]	23	15

[a]No probe; all deposition on body walls.
[b]Size range 1 to 20 μm.
[c]Size range 1 to 10 μm.

A consideration of the size retention characteristics of the cascade impactor indicates that it is poorly adapted for defining the size distribution of monodispersed particles (such as are commonly prepared for laboratory investigations in which the uniform size properties of dusts or mists play a major role). For example, in Fig. 2.16 the jets do not make a sharp size separation but remove varying percentages of all sizes present in the cloud. This is a characteristic of all types of jet impactors. If a hypothetical aerosol containing uniformly sized droplets of 1.5-μm diameter and unit specific gravity were to be sampled with the Casella cascade impactor, the distribution of collected particles would occur (in accordance with the penetration curves of Fig. 2.16) as shown in Table 2.5. This tabulation indicates that the median size is approximately 1.5 μm (which is correct), but it yields a false idea of the size distribution of the material. When the points are plotted on logarithmic probability paper (Fig. 2.19), the line of best fit shows a standard geometric deviation of approximately 1.6, when in fact it should be 1.00. This constitutes a very large discrepancy. These hypothetical, uniformly sized droplets represent an extreme case for May's cascade impactor, but the same factors are present to a lesser degree when dealing with aerosol particles having other diameters and a fairly uniform size. Other cascade impactors with six to ten stages that cover the same size range as May's four-stage instrument give less size spread on each stage and minimize the uncertainty in defining the standard deviation of monodisperse aerosol particles. Basically, not even an ideal impactor with a step function efficiency curve could identify a monodisperse aerosol with the entire sample deposited on a single stage. It would merely indicate that all particles had diameters between the upper and lower cutoff values. These difficulties do not occur with the majority of particle size distributions because they cover a wide size range.

TABLE 2.5
APPARENT SIZE DISTRIBUTION OF UNIFORMLY SIZED PARTICULATE CLOUD

Stage	E.D.S. (μm)	Stage efficiency for 1.5-μm droplets (%)	Number reaching stage (based on 1000 particles)	Number of particles retained on stage	Cumulative % equal to or less than stage size
1	—	0.2	1000	2	100.0
2	13.0	2.0	998	20	99.8
3	4.0	48.0	978	469	97.8
4	1.7	98.5	509	502	50.9
Absolute filter	—	100.0	7	7	0.7
			TOT.	1000	

Confusion in the interpretation of cascade impactor data can exist, since some investigators have expressed their calibration results in terms of "effective particles sizes of each stage" and others as "the particle median size of each stage." In the former method, the total material collected on each stage is represented by the "effective particle size" value, while in the other method half the weight on each slide is considered to be greater, and half less, than the stage median size. These two methods yield different results in practice, but this is not so serious a matter as calculating results by one method when the impactor has been calibrated by the other.

In spite of many non-ideal characteristics of jet impactors, they have

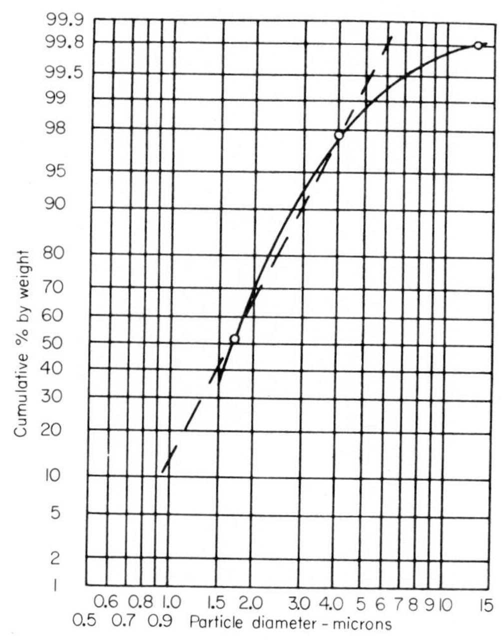

FIG. 2.19. Size distribution curve of uniformly sized droplets from cascade impactor. (From First, Ref. 42.)

proved to be valuable sampling and analytical instruments in practice. The theory of jet impactors is discussed in Chapter 1.

2-3.2 WINNOWING SAMPLERS USING GRAVITATIONAL AND CENTRIFUGAL FORCE

Sawyer and Walton[43] developed a centrifugal dust and mist sampling device called a Conifuge. The apparatus is designed for sampling airborne particles and is so arranged that the particles are deposited on the collecting surfaces in order of their settling velocities. For water droplets, the size deposited by the instrument ranges from 0.5 to 30 μm in diameter, whereas for materials of higher density there is a downward shift in the effective size range. With solid spherical particles, a complete size separation is obtained for all sizes up to 20 μm. With larger sizes, separation is less perfect. For liquid particles, the uniformity of the size separation is partially masked by uneven spreading and coalescene of drops which become superimposed during deposition. For irregularly shaped particles, separation is related to particle settling velocity and the microscopic appearance of the deposit is markedly less uniform.

Referring to Fig. 2.20, a diagrammatic section of the Conifuge, Sawyer and Walton[43] describe the operation of their instrument in the following manner:

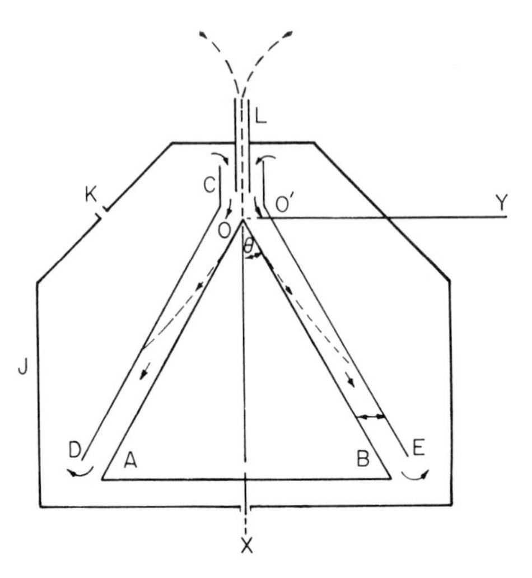

FIG. 2.20. Diagrammatic cross section of Conifuge. ⟶ clean air; - - → particulate cloud. (From Sawyer and Walton, Ref. 43.)

"(OAB) represents an inner solid cone, and (CDE) an outer conical shell, of the same vertical angle, which are rotated together about their common axis (OX). These are enclosed within a stationary outer container (J). Rotation of the cones causes a circulation of air through the annular space between them, air entering at the apex (C) and being expelled through the annulus (DA–BE) at the base. A controlled amount of the circulating air is allowed to leave the outer container through a small orifice (K), causing an equal amount of air to enter through the tube (L) onto the apex (O) of the inner cone, whence it streams over the surface of the latter in a symmetrical film separated from the outer cone by the larger quantity of clean air circulating through the system. Suspended particles in the sampled air, on arriving in the annular space between the cones, are subjected to a centrifugal force due to the rotation, which causes them to move radially outward across the streamlines of air flow at speeds proportional to (and greatly in excess of) their settling velocities under gravity. At the same time the particles are carried forward by the air flow so that each strikes the outer cone at a distance from the apex (O) characterized by its settling velocity."

Tillery[44] has investigated the use of a modified Conifuge for evaluating airborne radioactivity. The modification consists principally of a double instrument inlet tube which makes it possible to introduce the sampled aerosol into the centrifugal chamber in the form of a thin film separated from the deposition surface by a thick layer of dust-free air. This arrangement produces a sharper size separation, since all particles must traverse an equal radial distance before deposition.

Goetz and Kallai[45] developed a centrifugal particle separating and collecting instrument called an Aerosol Spectrometer, which operates on a principle similar to that of Sawyer and Walton's Conifuge. In this instrument, also, the collecting surface rotates with the winnowing air stream to maintain laminar flow conditions in the helical separating channel. The objective is to prevent the particles from experiencing a shearing action from the air stream prior to and after their precipitation, so as not to "disturb seriously the dynamic equilibrium of loosely sorbed or condensed components on the particle surface."

A schematic cross section of the Aerosol Spectrometer is shown in Fig. 2.21. The helical rotor functions as an air pump as well as a centrifugal particle separator. The maximum air flow rate through the helical channel depends on the pitch of the helix and its rotational speed. For lesser flows, a flow restriction may be placed at the channel exit. The authors describe the operation of their instrument in the following manner:

"The rotor functions not only as centrifugal separator, but effects also an impelling action by the helical channels in *downward* direction since the rotation is anti-clockwise and the helix represents a "right-hand" screw. This causes the induction of air through the *stationary* intake tube (E) into the tubular upper end of the rotor by ports (O) at the bottom of (E), which meet with the channel inlets (I, II) located at the upper end of the inner wall of each channel.

"The air flow is then propelled along the helical channels for two and one half turns (5π) which lead to the ring-shaped base of (R).

"This ring rotates freely, though at close tolerances, within the circular recess of the stationary base (P1). The free space between both parts connects with the outlet (A) through which the air escapes. The flow velocity through each channel is controlled independently by exchangeable orifices (D) before it leaves the rotor. Four pairs of such orifices 20-, 30-, 40-, 60-mil. i.d.) facilitate an adequate variation of the flow rates at different rotor velocities and thereby of the exposure to the centrifugal field. Each set of jets (D) is matched by a restricting orifice (H) inserted into (A), in order to balance the dynamic pressure conditions within the circular channel in the base with those prevailing in the space surrounding the rotor.

"At the upper end the rotor is guided by the air-lubricated "frictionless" bearing assembly (P).

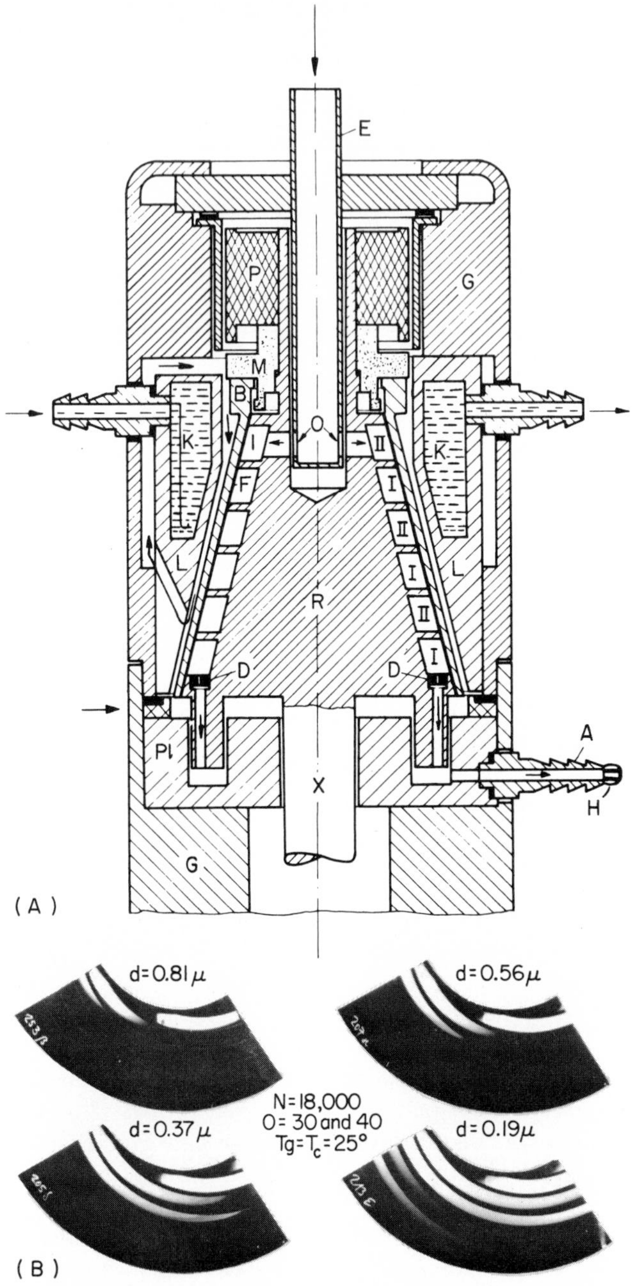

FIG. 2.21. Schematic cross section of Aerosol Spectrometer. (From Goetz and Kallai, Ref. 45.)

"The entire rotor assembly is contained in the massive, removable shell (G) which is rigidly attached to the base by bayonet fittings and is sealed on top by a lucite window for stroboscopic observation of the bearing (P).

"The motor (brush-type; AC, 135-V max, 0.25 hp), mounted in the lower part of the housing (G), supports and drives the rotor by the shaft (X) and also the armature of an electric tachometer.

"To minimize a change of the particle size and mass by thermal dissociation or volatilization of condensates a special refrigeration system maintains the rotor temperature within (±0.2°C) of that of the flow channels. Its function is briefly as follows:

"The conical insert (L) which represents the interior of the housing (G) surrounds the outer rotor surface (B) by a small clearance (1–2 mm). On the outside, (L) carries around its circumference 9 vertical channels which communicate with the gap between (L) and (B). This facilitates an air flow in downward direction along (B) which is effected by centrifugal action of the outer rotor surface while the recirculation in upward direction is facilitated through the channels between (L) and (G) (arrows in Fig. 2.21). (L) contains the cavity (K) to which a continuous flow of a coolant liquid from a separate refrigerator unit (cf.) is supplied at an adjustable rate."

The pressure relationships inside and outside the spinning cone are critical for proper operation of the instrument, and elaborate instrumentation is required to measure and adjust them. These are described in the article cited. It is the conclusion of most people who have attempted to use this instrument that it requires an unusual degree of skill. Raabe[46] has made a detailed study of calibration procedures for this instrument using polystyrene spheres. The exterior of the helical cone is covered with metal foil, on which the particles deposit.

The maximum rotational speed obtainable with this instrument is about 26,000 rpm, equivalent to 30,000 times the acceleration of gravity. The intense centrifugal field makes it possible to precipitate particles from 5 μm down to 0.01 μm and to distribute them on the spiral foil according to size. A typical foil with particle deposit is shown in the insert of Fig. 2.21. Deposits at precise locations progressively remote from the inlet may be measured with appropriate counting instruments for radioactive particles and by chemical or physical analytical methods (on known sections of the tape) for nonradioactive substances. The location of the deposit on the foil is indicative of the Stokes settling diameter of the particles. The equivalent Stokes diameter is calculated from the angular velocity of the helical cone and the air flow rate after suitable instrument calibration.

It appears possible to collect particles on membrane filters for microscopic observation or on suitable targets for electron microscope examination to check the calibration. The recommended method of calibration[47] is to use monodispersed aerosols of known size (e.g., polystyrene latex) and to note the central position and spread of the deposit on the collecting foil at various instrument settings. As the aerosol entirely fills the spiral air flow channel of the Aerosol Spectrometer, particles of identical size, shape, and density are deposited in a band depending on where they were located at entry relative to the depositing surface. Particles located near the outer surface on entry to the instrument deposit before those that enter at the same time but near the inner boundary of the spiral flow passage. For this reason, very fine particles, less responsive to radial movement from applied centrifugal force, deposit over a relatively wide band, whereas coarse particles are grouped more closely. Additionally, this produces considerable overlap of sizes at all points on the spiral deposit surface. The Conifuge largely overcomes this problem by drawing the aerosol into the instrument in a thin film along the inner wall and separated from the depositing surface by a layer of dust-free air. Thus, all particles start their outward migration from approximately the same radial distance away and hence particles of like size (or settling velocity, for nonspheres) tend to deposit in a narrow band at the characteristic distance down the cone.

The Conicycle (Fig. 2.22) described by Wolff and Roach,[48] is a simple centrifugal winnowing device with a cylindrical chamber (B) used as a selective sampler to simulate the human respiratory system. It has an inlet baffle (S) which prevents entry of particles above a preselected size depending on the air flow rate through the instrument. A flow rate of 10 lpm will permit particles less than 1 μm in Stokes diameter to pass through and prevent the entry of those larger than 7 μm. The air flow passages are noted by the letters A-A′, B, and O-O′ in Fig. 2.22. Because of its selective sampling properties, the device has little application for general particle sizing in its present form but may be used to determine the fraction of particles (by number or by weight) in the respirable size range.

A sampling instrument which uses gravitational instead of centrifugal force to deposit winnowed particles in a size array from 1 to 100 μm has been described by Timbrell.[49] An exploded diagram of the commercial instrument* is shown in Fig. 2.23A. The instrument is airtight except for the sampling orifice and operates in the following manner. The small air pump driven by a constant-speed electric motor (3) circulates clean winnowing air through the dust filter (1) and around the instrument at a constant rate in the directions indicated by the arrows. The flow rate (of the order of 100

*Fleming Instruments Ltd., Stevenage, England.

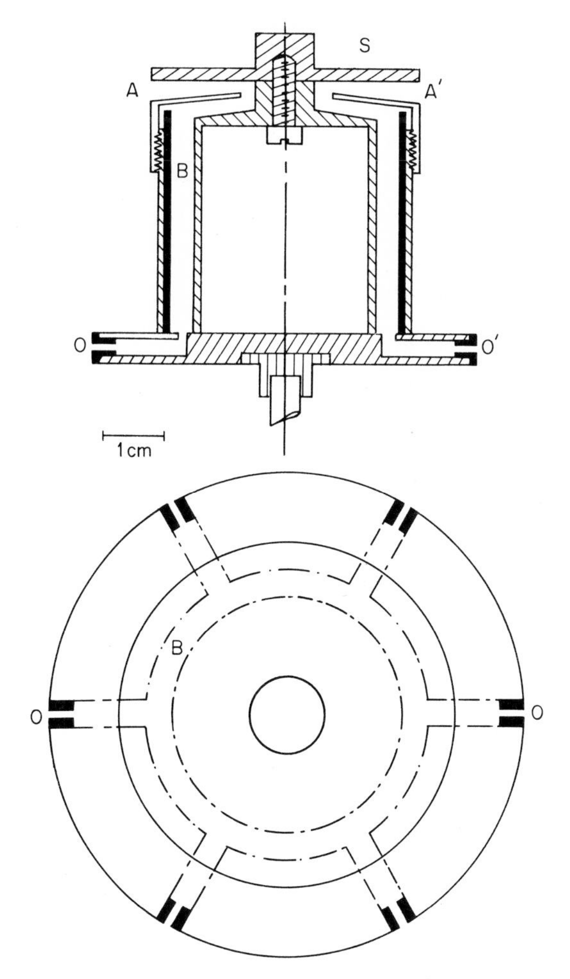

FIG. 2.22. Cross section and plan of Conicycle sampling head. (From Wolff and Roach, Ref. 48.)

ml/min) is measured by a rotameter and can be set to the desired value by means of a needle valve.

Fig. 2.23B is a rectangular fan-shaped aerosol passage milled into the lid of the instrument housing. This winnowing section is 45.7 cm long and 1.3 cm deep throughout, but it is 0.61 cm wide at the narrow, sample inlet end and 7.75 cm wide at the outlet end. Sampled air, at a controlled rate of 1 to 2 ml/min, is drawn into the instrument with the aid of a piston-cylinder aspirator pump driven by a manual, spring-operated time-switch clock mechanism. The aspirator provides constant sample flow for periods up to

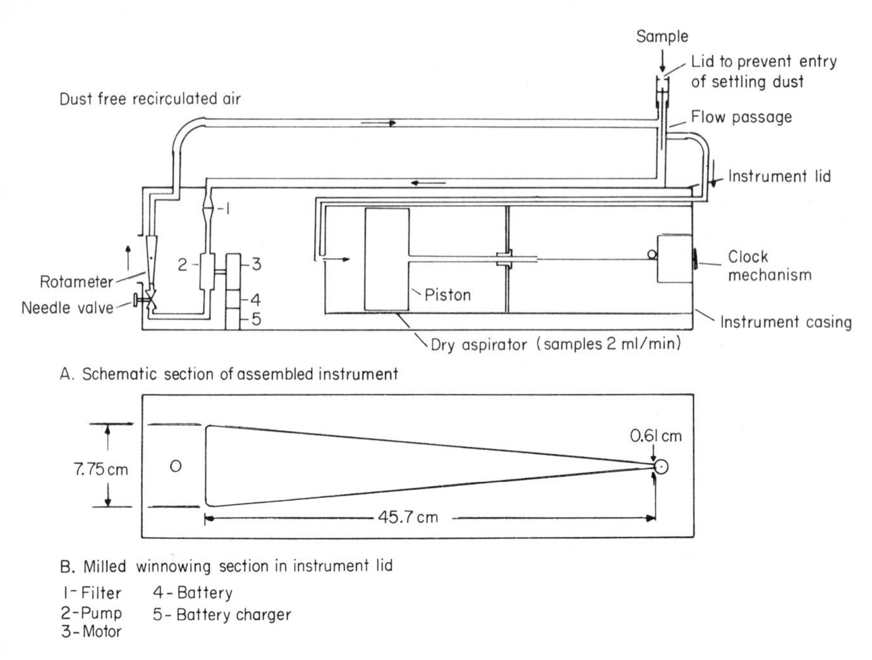

FIG. 2.23. Gravitational terminal velocity dust sampling instrument. (After Timbrell, Ref. 49.)

6 hr, and a scale on the exterior of the instrument casing indicates total volume aspirated. The two air streams (i.e., the sample flow at 1 ml/min and the recirculated dust-free winnowing air at 100 to 150 ml/min) mix before entry into the dust deposition chamber.

Before sampling, six 1 × 3-in. glass microscope slides are placed end-to-end on a rubber-covered recess in the top of the main instrument housing, which is directly below the centerline of the fan-shaped settling chamber when the lid is in sampling position. Strips of ¼-in.-wide adhesive plastic tape are placed along each side of the row of slides to attach them firmly in place before the lid is replaced.

To sample, the instrument is placed in a perfectly horizontal position, and the recirculating winnowing air pump and sample air aspirator are started. After operating for the required sampling period, the pumps are stopped and the slides on which dust has deposited are removed for examination.

All sampled dust particles, after entering the winnowing chamber, have a horizontal velocity equal to that of the winnowing air and settle under gravity at their individual terminal velocities. The lower the terminal velocity of a dust particle, the further it is carried along by the winnowing air before being deposited on a glass slide. Particles of the same terminal velocity lie on a line perpendicular to the length of the strip. The filter prevents particles of less than 1 μm, which are not deposited, from being recirculated.

The quality of the separation of particles according to terminal velocity has been demonstrated with solid spherical particles. Fig. 2.24 shows a photomicrograph at a site in the deposit where the diameter of collected glass spheres is 10 μm. The separation for other diameters is reported by Timbrell to be of similar quality. Once the instrument has been calibrated with homogeneous spheres of known density, the equivalent Stokes diameter of an irregular particle can be determined from the position of the particle in a deposit. Alternatively, calibration for nonspherical particles may be conducted by collecting under the same operating conditions solid spherical particles of known density in the same deposit as the irregular particles. When the deposit is scanned under the microscope, the average diameter of the spheres at each location can be used as an index of constant terminal velocity.

Even at the lowest size setting, few particles less than 1 μm will be precipitated by the gravitational field, and this limits the usefulness of this instrument for very small particle sampling.

Hatch and Hemeon[50] and Yaffe *et al.*[51] have described simple helical and spiral collecting devices for the separation of dusts into a continuous spectrum of overlapping graded sizes. One instrument (Fig. 2.25) contains an air passage in the form of a coil of glass tubing. As the dusty air advances through the coil, smaller and smaller dust particles are separated onto the

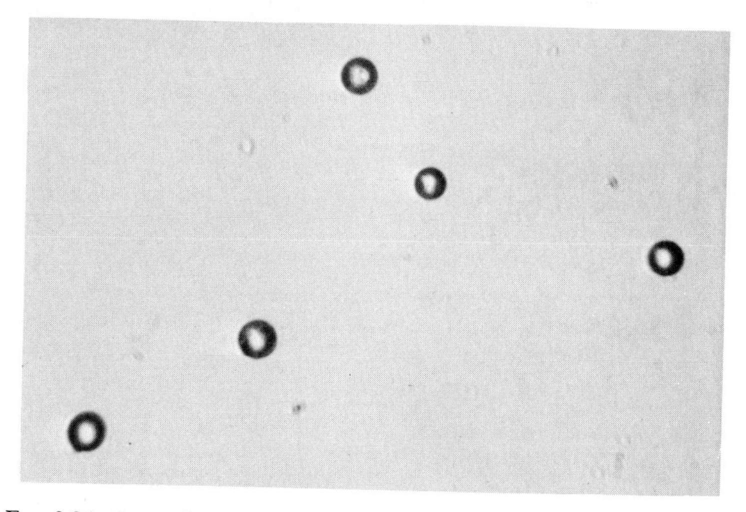

Fig. 2.24. Pyrex glass spheres of diameter 10 μm. (After Timbrell, Ref. 49.)

walls by centrifugal force. For analysis, the glass coil is broken into a number of segments, the particles are washed out, and the dust count is related to the location in the coil. The other device (Fig. 2.26) utilizes a helical passage lined with transparent film on which are deposited smaller and smaller particles as the helix tightens. The count density at selected locations on the film may be used to determine mean size and size distribution after the device is suitably calibrated. Just as for the cascade impactor, these types of devices may have advantages for the analysis of materials for which there are simple chemical or physical tests. The size separation of particles along the glass coil or plastic film is by no means sharp, however, and some question exists as to whether these methods offer any real advantages in the determination of size parameters by microscopic methods.

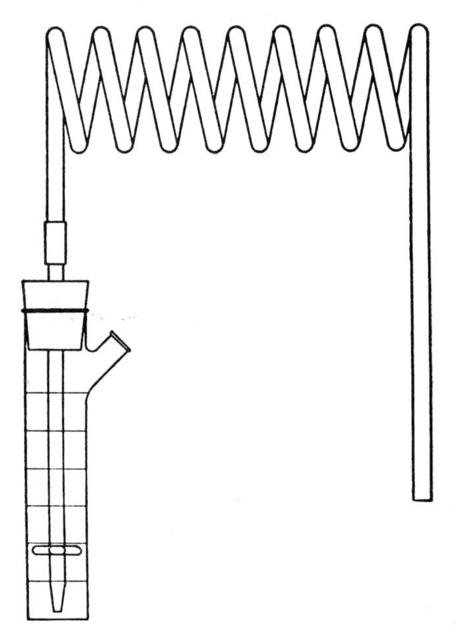

FIG. 2.25. Spiral inlet to impinger. (From Hatch and Hemeon, Ref. 50.)

Size-selective samplers which seek to duplicate the retention characteristics of the human lung are of great usefulness for assessing the hygienic risk associated with airborne dusts and mists, but, as was pointed out in the discussion of the Conicycle (p. 62), they are of limited value for collecting particles for general size analysis, since the upper and lower sizes are missing or inadequately represented. Collecting instruments of this type have been reviewed in considerable detail by Hatch and Gross.[52]

FIG. 2.26. Spiral sampler. (From Yaffe *et al.*, Ref. 51.)

Miniature cyclones may be used for collecting large amounts of dust from aerosols in a dry, easily recoverable state. The smallest commercial cyclone size (Aerotec, Design 2*) collects 100% of mineral particles (specific gravity 2.5) down to 5 μm in diameter. For smaller particles, retention decreases rapidly as diameter decreases.[53] Smaller, specially constructed cyclones (having cylinder diameters of ½ to 1 in.) may be utilized to capture particles as small as 1 μm in diameter.[54] Nevertheless, cyclones are not suitable for collecting a complete spectrum of particles in the smallest sizes (<1 μm) and therefore do not have universal application for particle sizing. Just as was noted for filter paper samplers, there is no way of determining whether airborne dust becomes deagglomerated by the shearing forces of the helical air stream, or agglomerated in the dust bin; consequently, no estimate can be made of the original state of the particles in the aerosol.

Whereas it appears possible to assemble cyclones in a series with decreasing diameter to remove progressively smaller fractions of airborne dust particles (a cyclone cascade), the particle separating characteristics of small cyclones have not proved sharp enough to make this idea of practical value.

2-4 Bulk Material Sampling

The methods customarily used for obtaining representative samples for analysis from bulk stores of finely divided dry powders are simple in concept, and at least the general outlines of the procedure are likely to be familiar to most engineers and chemists. In spite of the simplicity of the methods and their seeming crudity in comparison with the air sampling procedures already discussed, it is essential that each step be performed with care if satisfactory results are to be obtained, because the end product is a dust sample which

*UOP-Aerotec Division, Greenwich, Connecticut.

is only an infinitesimal fraction of the parent material, and it is difficult to make certain that the portion analyzed is indeed representative of the mass from which it was obtained.

Situations which require size analyses of bulk powders in industrial hygiene practice include the following:

1. Sizing of powders to be aerosolized for toxicological investigations or for testing and research on dust-collecting devices.
2. Rating performance of dust collectors on the basis of the size characteristics of the collected dust.
3. Examination of bulk materials to determine the nature of the dust control measures that will be required for bagging machines, conveyors, etc.

The process of sampling a bulk powder may be broken down into the following three steps.

2-4.1 Collection of a Gross Sample

The gross sample is composed of a number of sample units chosen in such a manner that each sample unit approximates the correct distribution of grain sizes, or the sum of the sample units (though each is known individually to be nonrepresentative) properly represents the overall size distribution. For example, a powder flowing from a grinding machine could be considered to be a reasonably uniform product. When small sample units are removed periodically from the discharge chute and pooled to form the gross sample, the moment-by-moment variations which occur in individual sample units because of small variations in feed, grinding efficiency, etc., are averaged, and there is a reasonable expectation that the smaller mass properly represents the material from which it was drawn. On the other hand, a truck or trainload of powdered material, or even a large container which has been on a long journey and subjected to intense vibration, will be size-segregated on arrival at its destination. Generally, the fines sift through the coarser grains to the bottom of the container. For this situation, sample units must be selected from specific locations in such a manner that when pooled to form the gross sample they will, *in toto,* be representative of the well-mixed mass.

Samples of fine powders may be taken with a spoon, a shovel, or a specially designed "thief" such as the one illustrated in Fig. 2.27, which is 43 in. long. It consists of two snugly fitting concentric thin-walled metal tubes with matching holes, or slots, in the sidewalls. The outer tube is pointed so that it may be slipped easily into a mass of powder. In use, the thief is pressed into the pile, container, or bin with the side holes open. After the powder has sifted into the tube, the inner tube is rotated to close the holes and the thief is withdrawn with its core of powder 40 in. long and 1.5 in. in diameter. From small piles, it is recommended[55] that one portion be taken with the thief pushed vertically downward from the center of the pile and one at each of ten points uniformly distributed along a line around the pile halfway between the peak and the edge of the pile. The eleven sample units are combined to form the gross sample.

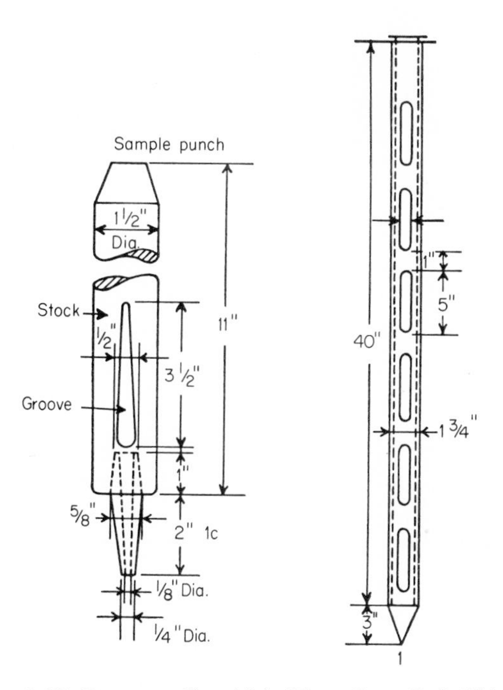

FIG. 2.27. Dust sampling thief. (From Scott, Ref. 55.)

2-4.2 REDUCTION OF GROSS SAMPLE TO A CONVENIENT SIZE

If the gross sample is small enough to be transported to the laboratory, it can there easily be mixed and reduced in size with an automatic proportioner such as the one illustrated in Fig. 2.28, which cuts out a representative one-sixteenth part of the material fed to it. If, however, the gross sample is too large, it must be reduced in the field by a process known as "coning and quartering." The procedure, based on a method recommended by the United States Bureau of Mines, is illustrated in Fig. 2.29. The steps are as follows:

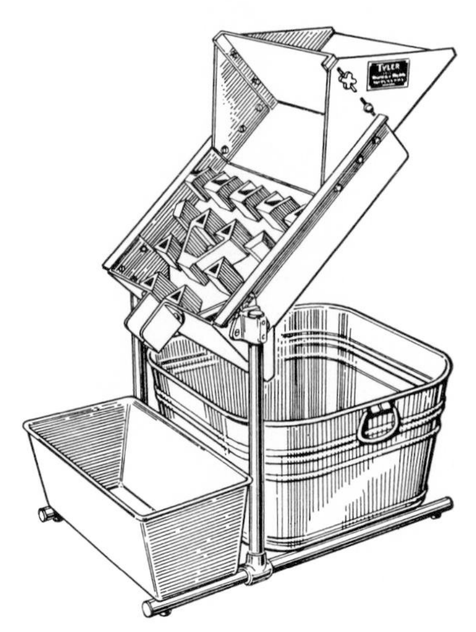

FIG. 2.28. Sample reducer. Courtesy of W. S. Tyler Company.

(1) mix the material and shovel it into a neat cone; (2) flatten the cone by pressing the top without further mixing; (3) divide the flat circular pile into equal quarters by cutting or scraping out two diameters at right angles; (4) discard two opposite quarters; (5) thoroughly mix the two remaining quarters, shovel them into a cone, and repeat the quartering and discarding procedures until the sample has been reduced to 5 to 10 lb. Samples likely to be affected by moisture or drying must be handled rapidly, preferably in an area with a controlled atmosphere, and sealed in a container to prevent further changes during transportation and storage. Care must be taken that the material is not contaminated by anything on the floor or that a portion is not lost through cracks or holes. Preferably, the coning and quartering operation should be conducted on a floor covered with clean paper. Coning and quartering is a simple procedure which is applicable to all powdered materials and to sample sizes ranging from a few grams to several hundred pounds.

FIG. 2.29. Coning and quartering. Courtesy of W. S. Tyler Company.

2-4.3 Preparation of Reduced Sample for Analysis

Small samples for size analysis are prepared by emptying the entire laboratory sample onto a large sheet of parchment, or other hard-surfaced paper, and thoroughly mixing the pile by giving the particles a rolling motion in opposite directions, as shown in Fig. 2.30. After the entire sample has been mixed, small samples are selected by successive quartering and remixing, or by passing the entire mixed pile through a laboratory sample splitter, similar to the one illustrated in Fig. 2.31, which mixes and divides the powder mechanically.

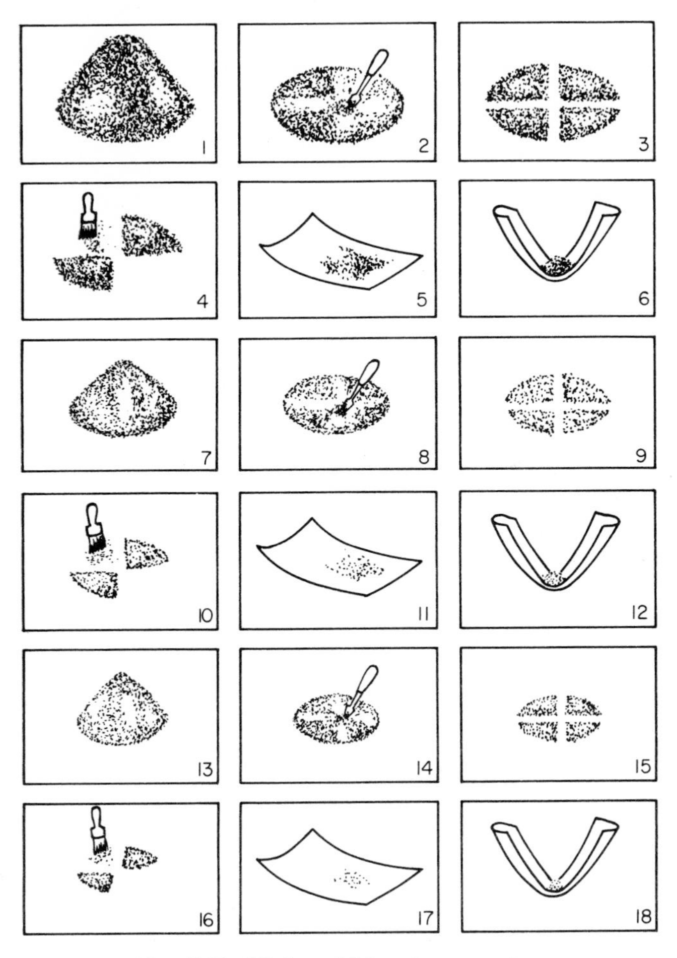

Fig. 2.30. Mixing of laboratory sample.

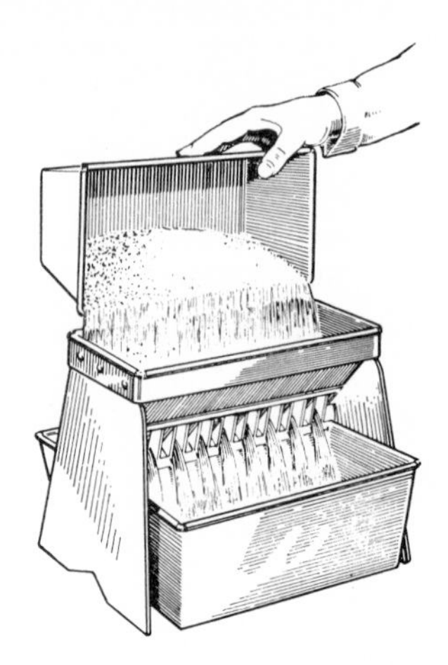

FIG. 2.31. Sample splitter. Courtesy of W. S. Tyler Company.

The size of the analytical sample will depend on the sizing method selected. For sieving, it should be 100 g; for sedimentation analysis, a few grams; and for microscopic sizing, a few milligrams. The sampling error of the selected procedure can be estimated by standard statistical techniques.[56] This is done by taking at least two portions from the gross sample for independent reduction and analysis.

References

1. L. Silverman, Sampling Instruments for Air Pollution Surveys, *Public Health Repts. (U.S.)*, 69-914 (1954).
2. P. H. Hardie, Résumé of Methods for Measuring Flue Dust, *Trans. Am. Soc. Mech. Engrs.*, *59*:355 (1937).
3. V. Vitols, Theoretical Limits of Errors Due to Anisokinetic Sampling of Particulate Matter, *J. Air Pollution Control Assoc.*, *16*:79 (1966).
4. K. J. Caplan, High Efficiency Collection of Radioactive Dust, *Heating and Ventilating*, *48*:79 (1951).
5. M. W. First, L. Silverman, R. Dennis, G. A. Johnson, A. T. Rossano, Jr., R. Moschella, C. Billings, E. Berly, S. Friedlander, and P. Drinker, *Air Cleaning Studies, Feb. 1, 1950-Jan. 31, 1951*, USAEC Report NYO-1581, Harvard University, April 1952.
6. C. J. Stairmand, Sampling of Dust-Laden Gases, *Trans. Inst. Chem. Engrs. (London)*, *29*:15 (1951).
7. P. Diamond, A. J. Kaimola, and R. G. Smith, The Importance of Sampling Velocity in Determining Particle Size Distribution, annual meeting of American Industrial Hygiene Association, Cincinnati, Ohio, Apr. 24, 1952.
8. L. Silverman, M. W. First, G. S. Reichenbach, Jr., and P. Drinker, *Air Cleaning Studies, Final Progress Report*, USAEC Report NYO-1527, Harvard University, Feb. 1, 1950.

9. K. R. May and H. A. Druett, The Pre-Impinger: A Selective Aerosol Sampler, *Brit. J. Ind. Med., 10*:142 (1953).
10. H. H. Watson, Errors Due to Anisokinetic Sampling of Aerosols, *Am. Ind. Hyg. Assoc. Quart., 15*:1 (1954).
11. B. M. Wright, A Size-Selecting Sampler for Airborne Dust, *Brit. J. Ind Med., 11*:284 (1954).
12. H. L. Green, Some Accurate Methods of Determining the Number and Size-frequency of Particles in Dust, *J. Ind. Hyg., 16*:29 (1934).
13. P. Drinker and T. Hatch, *Industrial Dust,* 2nd ed., McGraw-Hill Book Company, Inc., New York, 1954.
14. L. Greenburg and G. W. Smith, *A New Instrument for Sampling Aerial Dust,* United States Bureau of Mines, Report of Investigations No. 2392, 1922.
15. L. Silverman and W. Franklin, Shattering of Particles by the Impinger, *J. Ind. Hyg. Toxicol., 24*:80 (1942).
16. J. Aitken, *Collected Scientific Papers,* C. G. Knott (Ed.), Cambridge University Press, London, 1923.
17. H. L. Green and H. H. Watson, *Physical Methods for the Estimation of the Dust Hazard in Industry, with Special Reference to the Occupation of the Stonemason,* Medical Council of the Privy Council, Special Reprint 100, H. M. Stationery Office, London, 1935.
18. W. H. Walton, R. C. Faust, and W. J. Harris, *A Modified Thermal Precipitator for the Quantitative Sampling of Aerosols for Electron Microscopy,* Porton Technical Paper No. 1, Ser. 83, Chemical Defense Experimental Establishment, Porton, England, March 1947.
19. R. Wilson, S. Laskin, and D. Meier, *Design of an Oscillating Thermal Precipitator,* in University of Rochester Atomic Energy Project Quarterly Report, USAEC Report UR-199, December 1949.
20. J. R. Hodkinson, A. Critchlow, and N. Stanley, Effect of Ambient Airspeed on Efficiency of Thermal Precipitator, *J. Sci. Instr., 37*:182 (1960).
21. T. W. Kethley, M. T. Gordon, and C. O. Orr, A Thermal Precipitator for Aerobacteriology, *Science, 116*:368 (1952).
22. P. Drinker, Alternating Current Precipitator for Sanitary Air Analyses, I, An Inexpensive Precipitator Unit, *J. Ind. Hyg., 14*:364 (1932).
23. H. G. Bourne, Jr., and L. B. Fosdick, Collection of Mist and Dust for Particle Size Measurement: Electrostatic Precipitation on Hematocytometer, *Anal. Chem., 22*:1563 (1950).
24. R. Z. Bouton, R. W. Houston, and W. Grinter, *Semi-Annual Progress Report of Developmental Activities,* Health Physics Division, KAPL-492, General Electric Research Laboratory, Schenectady, N.Y., January 1951.
25. J. W. Baum, *Electrical Precipitator for Aerosol Collection on Electron-Microscope Screens,* Report HW-39129, Hanford Works, Richland, Wash., September 1955.
26. M. Arnold, P. E. Morrow, and W. Stöber, Vergleichende Untersuchung über die Bestimmung der Korngrossenverteilung fester Staube mit Hilfer eines Hochspannungsabscheiders und des Elektronenmikroskops, *Kolloid-Z., 181*:59 (1962).
27. P. E. Morrow and T. T. Mercer, A Point-to-Plane Electrostatic Precipitator for Particle Size Sampling, *Am. Ind. Hyg. Assoc. J., 25*:8 (1964).
28. C. E. Billings and L. Silverman, Aerosol Sampling for Electron Microscopy, *J. Air Pollution Control Assoc., 12*:586 (1962).

29. P. C. Reist, Size Distribution Sampling Errors Introduced by the Point-to-Plane Electrostatic Precipitator Sampling Device, in *Proceedings of 9th USAEC Air Cleaning Conference, Boston, Mass.,* USAEC Report CONF-660904, Vol. 2, p. 613, 1967.
30. F. J. Viles, Jr., personal communication, May 17, 1954.
31. M. W. First and L. Silverman, Air Sampling with Membrane Filters, *A.M.A. Arch. Ind. Health, 7*:1 (1953).
32. W. J. Smith and N. F. Suprenant, Properties of Various Filtering Media for Atmospheric Dust Sampling, *Proc. Am. Soc. Testing Mater., 53*:1122 (1953).
33. S. Posner, Air Sampling Filter Retention Studies Using Solid Particles, in *Proceedings of 7th USAEC Air Cleaning Conference, Brookhaven National Laboratory,* USAEC Report TID-7627, October 1961.
34. L. Silverman and F. J. Viles, Jr., A High Volume Air Sampling and Filter Weighing Method for Certain Aerosols, *J. Ind. Hyg. Toxicol., 30*:124 (1948).
35. United States Public Health Service, *Air Pollution Measurements of the National Air Sampling Network,* PHS Publication No. 978, United States Government Printing Office, Washington, D.C., 1962.
36. K. R. May, The Cascade Impactor: An Instrument for Sampling Coarse Aerosols, *J. Sci. Instr., 22*:187 (1945).
37. T. T. Mercer, On the Calibration of Cascade Impactors, *Ann. Occupat. Hyg., 6*:1 (1963).
38. J. J. Cohen and D. N. Montan, Theoretical Consideraions, Design, and Evaluation of a Cascade Impactor, *Am. Ind. Hyg. Assoc. J., 28*:95 (1967).
39. J. D. Couchman, Use of Cascade Impactors for Analyzing Airborne Particles of High Specific Gravity, in *International Symposium on Fission Product Release and Transport Under Accident Conditions, Oak Ridge, Tenn., April 5-7, 1965,* USAEC Report CONF-650407, Vol. 2, (1965).
40. M. W. First, L. Silverman, R. Dennis, A. T. Rossano, Jr., C. Billings, E. Conners, R. Moschella, E. Berly, S. Friedlander, and P. Drinker, *Air Cleaning Studies, Progress Report,* USAEC Report NYO-1586, Harvard University, 1952.
41. M. Lippmann, A Compact Cascade Impactor for Field Survey Sampling, *Am. Ind. Hyg. Asoc. J., 22*:348 (1961).
42. M. W. First, Instrumentation for Particle Sizing, in *Encyclopedia for Instrumentation and Industrial Hygiene,* University of Michigan, Ann Arbor, 1956.
43. K. F. Sawyer and W. H. Walton, The Conifuge, a Size-separating Sampling Device for Airborne Particles, *J. Sci. Instr., 27*:272 (1950).
44. M. I. Tillery, Design and Calibration of a Modified Conifuge, in *Proceedings of a Symposium on Assessment of Airborne Radioactivity,* International Atomic Energy Agency, Vienna, 1967, p. 405.
45. A. Goetz and T. Kallai, Instrumentation for Determining Size and Mass Distribution of Submicron Aerosols, *J. Air Pollution Control Assoc., 12*:479 (1962).
46. O. E. Raabe, Calibration and Use of the Goetz Aerosol Spectrometer, in *Proceedings of a Symposium on Assessment of Airborne Radioactivity,* International Atomic Energy Authority, Vienna, 1967, p. 417.
47. P. K. Mueller, H. L. Helwig, and C. L. Peterson, Some Performance Characteristics of the Goetz Aerosol Spectrometer, in *Proceedings of 52nd Annual Meeting of Air Pollution Control Association,* Los Angeles, Calif., June 1959.

48. H. S. Wolff and S. A. Roach, The Conicycle Selective Sampling System, in C. N. Davies (Ed.), *Inhaled Particles and Vapours,* Pergamon Press, London, 1961, p. 460.

49. V. Timbrell, The Terminal Velocity and Size of Airborne Dust Particles, *Brit. J. Appl. Phys.,* Suppl. 3, p. S86 (1954).

50. T. Hatch and W. C. L. Hemeon, Influence of Particle Size in Dust Exposure, *J. Ind. Hyg. Toxicol., 30*:172 (1948).

51. C. D. Yaffe, A. D. Hosey, and J. T. Chambers, Jr., The Spiral Sampler: A New Tool for Studying Particulate Matter, *A.M.A. Arch. Ind. Hyg. and Occupat. Med., 5*:62 (1952).

52. T. F. Hatch and P. Gross, *Pulmonary Deposition and Retention of Inhaled Particles,* Academic Press, New York, 1964.

53. R. Dennis, R. Coleman, L. Silverman, and M. W. First, *Particle Size Efficiency Studies on a Design 2 Aerotec Tube,* USAEC Report NYO-1583, Harvard University, April 1952.

54. R. Dennis, L. Silverman, C. E. Billings, D. M. Anderson, W. R. Samples, H. M. Donaldson, Jr., and P. Drinker, *Air Cleaning Studies, Progress Report,* USAEC Report NYO-4611, Harvard University, October 1956.

55. N. H. Furman (Ed.), *Scott's Standard Methods of Chemical Analysis,* 5th ed., p. 1305, D. Van Nostrand Company, New York, 1934.

56. *Symposium on Bulk Sampling,* Special Technical Publication No. 242, American Society for Testing and Materials, Philadelphia, Pa., 1959.

3

Direct Methods of Particle Size Analysis

3-1 Introduction

Analytical methods and instruments for performing particle size analyses are described in this chapter and the next two. Direct particle measuring methods are covered here, indirect methods in Chapter 4, and automatic counting and sizing instruments in Chapter 5. With direct particle measurement, the dimensions of individual grains are compared directly to a calibrated scale. Accuracy of measurement may be very great when an instrument of high sensitivity is used. With indirect measurement, a property of the dust which is a function of size is evaluated, and size is calculated from the relationship between the two, as when determining size by observation of particle settling velocity in still air. The accuracy of indirect methods depends on the correctness of the assumptions on which the size relationship is based. Direct methods include sieving and optical and electron microscopy.

As a further distinction, it may be noted that some sizing methods give information concerning both an average diameter and the manner in which particles are distributed around it, whereas others merely give an average diameter. Fig. 3.1 shows two dusts having the same median size and illustrates how misleading this description of size may be without added detail to describe size uniformity.[1] It seems natural, therefore, to classify sizing instruments according to whether or not they give information on size distribution as well as average diameter.

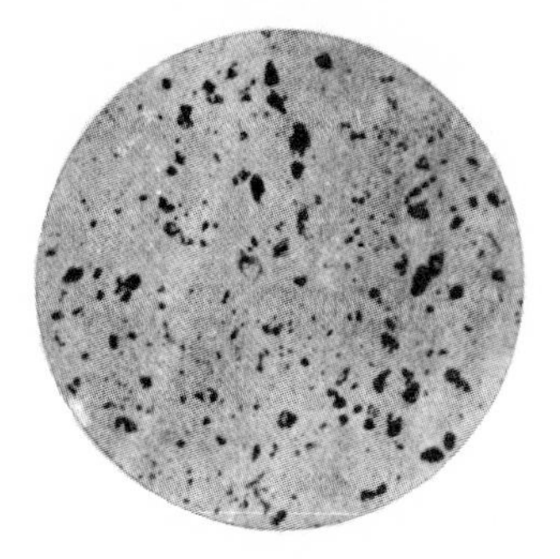

FIG. 3.1. Two powders with same average size but different size distributions. (From Drinker and Hatch, Ref. 1.)

It is convenient to divide particulate substances with respect to their measurement into three size groups:

1. Screen or sieve sizes, comprising particles greater than 37 μm, the size of the openings in the standard 400-mesh woven wire screen.
2. Subscreen or microscopic particles, from 0.25 to 37 μm.
3. Submicroscopic particles, those less than a quarter of a micron in size.

A sample of narrow size range may fall within one of these groups and require only a single method of measurement. Frequently, however, the size range extends into two or even three groups, necessitating the use of two or more methods for the measurement of the entire sample. The fact that different sizing methods often measure different size parameters, and that the size data so obtained may fit an assumed size distribution equation only as a first approximation, accounts for the frequency with which sizing data obtained by different techniques fail to agree. This should be kept in mind when attempting to check one method against another or when using data obtained partly by one method and partly by another.

Particles that are predominantly in the sieve size range are seldom measured by other methods, as a sieve analysis is rapid and easy to perform. Electrodeposited metal screens with openings smaller than the woven screen sizes are available, but their use as sieves requires techniques that have not yet been standardized.

In the microscopic range, particle dimensions may be measured directly with a linear scale incorporated into the microscope system, or indirectly by particle settling rate in air or liquid. This latter method may be extended to very small sizes by substituting centrifugal force for gravitational force. Other physical phenomena such as surface adsorption and light scattering are also utilized.

Particles in the submicroscopic range may be measured directly by scaling the particles shown on electron microscope photographs or indirectly by their electrical behavior, diffusion rate, and behavior as condensation nuclei.

3-2 Sieve Analysis

Particle size separations by sieving (square openings) or screening (round openings) are widely practiced for particles larger than 37 μm (the practical lower limit for conventionally woven wire testing sieves), and standard methods have been published by the American Society for Testing and Materials[2] and others[3] for a number of specific materials for which size and size distribution is a critical factor (e.g., Portland cement). Particles in the conventional sieve size range are seldom of significance in the practice of industrial hygiene, and for this reason no standard method having special relevance to the wide variety of very fine dusty materials encountered in this field of activity have been proposed. Nevertheless, dust samples which contain a sizable fraction of particles in the sieve range are sometimes encountered in practice, and it is often necessary to separate fine from coarse grains before undertaking an analysis such as microscopic sizing that is dependent to some degree on grain size uniformity for reliable results.

The subject of particle separation by sieving has recently acquired greater interest and importanc for industrial hygienists because of the introduction of new, and finer, woven wire sieves. For example, Work[4] indicated in 1958 that wire cloth manufacturers were able to make sieves with apertures of 25.2 μm and that they hoped soon to reach sizes of 18 μm. However, interest is more closely focused on the development of electrodeposited microscreens with openings as small as 5 μm in diameter.

The Tyler standard sieve scale is widely used in the United States for testing. The 200-mesh sieve (200 openings per inch) is the base for this series, and it has been adopted by the United States Bureau of Standards as well. It has square openings of 0.0029 in. (74 μm) and is woven from round wire with a diameter of 0.0021 in. The square sieve openings in the standard Tyler series increase in linear dimension by the square root of 2, so that each succeeding sieve of the standard series has openings twice the area of the preceding one in the numerical sequence. However, for closer size separations, Tyler sieves are also available in a series which increases by the fourth root of 2, as shown in Table 3.1, which reproduces only the lower end of the United States sieve series.[2] The International (ISO) Standard sieve series, in common use, is based on metric units and the square root of 2 progression in opening size. These sieves are designated by italic *b* in column 1 of Table 3.1. In addition, some non-standard sieve series have become established within certain industry groups by long-continued common usage. Frequently, the non-standard series is based on English units and a progression that doubles or halves the linear dimensions of the openings of adjacent numbers.

A sieve cannot make a perfectly sharp separation, especially of irregularly shaped particles. The fraction of particles passing through depends not only on the size of the mesh opening but also on the time and manner of shaking, the quantity of sample, moisture content, percentage of particles closely similar in size to the sieve openings, electrostatic attraction, and other factors of lesser importance, such as particle shape. It has been observed, also, that the average size, measured microscopically, of particles just passing a woven wire square-mesh sieve is considerably greater than the mean aperture.[5] This means that sieve specifications and sieving procedure must be standardized rigidly if accurate and reproducible results are to be obtained.

Using a single sieve, it is possible to obtain two fractions: one with its largest size smaller than the sieve size in use, and one with its smallest size larger.[6] However, it is generally more convenient to use a series of sieves, one below the other, with decreasing size of sieve openings, in order to separate a powder into a number of graduated size fractions in a single operation. Shaking may be done by hand. Small sieves, of 3-in. diameter, are most convenient for this purpose, but hand shaking is fatiguing, and it is virtually impossible to standardize the stroke and force of the vibrations. For accurate results, it is necessary to weigh and resieve the fractions repeatedly, until, on successive trials, weight changes on the sieves are less than 0.05%—an exceedingly laborious procedure. Therefore, most laboratories find it desirable to use a mechanical shaking device. The Ro-Tap Shaker* and the Syntron Test Sieve Shaker† (Fig. 3.2) are widely used for this purpose. Shakers consist of a frame to hold up to thirteen sieves at a time, a motor-driven shaker that imparts both a circular and a tapping

*W. S. Tyler Company, Cleveland, Ohio.
†Syntron Company, Homer City, Pennsylvania.

TABLE 3.1
U. S. SIEVE SERIES[a]

Sieve Designation		Sieve opening		Permissible variations in average opening %	Permissible variation for not more than 5% of openings %	Permissible maximum variation in individual openings %	Nominal wire diameter[c]	
Standard (1)	Alternate (2)	Millimeters (3)	Inches (approximate equivalents) (4)	(5)	(6)	(7)	Millimeters (8)	Inches (approximate equivalents) (9)
1.00 mm[b]	No. 18	1.00	0.0394	±5	+7½	+15	0.580	0.0228
841 μm	No. 20	0.841	0.0331	±5	+7½	+15	0.510	0.0201
707 μm[b]	No. 25	0.707	0.0278	±5	+7½	+15	0.450	0.0177
595 μm	No. 30	0.595	0.0234	±5	+7½	+15	0.390	0.0154
500 μm[b]	No. 35	0.500	0.0197	±5	+7½	+15	0.340	0.0134
420 μm	No. 40	0.420	0.0165	±5	+12½	+25	0.290	0.0114
354 μm[b]	No. 45	0.354	0.0139	±5	+12½	+25	0.247	0.0097
297 μm	No. 50	0.297	0.0117	±5	+12½	+25	0.215	0.0085
250 μm[b]	No. 60	0.250	0.0098	±5	+12½	+25	0.180	0.0071
210 μm	No. 70	0.210	0.0083	±5	+12½	+25	0.152	0.0060
177 μm[b]	No. 80	0.177	0.0070	±6	+20	+40	0.131	0.0052
149 μm	No. 100	0.149	0.0059	±6	+20	+40	0.110	0.0043
125 μm[b]	No. 120	0.125	0.0049	±6	+20	+40	0.091	0.0036
105 μm	No. 140	0.105	0.0041	±6	+20	+40	0.076	0.0030
88 μm[b]	No. 170	0.088	0.0035	±6	+20	+40	0.064	0.0025
74 μm	No. 200	0.074	0.0029	±7	+30	+60	0.053	0.0021
63 μm[b]	No. 230	0.063	0.0025	±7	+30	+60	0.044	0.0017
53 μm	No. 270	0.053	0.0021	±7	+30	+60	0.037	0.0015
44 μm[b]	No. 325	0.044	0.0017	±7	+30	+60	0.030	0.0012
37 μm	No. 400	0.037	0.0015	±7	+30	+60	0.025	0.0010

[a]From American Society for Testing and Materials, Ref. 2.

[b]These sieves correspond to those proposed as an International (ISO) Standard. It is recommended that wherever possible these sieves be included in all sieve analysis data or reports intended for international publication.

[c]The average diameter of the warp and of the shoot wires, taken separately, of the cloth of any sieve shall not deviate from the nominal values by more than the following: Sieves coarser than 595 μm: 5%; Sieves 595 μm to 125 μm: 10%; Sieves finer than 125 μm: 15%.

motion to the sieve bundle in the case of the Ro-Tap and a vibratory motion in the case of the Syntron, and a timing device that automatically regulates the sieving interval. Recommended sieving times range from 20 to 45 min; the finest dusts require the longest shaking periods.

The loss of fine dust from the sieves, which generally fit together poorly, can be appreciable unless all joints are carefully closed with pressure-

FIG. 3.2. Syntron sieve shaker. (Courtesy of Syntron Co.)

sensitive flexible tape. Even when this is done, there will be some losses because dust adheres to the exposed sticky tape surface between the pans, but these losses are likely to be minor, since the tape becomes nonadherent once a thin layer of dust builds up on the surface. Dust loss is likely to be greater when using the Ro-Tap machine, as the forceful tapping of the sieve bundle tends to shake the dust on the screens very violently. The Ro-Tap is exceedingly noisy.

There is a tendency for some powders to "ball up" when a large fraction of the particles are less than 37 μm in size. This tendency can be minimized by reducing the amount of material on the sieves; the best results are obtained when the material on each screen at the end of shaking is not more than one or two particles deep.[7] Schuhmann's calculations[5] show that, on this basis, there should, at the end, be no more than 30 g of quartz on a 100-mesh sieve 8 in. in diameter. This is equivalent to 4.5 g for the 3-in.-diameter sieve, and the allowable weight is, of course, decreased proportionately for the finer sieves. Thus, a half-gram of dust is close to being the optimum amount for a 325- or 400-mesh sieve.

Difficulties associated with fine powder losses and balling up can be overcome by wet sieving (i.e., washing the particles through the sieve openings), and many workers prefer it when hand methods must be used. However, for the industrial hygienist, there are practical objections, such as solubility, to using wet methods on heterogeneous dusts of imperfectly known composition.

Whatever method is adopted for depositing particles on the appropriate sieve or screen pan, the analytical procedure remains the same: the dust deposit on each pan is dried, if necessary, and carefully transferred with the aid of a brush to the pan of an analytical balance and weighed. The cumulative weight of all the fractions should, of course, agree with the quantity of weighed dust originally added to the sieve bundle, with the exception of the few percent that may adhere to the sticky sealing tape. If because of dust leakage, the deviation between dust added and dust recovered is greater than 5%, significant losses have occurred. When high accuracy is required, the analysis should be repeated, using improved methods for retaining the fine fractions. Fig. 3.3 shows the results of a typical sieve analysis and the calculations that are used to construct a smooth size-by-weight curve.

3-2.1 Sieve Calibration

A calibration method for sieves and screens that utilizes glass highway marking beads of graduated size has been proposed by Carpenter and

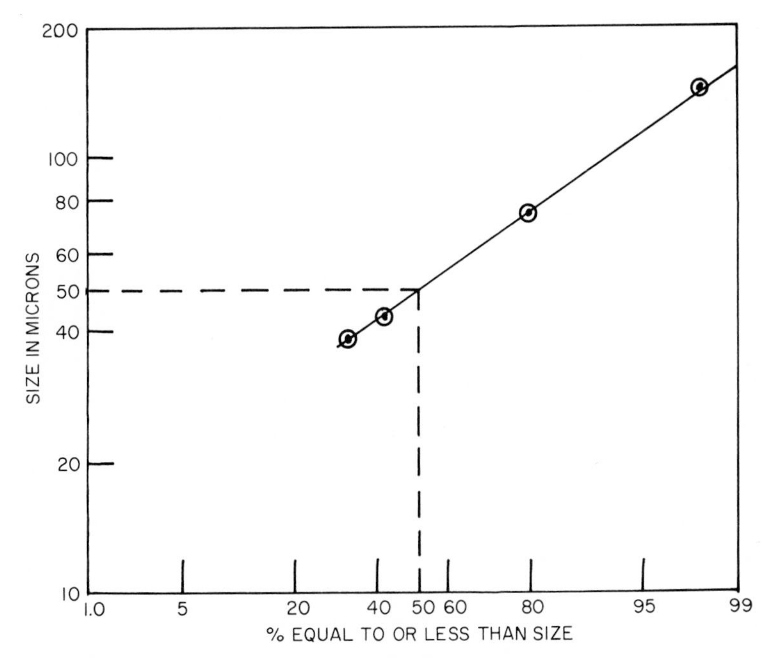

Sieve No.		Particle size (μm)	Weight (g)	Cum. wt. (g)	Cum. wt. % $\leq$ size
−400		−38	1.102	1.102	32.4
+400	−325	38–43	0.337	1.439	42.3
+325	−200	43–74	1.278	2.717	79.8
+200	−100	74–147	0.613	3.330	98.0
+100		+147	0.070	3.400	100

FIG. 3.3. Results of a typical sieve analysis: calculations and particle size curve.

Dietz[8] on the assumption that the spherical shape eliminates uncertainties over the average diameter of irregularly shaped particles, and over the orientation of such particles when they are passing through the openings. Effective sieve opening is determined by microscopically measuring the sizes of spheres that will just pass. Without suitable calibration, it has been found that size analyses with supposedly identical sieves vary significantly from laboratory to laboratory.

Calibration of small-mesh woven wire sieves and micro-mesh screens is most often done by direct microscopic measurement of a statistically significant number of openings. Parallel light rays are used to make it possible to focus on the narrowest point in the open area; this is generally midway between the top and bottom planes of a woven sieve screen and often reveals significant differences from the nominal size.

3-2.2 AIR-JET SIEVE

As an alternative to mechanically shaking and tapping the sieve screens, a continuous downward air flow may be used to "wash" the undersized particles through the mesh openings. Reverse, upward-flowing air jets can be used, as well, to keep the particles air-suspended and to prevent plugging or "blinding" of the fine-mesh sieve cloth. In the Alpine Air-Jet Sieve* this principle is utilized to make a separation on one sieve at a time. The undersizes are collected on a filter paper, and the oversizes are brushed out of the sieve pan for weighing. By starting with the finest sieve, the oversizes can be progressively transferred to coarser and coarser meshes until the entire size separation is completed, or a fresh sample can be placed on each sieve. Fig. 3.4 shows a diagram of the appartus, which uses a rotating reverse-flow air jet under the screen to air-suspend particles over the screen.

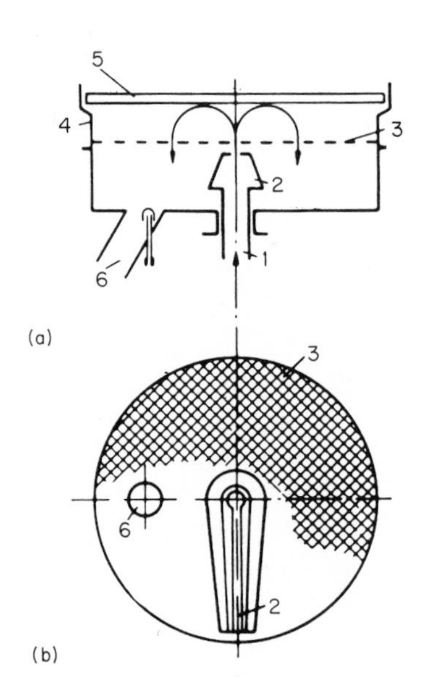

FIG. 3.4. Alpine air jet laboratory sieve: (a) schematic sectional view showing sieve pan (4), sieve screen (3), rotary reverse air jet (1 and 2), pan cover (5), and dust

*Alpine-American Corp., Saxonville, Massachusetts.

3-2.3 MICROMESH SCREENS

Electrodeposited precision micromesh sieve screens with square openings as small as 5 μm are available. Those manufactured by Buckbee-Mears* have been described by Daeschner *et al.*[9] as follows: A 3-μm-thick electrodeposited perforated film is mounted, for added strength, onto a separate grid that is 0.015 in. thick and has fourteen lines per inch. The two are bonded together by nickel plating for added strength. These strengthened

chute for undersizes (6). (b) Schematic plan view showing sieve screen (3), air jet (2), and dust chute (6). (c) Photograph of sieve. (Courtesy of Alpine-American Corp.)

*St. Paul, Minnesota.

perforated films are then mounted in conventional sieve pans 3 in. in diameter. Even though the film has been strengthened, it is exceedingly fragile and must be handled with extreme care at all times. Agglomeration, or balling up, of the fine powders used with these small-aperture sieves is severe. Daeschner *et al.*[9] attritbute this to electrostatic phenomena and indicate that considerable relief can be obtained by equilibrating dust and pans to 50 to 75% relative humidity prior to sieving. As a further aid, they suggest placing a small rubber O-ring on each sieve to break up the clumps mechanically by the gentle rubbing action of the ring. The procedure they suggest for micromesh sieving is as follows:

1. Weigh out sample (1 to 1.5 g).
2. Equilibrate to room conditions for 15 to 30 min.
3. Weigh sieves and bottom pan on an analytical balance to nearest milligram.
4. Assemble pack and add powder to the top.
5. Place in shaker and ground electrically.
6. Shake for 20 to 30 min.
7. Weigh.
8. Reshake for 5 min, reweigh, and repeat until constant weight is reached (i.e., less than 1 mg. change).

Sieving time is inversely proportional to the percent of open area of the screen and is apt to be long for the smaller openings, which have a low percent of open areas.

A sonic sifter* for electrodeposited screens with openings from 150 to 10 μm (Fig. 3.5), combines sonic vibration of the sieve pack with a mechanical pulse action to prevent blinding of the screens. Other methods for use of micromesh screens have been described.[10–12]

Multiple sieves, having even smaller openings than electrodeposited screens, may be fabricated from membrane filters of graduated size. Membranes are commercially available with uniform pores ranging from 0.05 to 10 μm.† Dry sieving is out of the question because of electrostatic effects, but the feasibility of sieving from liquid suspensions with multiple back washings to produce clean fractions has been explored[13] and appears to be practicable.

---

*Fisher Scientific Company, Pittsburgh, Pennsylvania.

†Millipore Filter Company, Bedford, Massachusetts; Gelman Instrument Company, Ann Arbor, Michigan.

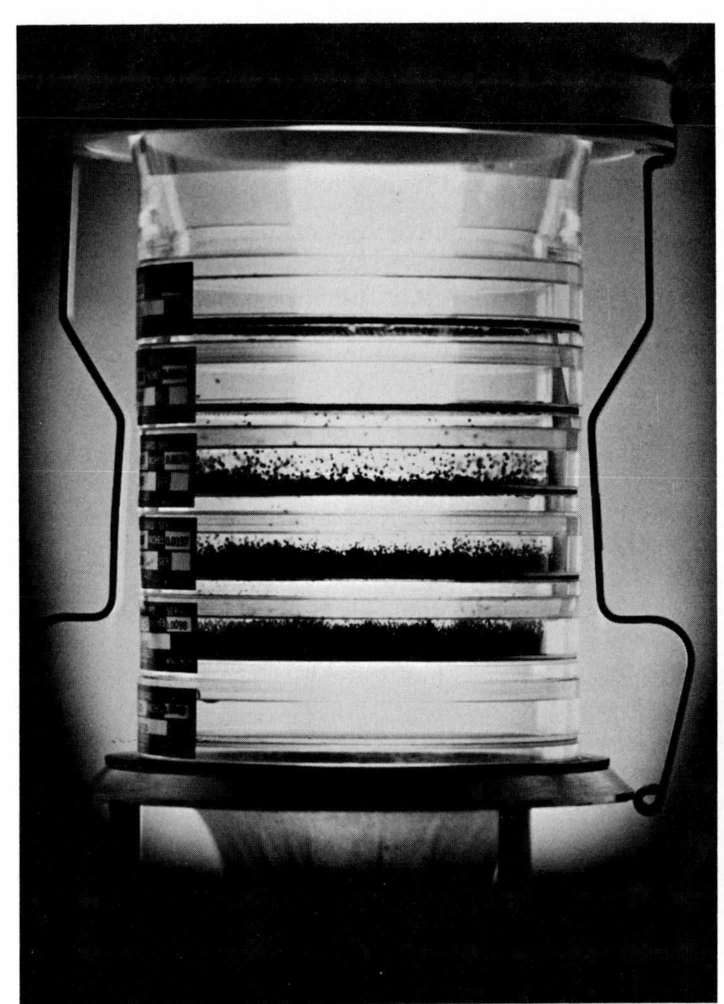

FIG. 3.5. Sonic sifter. (Courtesy of Allen-Bradley Co.)

3-3 Optical Microscopy

Visual measurement of particles is generally recommended for industrial hygiene purposes because (1) particles can be identified and size and shape determined directly, (2) only a calibration of a graticule is needed as the standard against which particles are compared, and (3) the amount of sample required for analysis is small and representative samples may be obtained with readily available industrial hygiene air sampling instruments.

Sizing with the light microscope is the basis for calibration of faster methods such as those involving sedimentation or turbidity. The light microscope is applied most efficiently to the range of sizes between approximately 0.25 and 20 μm. The lower limit is imposed by the resolving power of the optical microscope. The upper limit is determined by the thickness of the particle in relation to the depth of field of the optical system. For the particle size range of interest to the industrial hygienist (i.e., 10 μm and below), microscopic sizing is a fundamental method that requires no calibration by some other method.

3-3.1 THE OPTICAL MICROSCOPE

Optical microscopy is used in so many scientific disciplines that it is only necessary to describe here the specific optical systems and their uses which have special application to particle sizing. Comprehensive treatises on all phases of microscopy are given in several texts. *Photomicrography,* by Charles P. Shillaber,[14] is outstanding.

Because optical microscopes are used for other purposes in addition to particle sizing, it is not practicable to recommend a single microscope system (consisting of stand, lenses, condenser, stage, and lamp) that is uniquely designed for this purpose alone. Any professional-grade biological, chemical, research, or petrographic microscope with a mechanical stage can be used for particle sizing with a calibrated eyepiece graticule (see Section 3-3.2). Special microscope systems, such as Leitz's self-illuminating Ultropak objectives, make it possible to resolve and measure fine particles deposited on nonopaque substances by reflected surface illumination.

The factor that is important in microscopy for particle sizing, in addition to total magnification, is resolving power, or limit of resolution. No amount of magnification by the ocular lenses can enlarge an image that the objective lens of the optical system is unable to "see." The resolving power of an objective is nearly proportional to the reciprocal of its numerical aperture (N.A.), which is, in turn, the product of the refractive index *(n)* of the light-transmitting medium of least optical density times the sine of ½ the cone angle (θ) of the light rays that just fill the objective lens when properly focused on an object; i.e.,

$$\text{N.A.} = n \sin(\theta/2) \tag{3.1}$$

The numerical aperture of a microscope objective lens is also a measure of its light-gathering qualities, since the amount of light transmitted varies as the square of its N.A. For this reason, "an objective of higher power sometimes transmits more light, per unit area, than one of lower power, owing entirely to the increase of N.A."[14] Finally, field depth, or vertical resolution, is also controlled by the numerical aperture of the lens, increasing with decreasing aperture. Thus, it may be seen that large field depth, which requires small apertures and is a distinct advantage for the examination of particles with pronounced size differences, is antagonistic to the need for large apertures to preserve detail and maintain maximum resolving power; and these needs must be comprised in practical systems.

Resolution *(R)* equals the wavelength of light *(λ)* divided by twice the numerical aperture (N.A.); i.e.,

$$R = \lambda/2 \text{ N.A.} \tag{3.2}$$

Resolution is the least distance apart (expressed in the units used for λ) that two particles can be observed distinctly. The limit of resolution with the optical microscope when using blue light (i.e., the smallest value for λ) is about 0.2 μm. This limit is attainable only with excellent equipment and techniques. Table 3.2 lists the resolving power of objectives commonly used in professional-type microscopes. An excellent optical system for microscopic particle sizing consists of the three objectives marked with an italic *b* in Table 3.2. They are (1) a low-power 16-mm apochromat with N.A. = 0.30, (2) an 8-mm apochromat with N.A. = 0.65, and (3), of principal importance, an apochromatic oil immersion objective of high numerical aperture (N.A. = 1.40) and maximum resolving power. Apochromatic objectives are corrected chromatically for three colors and spherically for two colors, but are undercorrected for color to make possible especially fine spherical corrections. Therefore, compensating oculars that are overcorrected the proper amount must be used with apochromatic objectives to produce the best possible image. Ocular, or eyepiece, lenses which magnify the image projected by the objective lens 10 times are commonly used, although 20-times oculars matched to the objectives give a substantially larger image and are recommended for the examination of particles 1 μm or less in size. For best results, the microscope should contain a graduated, adjustable microscope draw tube so that tube length can be varied over at least a few millimeters (bear in mind that the image will deteriorate if tube length varies too widely from 160 mm).

A substage condenser of the three-lens aplanatic type is required for freedom from spherical aberration and coma when using high-power objectives to size submicron-sized particles. For best results, the ratio of condenser to objective numerical aperture should be between 0.8 and 1.0.

TABLE 3.2
RESOLUTION OF TYPICAL MICROSCOPE OBJECTIVES[a]

		Red light $\lambda = 691$ nm		Green light $\lambda = 546$ nm		Blue light $\lambda = 405$ nm		Ultraviolet $\lambda = 365$ nm		
Objective (1)	N.A. (2)	Lines per inch (3)	Separation for resolution, μm (4)	Lines per inch (5)	Separation for resolution, μm (6)	Lines per inch (7)	Separation for resolution, μm (8)	Lines per inch (9)	Separation for resolution, μm (10)	Magnification[c] required for a 1-mm image $\lambda = 546$ nm (11)
16-mm achr.	0.25	18,389	1.38	23,292	1.09	31,216	0.81	34,780	0.73	917
16-mm apo.[b]	0.30	22,073	1.15	27,889	0.91	37,896	0.67	41,666	0.61	1099
8-mm achr.	0.40	29,464	0.86	37,363	0.68	50,191	0.51	56,497	0.45	1471
8-mm apo.[b]	0.65	47,828	0.53	60,152	0.42	81,661	0.31	90,909	0.28	2381
4-mm achr.	0.85	62,500	0.41	79,375	0.32	106,617	0.24	120,773	0.21	3125
4-mm apo.	0.95	69,266	0.36	90,703	0.28	119,812	0.21	133,868	0.19	3571
3-mm achr.	1.25	92,024	0.27	116,510	0.22	157,759	0.16	169,491	0.15	4545
3-mm apo.	1.30	95,479	0.26	120,929	0.21	163,855	0.15	181,818	0.14	4761
3-mm apo.[b]	1.40	102,819	0.24	130,251	0.20	176,657	0.14	196,078	0.13	5000

[a]From Shillaber, Ref. 14.

[b]Excellent objectives for microscopic particle sizing.

[c]This column lists the magnification required to make the image of a particle at the threshold of resolution, 1 mm in diameter. It is based on the figures given for resolution in green light, column 6.

The illumination of small particles on the microscope stage is critical, particularly when the grains are translucent minerals such as quartz. The illumination level must reveal the outline of individual grains distinctly at the limit of resolution, and the microscope field must be free of glare. Insufficient light is as undesirable as too much, since either will mask detail. Ideal illumination may be achieved by using a diffusing plate at the microscope lamp house to act as a secondary light source on which the microscope mirror is focused. The diffusing plate should be located about 15 in. from the microscope mirror, as experience shows that this distance gives close to the minimum amount of glare. In practice, the image of the lamp filament is projected onto the center of the microscope mirror, and then the diffusing plate is dropped into place in the lamp housing. The field diaphragm at the lamp is closed to a small opening, and then the microscope is focused on a test slide. After this, the substage condenser is raised or lowered until the image of the spot of light is as sharp as possible. Next, the light spot is brought into the center of the field of view by turning the mirror, and finally the lamp diaphragm is opened to correspond to the field of view. If the "grain" of the diffusing plate is visible, the condenser should be raised slightly above focus. If the field is too brightly illuminated, light filters should be slipped into the lamp holder until the particles are clearly outlined against a soft, diffused background. For best results, the plane face of the microscope mirror should be used in combination with a substage condenser, as otherwise nonparallel light rays will pass to the microscope lens system. A number of illumination methods, including this one (which is believed to be the simplest), are described in detail by Shillaber.[14].

(a) *Focusing.* With a properly illuminated particle in the field of the microscope, it will be noted that the image changes its appearance as the microscope tube is moved a small distance up or down. This is because the microscope is alternately focused precisely on the top, the bottom, and intermediate points along the depth of the particle. It will be noted, in addition, that at certain focal positions a white zone, or band, known as the Becké line (and related to the relative refringency of the particle and its mounting medium), appears around, or just within, the edge of the particle; this tends to obscure the sharp outlines of the grain. It will be evident from this that the size of the microscope image can vary significantly depending on the position of focus, and therefore care must be exercised to focus sharply and exactly on the extreme outer edges of the grain if the true particle size is to be obtained. The microscope must therefore be carefully focused on each particle that is to be sized, and the entire depth of the slide must be examined systematically if the smallest grains (i.e., those close to the limit of resolution of the microscope) are not to be overlooked.

(b) *Measuring Scales.* The linear measurement of objects under the microscope is usually accomplished by comparison with a scale incorporated in the ocular of the microscope. This scale must be calibrated with a stage micrometer. The Patterson-Cawood eyepieces graticule (the first of the globe-and-circle graticules to receive extensive usage) has a series of disks graduated in an arithmetic series, as shown in Fig. 3.6a. Another graticule of similar type, called the Porton graticule (Fig. 3.6b), has the disks graduated in a series based on the square root of 2. This is a modification of Fairs' "globe-and-circle graticule,"[17] in which the two smallest circles did not follow the square root of 2 progression. A series based on the square root of 2 has certain advantages for irregular particles, as it is generally more satisfactory to compare areas than diameters, and successive disks double in area as they progress in size. Bosanquet[18] has shown that with the geometric series the error due to assigning a particle to a wrong class is only 0.18 times the counting error. Moreover, experience has shown that for most distributions

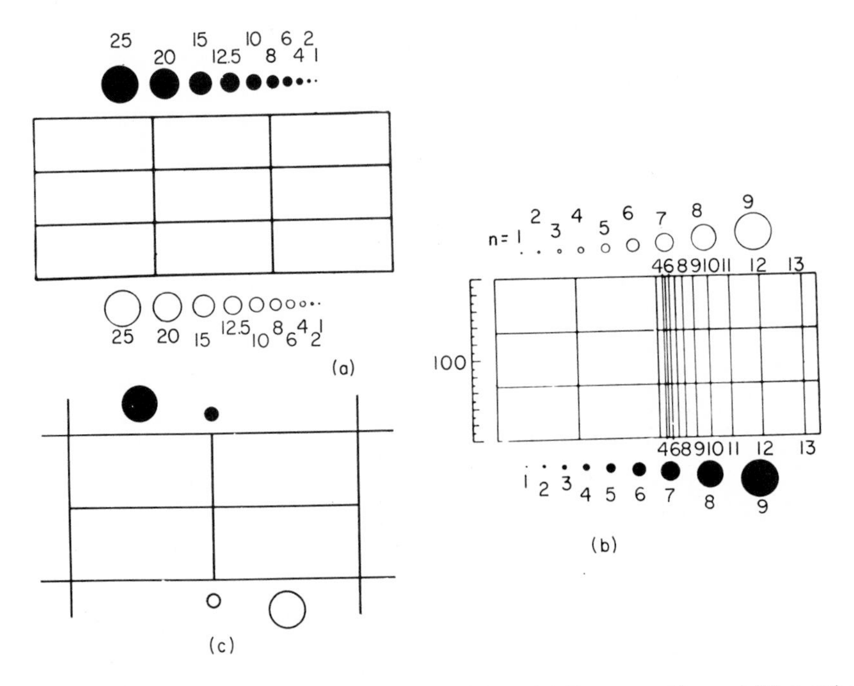

FIG. 3.6. Microscope graticules for particle sizing: (a) Patterson–Cawood (Ref. 15); (b) Porton (Ref. 16); (c) Watson (Ref. 21.)

the size range covered by an arithmetic series of graduated disks (e.g., the Patterson-Cawood) is too narrow.

The rectangular grid spaces of each of the three graticules shown in Fig. 3.6 are used to define precise areas of the microscope slide in which all particles are to be sized for requirements of statistical significance. The linear scales of a Porton graticule may be used conveniently to measure extreme dimensions of irregular particles, but the commonest way of using these graticules is to employ the graduated circles to estimate the projected area of particles. This may be done by actually superimposing the graduated circular images on a particle until one circle is found which is just equal to or slightly larger than the particle. This is laborious, but one rapidly learns to estimate the correct matching circle without physically superimposing the images, and then particle sizing can be accomplished much more rapidly.

For spherical particles, size is defined rigorously and unambiguously by a measurement of diameter. This single value may be utilized to calculate many other important parameters, such as surface area and, with particle density, mass and aerodynamic behavior. For nonspherical particles it is highly desirable to express size by a single number, as this permits simpler methods of data recording and analysis to be utilized. For particles that are roughly spheroidal, such as quartz dust, a frequently used size designation is the diameter of a perfect sphere having the same volume as the spheroid. This is convenient because (1) this length will approximate the average of maximum and minimum particle dimensions and will be close to the mean distance between all particle surfaces when measured through the center, and (2) this diameter can be approximated by a simple measurement of the diameter of a graticule circle which has the same area as the projected area of the particle under the microscope. Imperfect sphericity affects the dynamic behavior of spheroidal particles, and it has been observed that quartz particles settle at a rate two-thirds that of perfect spheres having the same volume.[1]

When particle shape diverges more markedly from spherical, a single-length parameter describes less well the multiple characteristics of a particle (e.g., geometry, settling rate). For needle-shaped particles (e.g., talc), platelets (e.g., mica), or fibers (e.g., asbestos), the diameter of a sphere of equal volume is unlikely to reflect relevant geometric, dynamic, or biological properties of these particles to a sufficiently close degree to be an accurate representation of particle properties related to size. This is especially likely to be true when the equivalent diameter of a sphere of equal volume is determined from the projected area of a particle on a microscope slide, because an extreme length dimension or actual diameter may be of greater interest for needle shapes or fibers than the average diameter, because they characterize the size of these particles more closely with respect to their dynamic and biological behavior. For example, Corn[19] has shown that small particles in motion (e.g., settling under the influence of gravity) have an orientation relative to the direction of motion that is different from that assumed when they are at rest. This means that the dynamic properties of grossly nonspherical particles cannot be estimated accurately from microscopic sizing of settled dusts performed with an area eyepiece graticule.

Numerous characteristic diameters of nonspherical particles have been suggested by a number of investigators to represent accurately one or another important particle characteristic. The various diameters which can be determined by observing a silhouette in the microscope have been reviewed and summarized by Davies[20] and given mathematical expression. They include the projected area diameter, already mentioned; Feret's statistical diameter, the distance between parallel tangents for a number of particles; Crofton's diameter, the mean length of a large number of chords drawn at random through the outline of a particle; and Martin's diameter, the mean of chords through the centroids of a group of particles. Various diameters are illustrated in Fig. 3.7.

In practice, the sizes of projected areas of particles are designated by the disk or globe numbers engraved on the graticule and are customarily enumerated by disk number on tabular forms as they are measured, or a

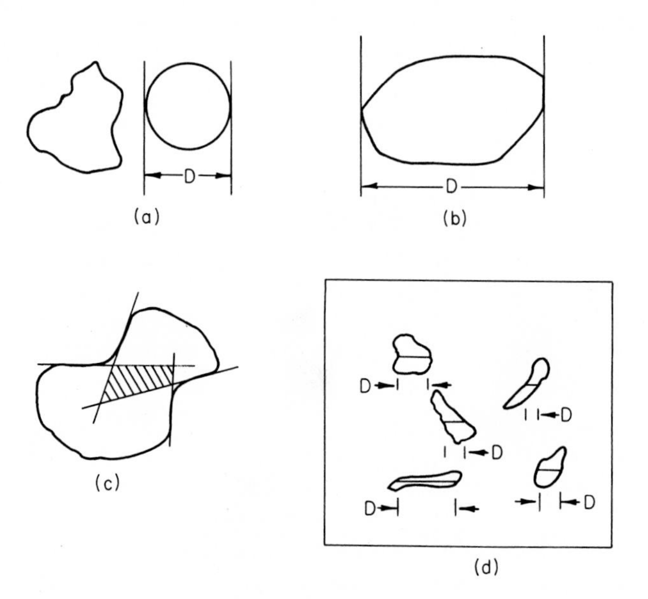

FIG. 3.7. Particle diameters. (a) Projected area diameter. (b) Feret's diameter. (c) Crofton's diameter. (d) Martin's diameter.

bank of hand tallies may be used to register the total number of particles sized plus the number of particles in each size interval delineated by the graticule disks. Another way of recording sizing data without lifting the eyes from the microscope field is to dictate it to an assistant or into a dictating machine.

When only limited information on size distribution is required, the graticule may be considerably simplified by omitting a number of intermediate circles. For example, Watson[21] has proposed a three-spot graticule (Fig. 3.6c) for the investigation of pneumoconiosis-producing dusts that can be used to size-classify particles into the following groups: < 0.5, 0.5 to 2, 2 to 5, and >5 μm. Simple scales of this type accelerate microscope size analyses when information is needed only on the proportion of particles within a specified (e.g., respirable) size range.

The filar micrometer is a more accurate eyepiece-measuring device than those already described. It has a movable cross-hair built into a standard ocular and is actuated by a calibrated micrometer screw. To make a size measurement, the cross-hair is aligned with one edge of the particle, and the micrometer setting is read and recorded. Next, the cross-hair is moved to the opposite side of the particle along a horizontal traverse, and the micrometer setting is reread and recorded. The difference in micrometer readings is proportional to the measured particle dimension and may be evaluated exactly after calibration of the filar micrometer with a stage micrometer, as explained in Section 3-3. As there may be backlash in the filar drives, highest accuracy is obtained by making at least one duplicate measurement traverse in the opposite direction. It is important to have the same side of the cross-hair exactly touching the edge of the particle for each pair of measurements. The filar micrometer is employed best to measure the extreme dimensions of a particle (which will be larger than the so-called average diameter of irregular particles). The particle or the micrometer may be rotated to measure the largest and/or the smallest dimensions of particles. Alternatively, only the horizontal lengths of particles distributed randomly in the microscope field may be measured by sweeping the cross-hair from one side of the field to the other.

The alternate viewing of the microscope field and reading of the linear scale that is required with this apparatus is exceedingly laborious and trying on the eyes, and for this reason direct comparison with a fixed graticule is preferred. The filar micrometer has been recommended for the most exact work, but Heywood[22] and Fairs[17] both demonstrated that, for particles in the range of interest of the industrial hygienist, the accuracy of the globe-and-circle type of reference scale, when used for the measurement of the projected area of irregular particles, provides a satisfactory degree of precision.

The image-splitting microscope[23] (Fig. 3.8) may also be used for particle measurement, although it possesses the same disadvantages as the filar micrometer in that a reading of an external scale is required for each particle measured. A special prism assembly may be mounted in any conventional compound microscope to make an image-splitting microscope. The prisms are linked to a micrometer screw by means of which their angular relation to each other can be varied. When the prism faces are parallel, two images of the object, exactly superimposed and appearing as one, are visible, as shown in Fig. 3.8a. As the micrometer screw is turned, the images move across each other. When they lie exactly side by side, touching but not overlapping, as shown in Fig. 3.8c, a reading of the micrometer screw is noted. The images are then reversed by changing the position of the prisms, and the total amount of micrometer run is read and referred to a calibration chart for particle size. Calibration of the micrometer screw is accomplished with a microscope stage micrometer in the manner described below. Comparative measurements can be made fairly simply with the image-splitting microscope by presetting the micrometer screw to some specific size, such as 5 μm, and then examining the particles in the microscope field. Those exactly 5 μm in size will appear as shown in Fig. 3.8c; those larger as in Fig. 3.8b; and those smaller as in Fig. 3.8d. For irregular shapes, the image-splitting microscope measures the extreme horizontal dimension of particles randomly distributed on a slide.

(c) *Calibration of Measuring Scales.* Stage micrometers with scale intervals of 10 μm are suitable for calibrating ocular scales to be used for the measurement of small particles. These scales usually contain 100 intervals, as shown in Fig. 3.9a, and are permanently mounted on 1 $\times$ 3-in. glass microscope slides. Micrometer scales are made (1) by a facsimile process (photographing the rulings onto the sensitized slide), or (2) by engraving the lines directly onto the glass. Facsimile types are inexpensive (only a few dollars) and are suitable for low microscope magnifications (e.g., $\times$100).

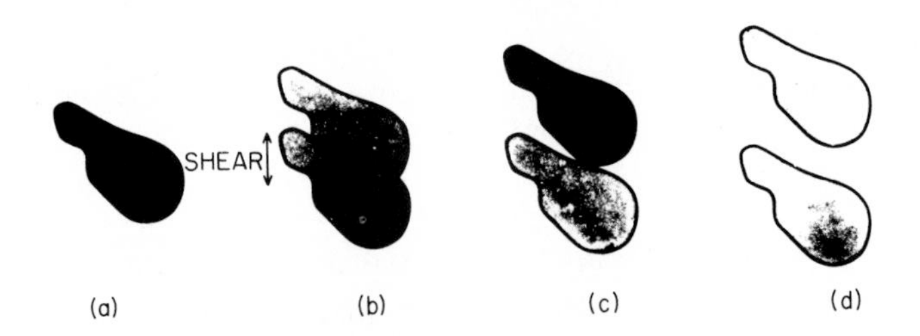

FIG. 3.8. Image-splitting microscope fields for particle measurement.

At high magnifications, however, the lines appear wide and indefinite, as shown in Fig. 3.9b. An engraved stage micrometer costs about $30, but the lines are sharp and clear even at 1000 diameters, as shown in Fig. 3.9c.

Ocular measuring disks are calibrated with a stage micrometer placed on the microscope stage. After the scale is brought into sharp focus, the stage micrometer is moved by the mechanical stage until its ruled lines exactly coincide with an ocular disk, as shown in Fig. 3.10. At high magnifications, the thickness of the micrometer lines is appreciable, and care must be taken to make measurements from the same side of the ruled lines each time. With

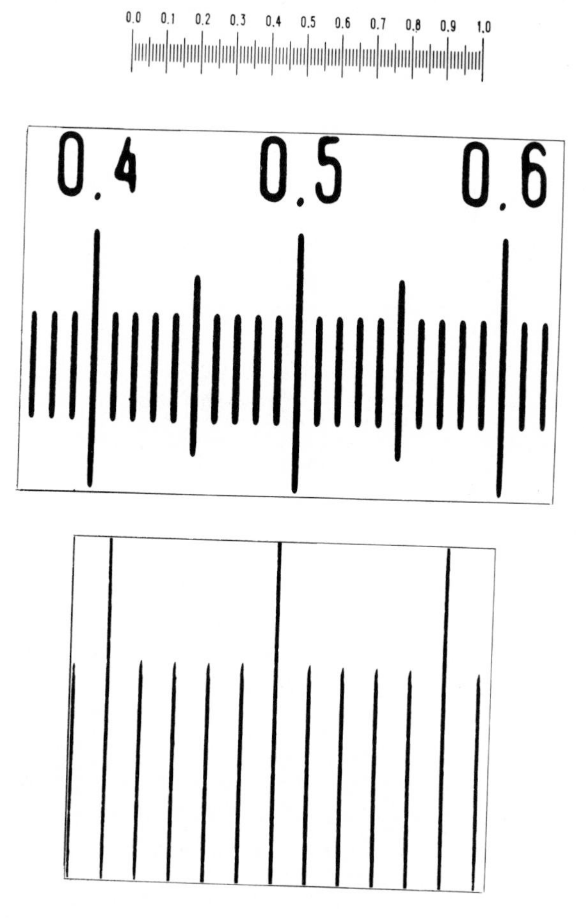

FIG. 3.9. Stage micrometers: (a) photomicrograph of photoprocess scale, ×47, (b) photomicrograph of photoprocess scale, ×275, (c) photomicrograph of engraved scale, ×455. (From Shillaber, Ref. 14.)

ocular scales that have a known relationship between graduated circles, it is necessary to measure only a single circle. With the Porton eyepiece graticule, either the circles or the exactly equal rectangular grid intervals may be measured for calibration purposes, and the sizes of the remainder may be calculated. For example, in Fig. 3.10, the diameter of the No. 9 circle of the Porton graticule is exactly equal to three stage micrometer intervals, or 30 μm. Since the Porton size intervals are based on a square root of 2 series, each circle is 1.414 times as large in diameter as the one lower in the series, and therefore the next smaller circle, No. 8 is $30/\sqrt{2}$ or 21.2 μm. The size of each of the remaining circles may be calculated in a similar manner.

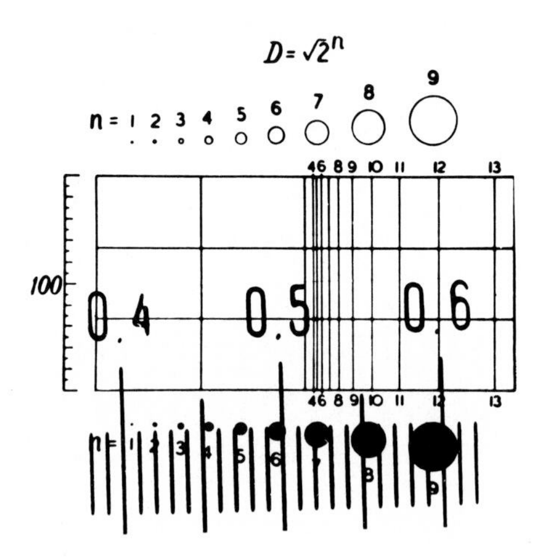

FIG. 3.10. Calibration of Porton graticule with stage micrometer.

If no circle on the ocular disk corresponds exactly with one or more entire intervals on the stage micrometer scale, the distance may be estimated or the microscope draw tube may be extended to increase the magnification slightly so as to achieve exact coincidence. If the second method is used, the draw tube must be extended to the same mark each time the microscope is used for particle sizing. An individual calibration is required for each combination of objective and ocular lenses that is to be employed for particle sizing. The actual magnifications obtained will not coincide exactly with the product of the nominal magnifications of objective and ocular lenses.

(d) *Projection and Photographic Methods of Direct Microscopic Observations.* Microscopic sizing may be less eye-fatiguing with microprojectors. The image of the particles is projected onto a screen, and particle diameters are measured with a scale. This method is not recommended for dusts with a preponderance of particles below 2 μm. The apparent size tends to be larger than that obtained by direct vision, and resolution is poorer (leading to very fuzzy edges); hence, measurement is not exact. Well-designed remote controls for microscope focusing and field searching are essential for this method.

Methods described by Loveland[24] as useful for determining projected areas of particles on a screen are as follows:

1. The use of templates containing a series of idealized particle shape outlines of graduated size to permit both shape and size classification of each particle. When the templates are on glass and projected from below, no parallax occurs. Grids, scales, and templates, identical with those described for use in making direct microscopic observations (e.g., the Porton graticule), are used also for projection sizing.

2. Graph-like cross-hatching, superimposed on the images. The squares intercepted by the particle are counted. This method is slow and tedious.

3. Particle outlines traced on transparent paper held against a ground-glass screen containing the projected image. The outlines can then be cut out and the paper weighed. Alternatively, the outlines can be traced through very thin Kodaloid* film with a hotpoint etching tool, thereby automatically making the cutouts as the particles are traced.

4. Areas of projected particles traced by planimeter. Although too slow for routine use, this method is accurate and may be used to calibrate other methods.

5. Automatic measurement of projected areas of particle images by a photoelectric cell. The shadow is the projected area, regardless of the intricacy of the outline or the shape of the particle. A null instrument for performing this measurement has the following features (quoted from Loveland[24]):

> Light beams from two sides of a concentrated source are sent on diverse and independent paths but meet again to form an image of the source superposed accurately and at identical magnification. A rotating-sector mirror cuts accurately through this image and sends a light from each beam alternately to a vacuum-tube photocell. The latter, in turn, is connected through an a-c amplifier to a servomotor that either opens or closes a sector diaphragm in one of the two beams. If the two beams produce equal illumination at the photocell, the a-c amplifier and the motor are unaffected, and nothing happens. However, when any obstruction is introduced into the measuring beam, an a-c current is produced

*Eastman Kodak Company, Rochester, New York.

and the sector wheel closes down until the two beams are again equal; the turning of the diaphragm is a measure of the size of the obstruction. It is most simply measured on a Veeder counter or rotating scale and more elaborately on a series of Veeder counters denoting size ranges. With this instrument, a series of apertures can be measured as readily as a series of area obstructions.

Another method of reducing fatigue from direct observations involves photographing each microscope field and measuring the particles on an enlarged print. This method allows the employment of illumination of a shorter wavelength than can be used for direct visual observation, with consequent improvement in resolving power. In addition, the photomicrographic method makes it possible to increase the contrast between particles and their background. It also gives a permanent record of the sample. The size range of particles that can be measured by the photomicrographic method depends essentially on the thickness range of the particles in relation to the depth of field. For example, an oil immersion objective having a numerical aperture of 1.4 will have a field depth of approximately 0.2 μm. If the particles are plates or needles lying in a plane, the size range may be quite large without appreciable change in thickness. For particles covering a greater size range, negatives at more than one focus or magnification may be used, but a system of checks must be maintained to ensure that all particles in the fields involved are measured once, and only once. The apparent size of a photographic image is primarily dependent on resolving power, but it can be altered substantially (particularly with images of small particles) by variations in exposure, development time, film grain, and printing. The same disadvantage applies here as to microprojection in that small particles near the limit of resolution become extremely fuzzy and exact measurement of their diameter is somewhat uncertain.

A semiautomatic instrument for particle size analysis of photomicrographic particle images was described by Endter and Gebauer[25] and is manufactured by Carl Zeiss. This instrument was described excellently by the manufacturer in the following manner:

"The smallest particle in a photomicrograph or electronmicrograph must be at least 1 mm in diameter. Particles, which often are marked off only by their contours, may be in contact, overlap, or even lie superimposed. Greatly elongated particles can be measured in length or width. Mixtures of different kinds of particles can have their components evaluated separately.

"The image of an iris diaphragm is projected by a lens in the plane of a Plexiglass plate (as shown in Fig. 3.11b). An enlargement of the photomicrograph or electronmicrograph is laid on this plate (as shown in Fig. 3.11a). By adjusting the iris diaphragm the diameter of the sharply defined circular light spot appearing on the enlargement can be changed and its area made to coincide with that of the individual particles.

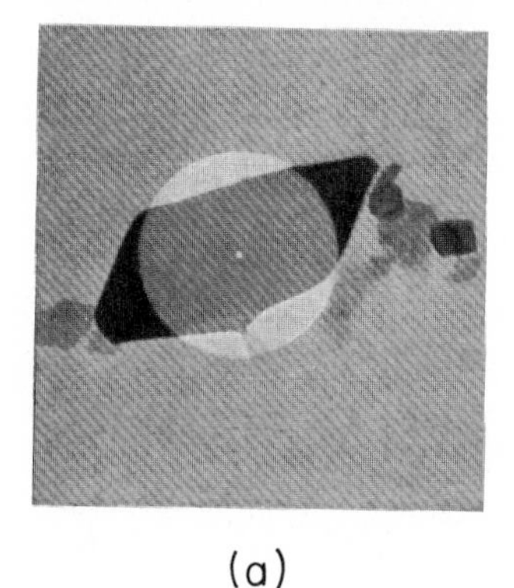

(a)

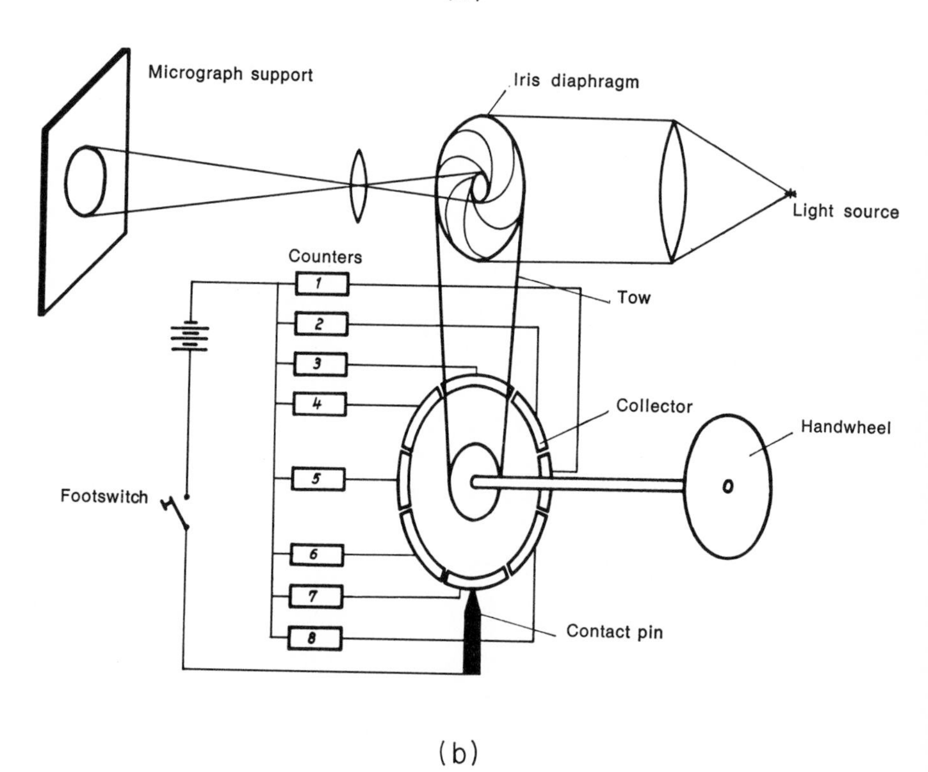

(b)

FIG. 3.11. Particle size analyzer. (a) Image of particle in circular light spot. (b) Image of iris diaphragm on Plexiglass plate and telephone counters covering a range of iris diaphragm openings.

"If particles deviate from a circular shape, this circular light spot must be so adjusted that it is approximately equal in area to that of the particle. Experience has shown this is readily estimated.

"Diameters of the iris diaphragm are correlated, via a commutator, with a number of telephone counters (as shown in Fig. 3.11b), each counter covering a certain range of diameters.

"When the light spot is equalized with a particle in the photomicrograph or electronmicrograph, a foot switch is depressed. That activates the correlated counter, and a puncher marks the counted particle. The photomicrograph or electronmicrograph is then shifted until the next unmarked particle is over the light spot.

"About 15 minutes are required for analyzing 1000 particles. Since the eye participates in the measuring process, the image diameter of the smallest particle to be measured in the photomicrograph or electronmicrograph should be at least 1 mm. The particle sizes are divided into 48 continuous categories. In addition to the individual counters, the instrument is equipped with a counter which registers the total counted particles."[25]

Photographic methods of recording particle images for size analysis have special application for very fine particles suspended in liquid media, since a flash exposure is capable of stopping motion even when the particles are in Brownian motion or in movement because of convective currents in the liquid.

3-3.2 Preparation of Slides for Microscopic Sizing

For samples collected in a filter thimble or other dry dust collector or obtained from a bulk dust, it is essential that the particles be dispersed randomly on a slide without reference to size or shape. Agglomerates must be deflocculated, but in the process individual particles must not be shattered.

Green's technique[26] for the dispersion of paint and rubber pigments in preparation for microscopic examination has been found ill-suited to crystalline materials. According to Silverman and Franklin,[27] the crushing forces exerted by a rigid cylindrical glass rod on a sector of the surface of a small particle resting on a rigid glass slide are tremendous, and shattering occurs. In addition, the recommended wedge-shaped dispersion (although it may be of value for photography in which one needs only a single suitable field) has proved of no value in the examination of slides for particle sizing.

As a substitute for Greeen's method, the rubbing-out process (required for producing an even dust dispersion free of agglomerates) may be performed with the aid of a wooden toothpick. Because the wood is soft and flexible, shattering is minimal. However, since less force is applied to the particles, longer treatment is required to achieve satisfactory dispersion. The shearing forces may be increased materially by the use of viscous mounting media such as gum dammar or Canada balsam dissolved in xylene (in place of the turpentine called for in Green's method). These media have the advantage of producing a permanent mount. Whether temporary or permanent mounts are made, the use of a viscous medium is highly desirable to stop or retard the Brownian motion associated with very small particles (those less than 1.0 μm) and to prevent rapid reagglomeration of the dispersed dust. Glycerin and mineral oil are suitable viscous substitutes for water, the medium most commonly used for preparing temporary dust mounts.

The steps in the preparation of a representative dust slide with the use of permanent viscous mounting media are as follows.[27,28] A drop of mounting medium, such as Canada balsam dissolved in sufficient xylene to make a somewhat fluid solution (80 to 90% solvent), is placed on a clean glass slide, and a small quantity of dust is added from the end of a toothpick. The dust is dispersed in the drop by rubbing, and the suspension is spread over the glass in the form of a thin film. Most of the solvent evaporates during this treatment, and the well-dispersed dust particles become embedded in the viscous residue. A clean coverglass completes the preparation.

The slide is usually put aside for a few hours to permit the Canada balsam to harden, but it may be examined immediately. For temporary mountings in which glycerin or mineral oil is used, drying is not necessary. This type of mounting does not dry, as water preparations do, and has excellent optical properties for use with the oil immersion microscope objective.

If the particles are of such nature that the rubbing-out process (even with a toothpick) is likely to cause shattering or crushing, a small quantity of dust may be dispersed by hand or machine shaking in a few milliliters of a very dilute resin (such as gum dammar in turpentine). A drop of this mixture is placed on a glass slide, covered with a coverslip, and allowed to harden. This method is least subject to criticism so far as shattering of particles is concerned, but the shaking must be very thorough and prolonged to produce a satisfactory dispersion. Because the resin must be greatly diluted with solvent to produce a nonviscous mixture suitable for shaking, there is some tendency toward reagglomeration when a drop is placed on the microscope slide, but hardening can be hastened (and reagglomeration minimized) by drying the slide in a warm place.

Foster and Schrenk[29] have shown that the visibility of small crystalline particles decreases as the relative index (the ratio of refractive indices of particle and mounting medium) approaches 1.0. They note, in addition, that the smaller the particle size, the larger is the range of the refractive index of the liquid which will cause it to become invisible. This indicates that it is impractical to use a mounting medium having an index of refraction close to that of the particles under investigation. For example, Canada balsam has a refractive index (1.535) which is very close to those of quartz (1.544 and 1.553); consequently such a medium should not be used when this mineral is a principal constituent of the dust to be sized. Fortunately, there are other gums and resins which have refractive indices sensibly different from that of Canada balsam. These may be substituted when the need arises. Phase-contrast objectives may be used also to overcome this difficulty. A dust should be examined in more than one mounting medium if there is any doubt as to the refractive index. Table 3.3, largely taken from Shillaber[14] and Loveland,[24] indicates the refractive index of some of the common gums and resins useful for temporary and permanent mountings. Some viscous

TABLE 3.3

REFRACTIVE INDICES OF MOUNTING MEDIA SUITABLE FOR DUST EXAMINATION[a]

Name	Common solvent	Refractive index[b] (25°C)
Water	—	1.333
Silicone oil (30,000 centistokes)	Carbon tetrachloride	1.404
Glycerin	Water	1.463
Isobutyl methacrylate	Xylene	1.47
Mineral oil	Kerosene	1.48
Cedarwood oil (thick)	Xylene	1.515
Gum dammar	Turpentine	1.521
Canada balsam	Xylene	1.535
Colophony	Turpentine	1.545
Styrax	Xylene	1.62+
Aroclor (chlorinated diphenyl)	Xylene	1.63
Balsam of Tolu	Xylene	1.64
Polyvinyl alcohol[c]	Water	~1.54
Gelatin	Water	1.516–1.534
Glucose–pectin	Water	1.43
Glycerin boriphosphate (Abopon)	Water	1.44
Polyvinl acetate[c]	—	1.47
Polystyrene[c]	—	1.6

[a]From Shillaber, Ref. 14, and Loveland, Ref. 24.

[b]Refractive index of resins refers to dried materials. In liquid form these resins have a lower index, since most solvents have an index considerably lower than that of the resin.

[c]These materials form a hard film that may be used with an oil immersion objective without a coverglass.

materials which may be used for temporary mountings are also given. Water is included in the table as reference.

Loveland[24] observed that, when a powdered sample is dispersed in a solution of a resin and mounted between a glass slide and its coverslip, the center of the preparation may contain unevaporated solvent for many days and permit extensive clumping of particles. Therefore, he recommends the use of resins such as polyvinyl acetate and polystyrene, which can be hardened rapidly by polymerization, in preference to mounting liquids that harden slowly by evaporation. He also recommends the addition of a small quantity of a wetting agent to aid dispersion of samples prepared in aqueous mounting media such as polyvinyl alcohol, gelating, and pectin, and the addition of 2 to 5% of a plasticizer such as castor oil or diethyl phthalate to nonaqueous polymer mounting media for the same purpose.

Collecting devices which depend on jet impaction (e.g., Owens jet and Bausch and Lomb dust counters, cascade impactor), electrical and thermal precipitation, settling, and volatile or transparent (membrane) filters can be used to collect airborne dust on a glass slide or film suitable for light or electron microscope examination without further dispersion. Methods of preparation with suitable mounting media have been noted.

Natural gums and resins are apt to be rather dirty. Prior to use, a very dilute solution should be prepared and filtered through a fine paper filter to remove suspended dust and dirt. After filtration, the solution can be made more viscous by evaporating the excess solvent.

Laskin[30] described a method of coating dusts mounted for microscopic examination with a surface layer of selenium or stibnite (or other materials possessing refractive indices greater than 2) by means of a vacuum evaporation technique similar to that used for the production of front-surface aluminum mirrors. He states that by this method it is possible to extend the resolution limit of a particular microscope system from about 0.8 to 0.2 μm or less. The value of the coating procedure is illustrated by results of size measurements on identical slides before and after coating. These show, for a typical specimen, a shift from 0.89 to 0.25 μm in median size after coating.

3-3.3 Size Analysis

To assure statistical validity for the results of the size analysis, the particles selected for sizing must be chosen in a completely random manner. The most serious criticism that may be advanced against the microscopic method of size analysis is that the number of particles actually measured (a few hundred grains) is a minute fraction of the total sample. This factor makes it essential to follow approved techniques throughout all phases of the analysis.

With bulk samples of dust, such as those from product bins, dust collector hoppers, and high-volume air samplers, it is necessary to mix the sample thoroughly. This is especially important for dust samples that have been shipped or transported over long distances. Vibration has a tendency to segregate the dust by particle size, the fines sifting to the bottom of the container through the spaces between the larger particles. If the sample is unusually large, it should first be reduced, mechanically or by hand, using the standard techniques described in Chapter 2 (p. 69). A dust specimen is then taken from this small, freshly mixed bulk sample and used for preparing a microscope slide in the manner described above.

Size segregation may occur on the microscope slide during preparation because the smallest particles tend to stream outward to the edges of the drop of mounting fluid. There is also a tendency for the largest particles to settle to the bottom of the viscous mounting medium. Representative microscopic fields must therefore be selected at the edges of the mount as well as at the center, and a search for particles must be made throughout the entire depth of the dust deposit on the slide. This may be accomplished by selecting the fields to be examined on some pre-established pattern, such as one which makes at least one complete horizontal sweep of the entire slide in the upper, middle, and lower thirds of the mounted specimen. All particles falling entirely within this area are sized, and all others are strictly omitted. Additional fields may be selected on a similar basis when three sweeps of the slide expose an insufficient number of particles to satisfy the statistical counting requirements explained in Chapter 6 (p. 247).

There is an optimum number of particles per microscope field. When particles are crowded together, (1) the largest grains may cover or otherwise obscure the smallest; (2) grains may rest against or atop one another to such an extent that clumps are mistaken for single particles; and (3) some particles may be sized twice, whereas others may be overlooked because of the difficulty in making an orderly enumeration in an overcrowded field. When there are too few particles, (1) a long time is required to locate them; and (2) some of the smallest particles may be overlooked because they are out of focus as the eye sweeps the field.

Although the optimum number per field may vary somewhat from operator to operator and from dust to dust, depending on the size distribution, a population of 100 to 200 particles in an entire microscope field is preferred by many investigators. In most cases only a small area in the very center of the microscope field will be used for actual particle measurements, so that the number of grains sized in each field will be a much smaller number.

When analyzing bulk dust samples, the objective is always to determine the size of individual grains, since the presence of agglomerates merely indicates inadequate dispersion of the sample during the preparation of the slide. No useful information can be obtained regarding the nature and number of agglomerates in a dust cloud from an examination of a bulk dust sample. Nevertheless, it is often necessary to examine and size airborne dust in its original state, taking into account the number and size of agglomerates, as well as unitary particles. The membrane filter is unexcelled for this purpose. Sampling with membrane filters has been described in Chapter 2 (p. 44). For successful results, the density of particles deposited on the filter must be kept sufficiently low so that the probability of superposition (i.e., one particle landing on or adjacent to another) is remote.

Procedures for preparing membrane filter samples for *in situ* examination of the undisturbed dust deposit representing the state of particle agglomeration existing in the aerosol are as follows:

1. Using forceps, place the membrane, dust side down, on a dust-free glass microscope slide and add immersion oil of the correct index of refraction (1.515), drop by drop, very slowly and carefully, to the clean upper surface. The membrane will become translucent and then completely transparent as immersion oil is added. If the oil is added rapidly, however, the dust deposit may be washed out of the filter and float to the edges of the oil drop.[13]

2. The Millipore Filter Corporation[31] suggests gently floating the filter, dust side up, on a film of immersion oil in a petri dish until saturated and transparent. The film is drawn over the lip of the dish to remove excess oil from the bottom side and then placed on a clean glass slide with a rolling motion so as to exclude air bubbles.

3. Viles[32] developed the following technique which gives permanent membrane filter mounts and is least likely to disturb the dust deposit. The membrane, or a section of it, is placed, dust side up, on plastic coverslip supports that have been positioned on a microscope slide so as to form channels under the membrane, as shown in Fig. 3.12. Fixation of the particle to the membrane surface is accomplished with a volatile fixation liquid, and then the regular mounting procedure using Shillaber's immersion oil* is employed. The fixation liquid is 0.5 to 1.0% Lucite or Plexiglas (index of

*Bausch and Lomb Optical Company, Rochester, New York.

refraction 1.49 to 1.51) in benzene. It is made by adding the proper amount of benzene to a weighed block of the plastic in a tightly stoppered bottle and allowing a few days for the plastic to dissolve. The liquid is added slowly to the channels with a fine capillary dropper until it completely fills them or completely wets the membrane. After drying for half an hour the particles will be fixed to the membrane in their original positions and the entire mount cemented to the slide. Then Shillaber's immersion oil is added to the channel with an applicator rod until the channel sections of the membrane are completely transparent, and next oil is added to the top of the stain to assure covering the cemented particles. This procedure permits use of the microscope with or without coverslips for all microscope objectives.

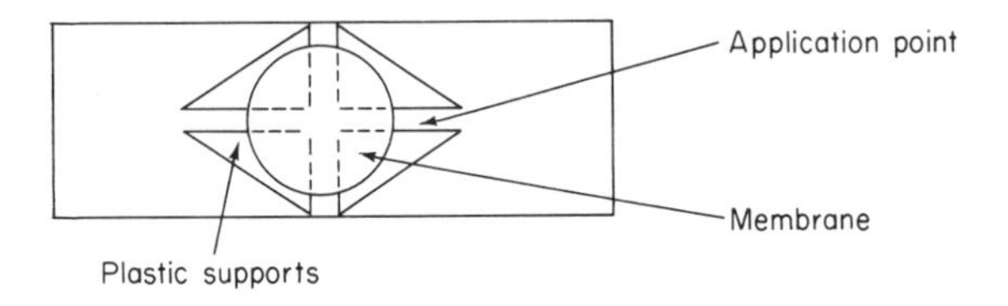

FIG. 3.12. Method of mounting membrane filters. (From Viles, Ref. 32.)

Because of the rapid volatilization of benzene, the application of the fixation fluid on humid days should be done in a confined atmosphere, such as under a watchglass, to eliminate condensation of water. For hygroscopic particles, fixation should be carried out in a thoroughly dry atmosphere, which can be obtained by utilizing plastic-covered desiccators containing Drierite or adding compressed air which has been dried and filtered through appropriate desiccant-filled towers. After the prepared slides have been set in the desiccator and allowed to dry for a period of time, the fixation liquid is applied with a long capillary dropper through an access hole in the cover.

The number of particles that must be measured in each size grouping to give statistically significant results is discussed and illustrated in Chapter 6 (p. 247).

Fig. 3.13 illustrates the type of data, calculations, and results associated with a typical microscopic size analysis. Microscopic sizing gives a size distribution by particle count. The size distribution by weight or volume may be calculated from the size-by-count results with the aid of the conversion equations discussed in Chapter 6 (p. 244). In this figure the mass (weight) median diameter, calculated from the count median diameter and the geometric standard deviation of the size distribution, has been shown.

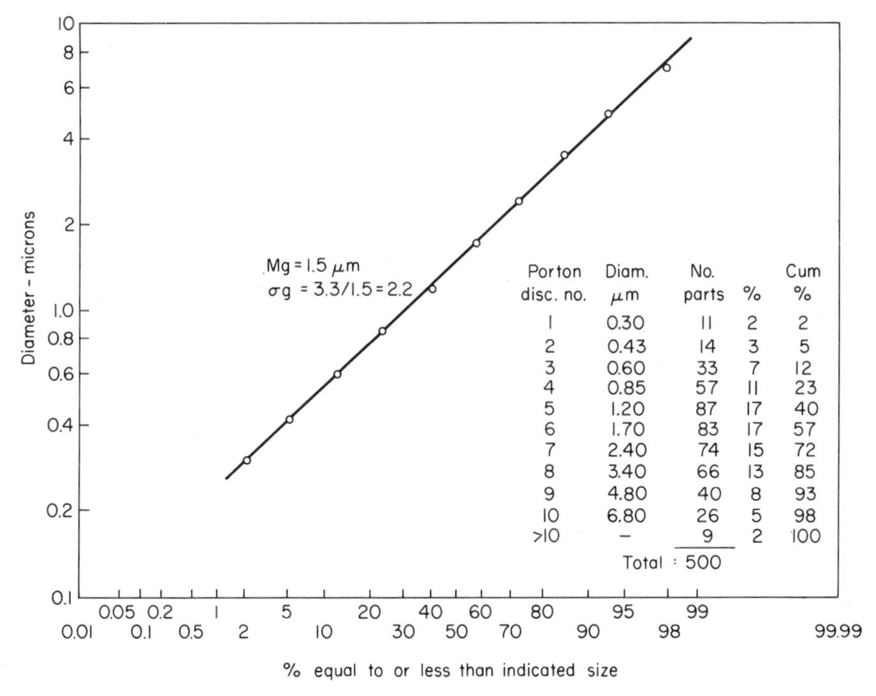

Porton disc. no.	Diam. μm	No. parts	%	Cum %
1	0.30	11	2	2
2	0.43	14	3	5
3	0.60	33	7	12
4	0.85	57	11	23
5	1.20	87	17	40
6	1.70	83	17	57
7	2.40	74	15	72
8	3.40	66	13	85
9	4.80	40	8	93
10	6.80	26	5	98
>10	–	9	2	100
	Total :	500		

FIG. 3.13. Microscopic size analysis data and size curve.

3-4 Electron Microscopy

3-4.1 THE ELECTRON MICROSCOPE

The electron microscope is a powerful tool for the study of submicron dusts and fumes of concern in industrial hygiene, air pollution control, radiological health, and general small-particle technology. It permits observation, counting, and sizing of particles down to about 0.001 μm, as contrasted to the limit of resolution for light optical systems of about 0.25 μm.

Figs. 3.14a and b are photomicrographs of iron oxide fumes and sodium chloride particles, respectively, collected on membrane filters made transparent with cedarwood immersion oil and photographed at maximum magnification in a light microscope (×1200; actual magnification shown is ×900). Figs. 3.14c and d are electron photomicrographs of the same aerosols (collected directly on a specimen support with an electrostatic precipitator) and enlarged to a magnification of ×25,000 (actual magnification shown in photograph is ×15,400). The material just discernible in the light photomicrograph is seen to consist of chains and agglomerates of spheres (Fe_2O_3) and cubes (NaCl) of very much smaller size, of the order of 0.05 μm.

The user of the electron microscope should consult one or more of the standard reference texts in this field[33,34] for detailed information on the construction and operation of these instruments; that of Hall[33] is of most

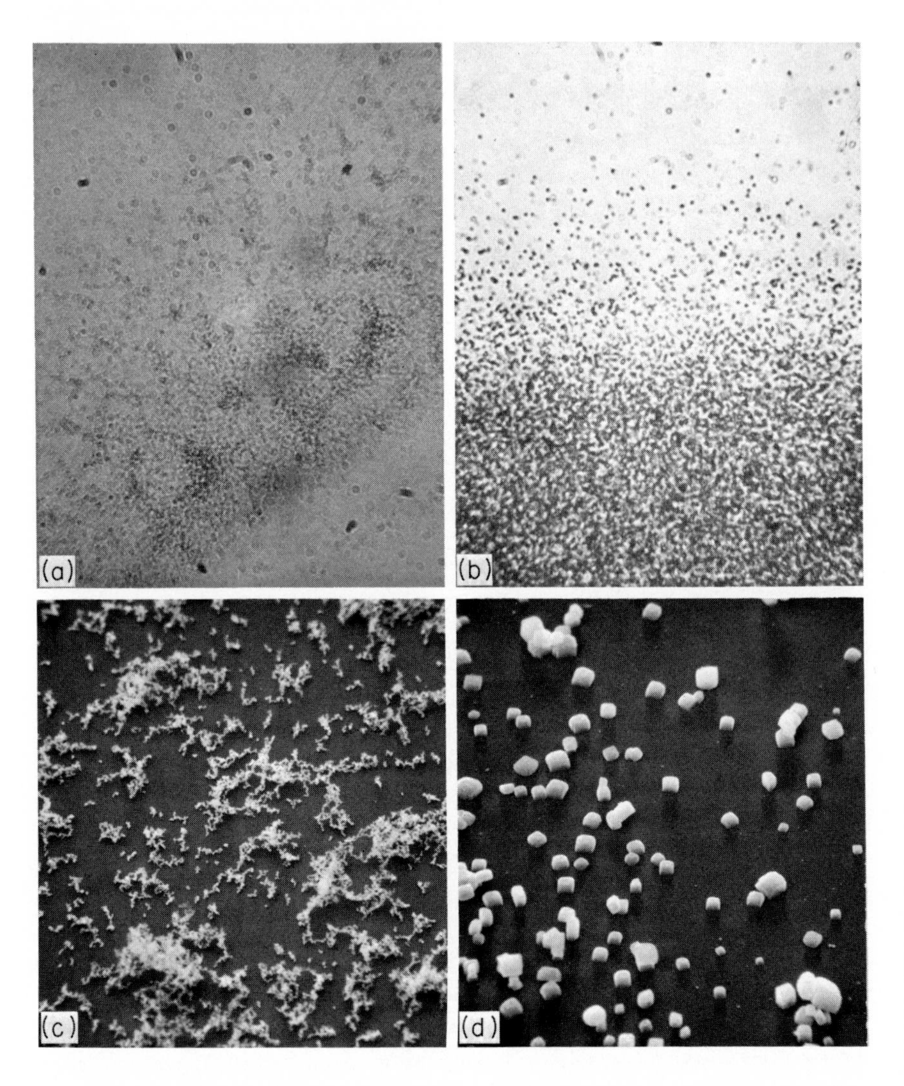

FIG. 3.14. Photomicrographs of iron oxide fumes and sodium chloride particles: (a) and (b) light microscope; (c) and (d) electron microscope.

general utility in explaining the theoretical and practical aspects of function and use.

The electron microscope has components whose function is analogous to those in the compound light microscope, as shown in Fig. 3.15. Illumination is provided by a beam of electrons of short wavelength (<0.05 Å) thermionically emitted from a heated tungsten cathode, accelerated, and initially collimated by an aperture and electromagnetic condenser lens. The specimen is place in the path of the collimated beam on a transparent carrier (the stage), and electrons are absorbed or scattered by the various parts of the specimen. An image is formed of the specimen with varying internal structural details, depending on its thickness and atomic density.

An electromagnetic objective lens magnifies the image 10 to 200 times into the object plane of a magnetic projector lens (equivalent to the light microscope ocular, or eyepiece, and eye–lens combination). The image is magnified further by the projector lens (50 to 400 times) and directed onto a fluorescent screen for viewing (corresponding to the retina of the eye, or a translucent screen of a microprojector). Focusing is accomplished by varying the current to the electromagnetic lenses and viewing the magnified image on the fluorescent screen. Finally, a photograph is taken with standard fine-grain photographic emulsions sensitive to the electron beam, and the negative is enlarged five or ten times to provide a maximum total magnification of the specimen of several hundred thousand times.

A minimum separation of 0.2 mm is required for resolution of two separate points by the human eye. Useful magnifications in the electron microscope range from about 1000 to over 100,000 times, with subsequent photographic enlargement to about five times more. Thus, images just able to be resolved by the human eye (0.2 mm) on an electron micrograph correspond to particles sizes from 0.2 μm, down to less than 0.002 μm, depending on total magnification. The practical limit of resolution of existing electron microscopes is about 0.0003 μm. Because of inherent aberrations in the lenses, the theoretical resolving power of the microscope (0.0001 μm, based on an electron wavelength of 0.000005 μm) has not been realized. The magnification and resolving power attainable in both light and electron microscopes are summarized in Fig. 3.16.

Differences between light and electron microscopes include the high vacuum (10^{-4} torr) necessary for linear transmission of the electron beam, and methods of image formation by electron scattering and absorption rather than by light absorption and reflection as in a light microscope. Although heating is occasionally a problem in light microscopy with certain specimens, the high-power densities applied to specimens in the electron

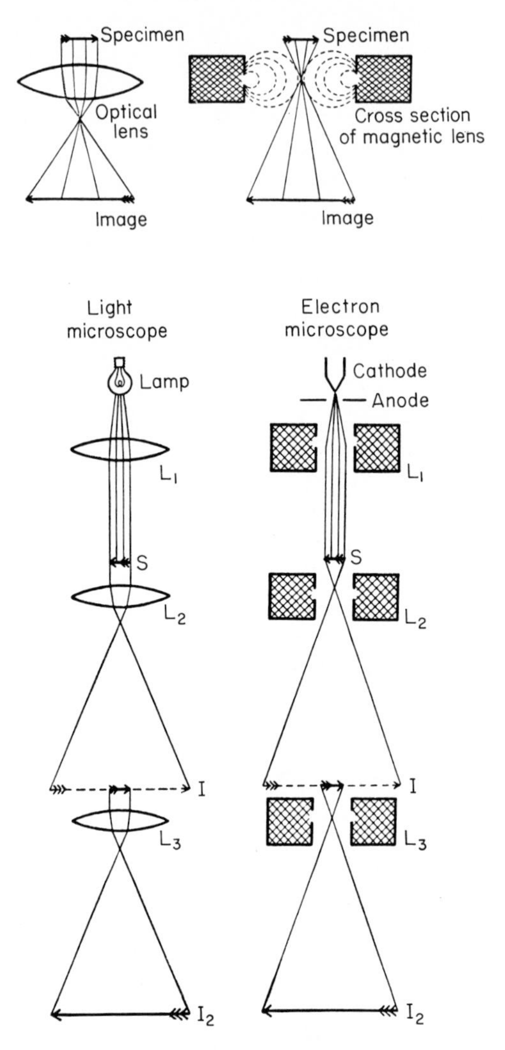

FIG. 3.15. Electron and light microscopes. (Courtesy of Radio Corporation of America.) *Top:* A cross section showing how the field (dotted lines) of a doughnut-shaped magnetic coil can be used to bend the paths of electron rays, thus bringing the rays to a focus and causing the projected image of the specimen to be enlarged in exactly the same way that an optical lens focuses and enlarges a beam of light. *Bottom:* The magnetic lenses in the electron microscope are arranged in the same manner and perform the same functions as the optical lenses in the light microscope. The magnetic lenses, however, are not moved as are the optical lenses; the same effect is more easily obtained simply by varying the current flowing in the coils. This varies the intensity of the field acting on the electron rays and is equivalent to producing a continuously variable focal length lens.

beam (of the order of 1 watt/cm²) introduce serious problems of over-heating. Temperatures in excess of several hundred degrees centigrade would occur in a few seconds if the heat were not conducted away efficiently. The combination of vacuum, high local temperature, and electrostatic charge induced by electron bombardment often causes evaporation, degradation, and movement of the sample while in the beam. In many instances, difficulties can be minimized by suitable specimen preparation. Degradation effects are likely to be major problems for the examination of biological specimens, and the biological literature contains descriptions of many useful techniques designed to prepare specimens for viewing in the electron microscope. Many of these can be applied for small-particle studies (e.g., plastic embedding, microtomy, and replication).

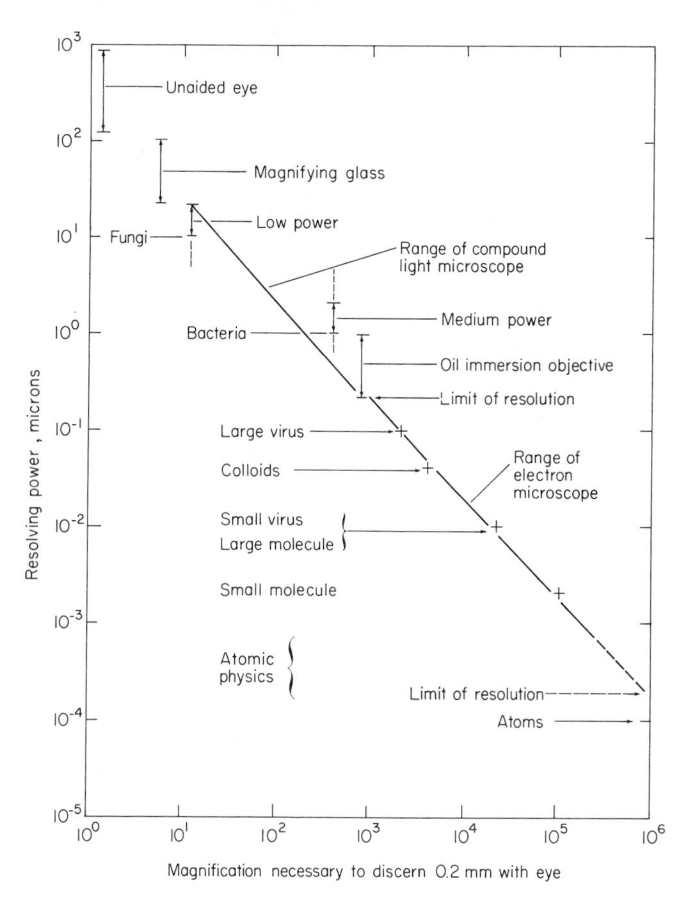

FIG. 3.16. Resolution and magnification. (From Zworykin *et al.*, Ref. 34.)

There are no standard methods that have been universally accepted for sampling airborne particulate material for electron microscopy. The American Society for Testing and Materials has prepared and circulated a tentative standard method, which is summarized below. In addition, several texts on microscopy,[33-35] particle size measurement,[36,37] and small-particle technology[38] deal with the subject of sample collection and preparation in a general way.

3-4.2 Preparation of Samples for Electron Microscopy

(a) *Specimen Mounting.* The type of support used to hold the particle sample must be evaluated for its chemical compatability with the dust under examination. The sample is placed on a thin membrane, or substrate, which is 0.01 μm thick and transparent to the electron beam. It is usually held on a piece of 200-mesh electrodeposited screen grid ⅛ in. in diameter. In some instances it is possible to deposit the aerosol particles directly onto a substrate or an uncoated grid, whereas in others various transfer or replication procedures are applied to the original collected sample to prepare it for microscopy. Structureless substrate materials in common use include Parlodion [1 to 4% nitrocellulose (collodion) in amyl acetate], Formvar (0.5 to 3% polyvinyl formal in ethylene dichloride), and thin carbon films formed by vacuum evaporation and deposition.

The plastic films are prepared and floated on a water surface for transfer to the grid. Hall[33] outlines the following steps for the preparation of plastic-coated electron microscope specimen holders.

"The items needed for producing filmed grids are a shallow circular dish filled to overflowing with distilled water, a clean chromium-plated metal bar for sweeping the surface clean of dust, a grid spade, and 3% collodion (nitrocellulose) in amyl or butyl acetate. Clean the chromium bar by rubbing vigorously with a lintless cloth before and after using. The grid spade consists of a piece of Lektromesh screen about 2 by 3 in., welded or soldered to a piece of heavy nickel wire bent to form a handle and a frame around the edge of the mesh. Place the grid spade in the dish as shown in Fig. 3.17 and deposit the grids on the surface with a long forceps. (The grids cannot be placed on the spade before immersion since they will float on the surface owing to surface tension.) Sweep the surface clean, starting at the point where the handle comes over the edge of the dish. Dip the end of the glass rod drawn down to a round tip about 1 to 2 mm in diameter into the collodion solution, and bring the drop gently into contact with the water surface near the center of the dish. The drop spreads rapidly, forming a circular patch nearly out to the edge of the dish. Broad, circular interference colors appear during drying, but the dry film is colorless and visible only through a difference in reflectivity at the sharply bounded circular edge. Lift the grid slowly upward, bringing the flat surface against the underside of the film, and then lift it clear of the surface at a slight angle. Blot excess water from the underside with filter paper, and allow to dry thoroughly. Lift the grids from the spade with a pair of forceps whose bottom point has been sharpened to a knife-edge to slide under the grids, and transfer them to a glass slide for later use. It is not necessary to cut the film around the edge of the grids before lifting them from the spade. Wash the spade with acetone before using again. About 30 to 100 grids can be prepared in one operation, depending on the size of the grid spade."

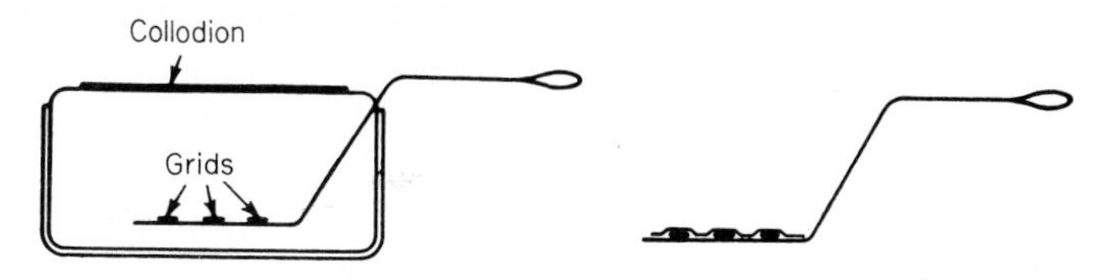

FIG. 3.17. Preparation of filmed grids: (left) dish, (right) grid spade. (From Hall, Ref. 33.)

A carbon film can be prepared by high vacuum evaporation of a spectrographic carbon rod onto a 1 × 3-in. glass slide. The film is stripped from the slide by flotation on a clean water surface, and the coating of individual grids proceeds as described above.

The American Society for Testing and Materials[39] has proposed a standard method for the preparation of aerosol samples for electron microscopy based on the method presented by Posner and Miner.[40] The steps in the preparation of carbon specimen holders given there are as follows:

"(a) Prepare carbon electrodes. Use ⅛-in. diameter spectrographic electrodes. One electrode should have a sharpened point and the other should have a blunt end. Insert them in the electrodes under the bell jar (see Fig. 3.18).

"(b) Cleave mica into a very thin sheet and immediately insert it in the vacuum evaporatory system. Extreme care should be taken to avoid handling the freshly cleaved surfaces.

"(c) Evacuate the system to a pressure of 10^{-5} millimeters of mercury. If the system allows, better results may be obtained at 10^{-6} millimeters of mercury.

"(d) Sublime the carbon electrodes at 30 to 50 amperes for approximately 30 sec to obtain the proper thickness of film on the mica. Individual

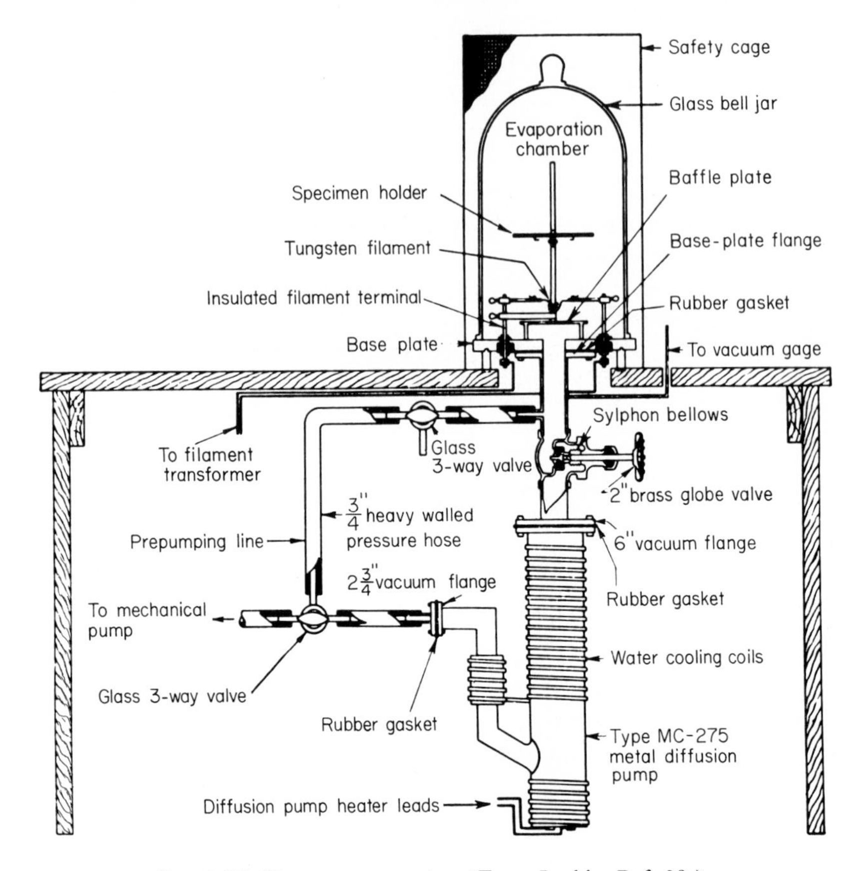

FIG. 3.18. Vacuum evaporator. (From Laskin, Ref. 30.)

adjustments should be made to allow for the geometry of the system being used. A simple method for estimating the evaporated film thickness uses a small piece of white porcelain mounted below the point of evaporation. On it is placed a small drop of low vapor pressure vacuum oil. The carbon evaporates onto the porcelain and visibly darkens it except for the region of the oil drop, where no darkening occurs. When the darkening is just visible, the thickness of the vertical film is about 100 angstrom units at about 10 cm from the source. Thinner film can be obtained by keeping the indicating oil droplet in a region where reasonable darkening occurs, and placing the target further away and at an angle to the source.

"(e) Bring the vacuum system back to atmospheric pressure slowly to avoid destruction of the carbon film in the decompression process.

"(f) Remove the carbon-coated (mica) surface from the bell jar.

"(g) Float the carbon film off the (mica) surface in a bath of double distilled water.

"(h) Fit the desired number of copper grids into the holes in the grid holder (see Fig. 3.19).

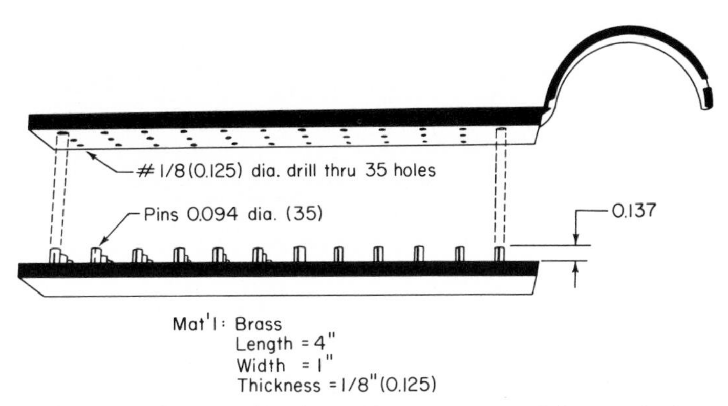

FIG. 3.19. Copper grid holder. (From Posner and Miner, Ref. 40.)

"(i) Immerse the loaded grid holder in the distilled water bath and then lift the carbon film with one single swift motion.

"(j) Wipe off excess water.

"(k) Score the carbon film around each grid. This should be done when the carbon is partially dry. If the carbon film is scored while too wet, it will fly off the grids. If scored when fully dry, the strain on the film will shatter it or detach it from the outer periphery of the grid. (Note: An adhesive can be used to bind the film to the grid. This may introduce undesirable contamination which the above method avoids.)

"(l) Allow the carbon-coated grid to dry completely, e.g., overnight in a desiccator. Then place the grid holder on the specially designed grid lifter to raise the prepared grids from their beds in the holder (see Fig. 3.19).

"(m) Pick up individual grids with fine tweezers, and place immediately in sterile #5 gelatin capsules for safe keeping until needed."

The above technique is frequently modified by replacing the mica by a scrupulously clean glass slide.

Evaporated metals and metal oxides (Si, SiO, A_2O_3, Be, BeO, LiF), as well as graphite, mica crystals, and plastic materials, are used as substrates when special properties are required, or when it is necessary to perform specimen manipulations in the electron beam (e.g., heating, cooling, stretching, micromanipulation, and chemical treatment).

After aerosol particles have been deposited on a suitable substrate it is often desirable to improve three-dimensional visualization by vacuum evaporation of a thin film of metal obliquely onto the object and substrate in the apparatus shown in Fig. 3.18. For vacuum evaporation, a metal (chips or ribbon) is placed in a tungsten carrier and heated to incandescence. Metal atoms are emitted in all directions in essentially straight lines, and when the object is placed some distance away and below the heated metal source, atoms strike one side of the object (and the substrate) and leave "shadows" of uncoated substrate immediately behind the object. The shadow shape bears a definite relation to the shape of the object, and its length is inversely proportional to the tangent of one-half of the shadowing angle. Thus, a spherical aerosol particle, which appears as a two-dimensional circle in a direct electron image on the fluorescent screen of the microscope, takes on a three-dimensional appearance when vacuum-metallized at an angle because of the distinctive appearance of its projected shadow. The shadow can be measured and its form used to determine whether the particle is a sphere or some other shape. Fig. 3.20 illustrates how shadowing is employed to assist in the three-dimensional visualization of particles. This preparative procedure is also useful when electron stereomicrographs are to be used for three-dimensional presentation. Metals frequently used for shadowing includ chromium, germanium, platinum, palladium, silicon monoxide, uranium, gold, nickel, and combinations such as platinum–palladium, 80–20%, palladium–gold, 40–60%, and platinum–carbon. Grain size and ease of preparation are important factors in selecting a shadowing method. The proposed ASTM tentative method for shadow casting[39] is as follows:

"(a) Place specimen or replica of specimen already on support grid

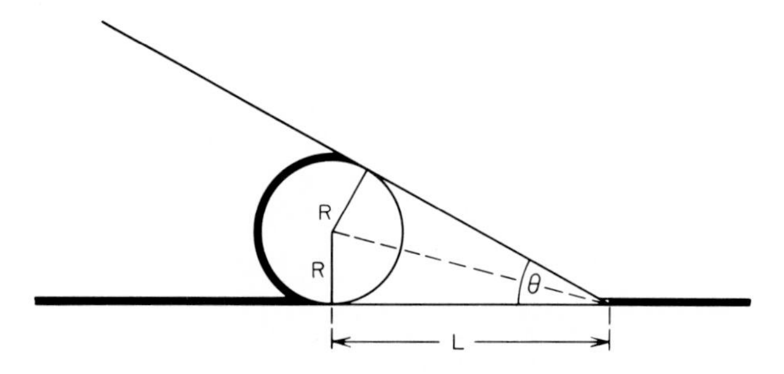

FIG. 3.20. Shadowing of a spherical particle to assist interpretation of shape.

in vacuum evaporator bell jar at a distance of 15–20 cm from evaporating source.

"(b) Place a small piece (200 mg) of chromium on a V-shaped filament or tungsten boat (cup) or basket which forms a resistance element in an electrical circuit. Other heavy metals such as gold or uranium can also be used.

"(c) Evacuate system to 10^{-5} mm Hg and sublime metal (e.g., 1 min at 50 amp for Cr). (Note: It is important not to evaporate excessive metal which might obscure details of the specimen. In making calculation of how much metal to evaporate in order to obtain the desired thickness, the inverse square law may be used [considering] the angle between specimen and source.)

"(d) Note the angle of deposition. The height of the specimen's detail can be roughly estimated from the length of shadows they cast.

"(e) Bring evaporator bell back to atmospheric pressure. Samples are now ready for study in the electron microscope. (Note: In high resolution microscopy it is found that almost all metals aggregate somewhat. This defect is accentuated if the specimen is overheated in the electron beam.)"

Some aerosol investigators have used replication tchniques on particles collected on a flat surface (e.g., a membrane filter) by first vacuum-evaporating a metal perpendicularly (90-degree shadowing) onto the surface. Next, the collecting surface and particulate material are dissolved away, leaving finally the metal replica for examination. The lack of contrast between particles and the collecting surface (particularly in the case of membrane filters) makes subsequent analysis difficult. Other replica techniques use carbon or plastic films, metal-shadowed in some cases, and rereplication of the original replica (which is a "negative") to obtain a "positive" copy of the original material. These techniques are used frequently in metallography for examination of metal surfaces and crystal structure but have severe limitations for aerosol investigations. Fig. 3.21 shows a schematic representation of negative and positive replicas. The proposed ASTM tentative method for preparation of sample replicas[39] is as follows:

"(a) Evaporate 100 Å carbon film onto segment of membrane filter under vacuum using same technique as described above for preparation of carbon film specimen holders, steps (c) through (f).

"(b) Place carbon-coated filter segment on a bare electron microscope grid, carbon film up. The grid should be supported on a screen or glass filter stick in a small petri dish.

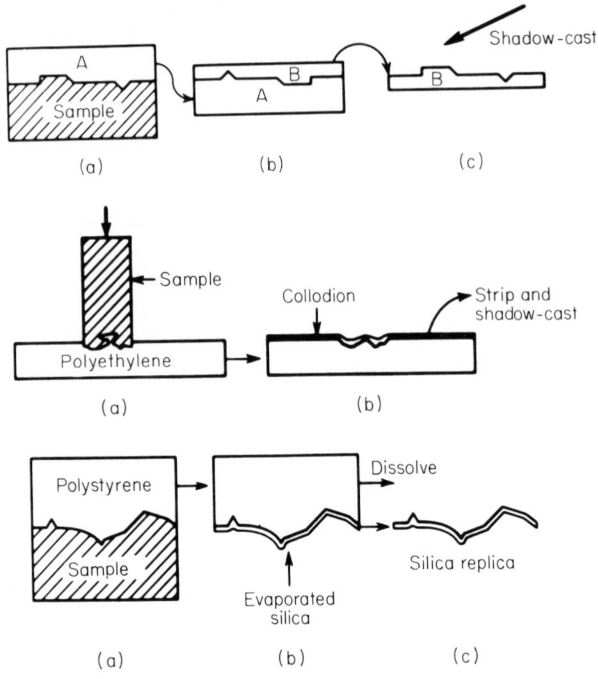

FIG. 3.21. Replication of electron microscope specimens. *Top:* Positive replica technique: Top (a) negative replication (A); (b) positive replication (B); (c) shadow casting stripped positive replica. *Center:* Polyethylene method of positive replication: (a) sample cast in polyethylene; (b) collodion positive replica. *Bottom:* Silica replica method: (a) negative replication with polystyrene; (b) silica evaporated onto replica surface in vacuum; (c) polystyrene dissolved leaving positive silica replica. (From Hall, Ref. 33.)

"(c) Strip membrane filter and sample from replica with ethyl acetate by vapor dissolution.

"(d) Remove ethyl acetate and allow film to dry on grid."

Step (c) above can be improved by using the Tracerlab*[41] Excited Oxygen Asher to oxidize the membrane at low temperature.

For best results, it is advisable to vacuum-evaporate a final thin carbon layer onto the prepared specimen (be it direct, shadowed, or replicated) to provide a conducting layer to dissipate charge and to help protect the specimen preparation from the detrimental effects of intensive electron bombardment.

---

*Trapelo Division-West, LFE, Richmond, California, Model 600 Low Temperature Asher.

(b) *Aerosol Sampling for Electron Microscopy.* Devices generally useful for the collection of submicron aerosol particles for electron microscopy include the thermal precipitator, the electrostatic precipitator, and the membrane filter, although many other methods have been reported.[42]

When using the thermal precipitator, the sample is collected on glass microscope coverslips held ajacent to the heated wire. The glass slides are precoated with a vacuum-deposited carbon film to act as a substrate for subsequent transfer to an electron microscope sample grid. Coated grids should not be used directly in the thermal precipitator, because the thermal field in the vicinity of the grid wires is quite different from that adjacent to grid openings covered by the carbon substrate. Since heat flows more readily to the grid wires, this leads to preferential deposition on the wires and a sparse deposit on the film.[43] After sampling is completed, the coverslip is removed from the instrument and shadowed with a metal to improve contrast and three-dimensional visualization; a final thin carbon film is added to the surface of the sample to preserve and protect it during handling and electron bombardment. Next, the prepared sample is removed from the glass slide to water surface by flotation and transferred to a grid for observation in an electron microscope.

Electrostatic precipitators designed primarily for the collection of particles on a carbon-coated grid for direct insertion in the electron microscope have been reported by several investigators, including Bouton *et al.* (Chapter 2, reference 23), Baum (cited in reference 42), Billings and Silverman,[42] and Morrow and Mercer.[44] These devices generally consist of a point electrode disposed a short distance from a plane carrier supporting the coated grid, as the opposing electrode, and are called point-to-plane precipitators. The simple, inexpensive device described by Billings and Silverman[42] for field stack sampling and laboratory aerosol research is shown in Fig. 3.22. It consists of a glass tubing cross and two electrodes. One electrode comprises two concentric threaded tubes and a central screw to hold the grid in position for sampling. This is placed in a rubber stopper in one arm of the cross opposite a second electrode (a needle) in the other arm. The aerosol sample is drawn through the larger tube of the cross by means of a suction source at a velocity of 14 ft/min (5 lpm). After air flow conditions have been established, the precipitator is energized with a luminous tube transformer at 5 to 15 kV until a deposit is just barely visible on the inside of the glass tube near the electrodes; usually a few seconds are sufficient for industrial dust or stack concentrations.

The more elaborate instrument of Morrow and Mercer,[44] shown in Fig. 3.23, has a sampling rate of 70 ml/min. It was developed for laboratory

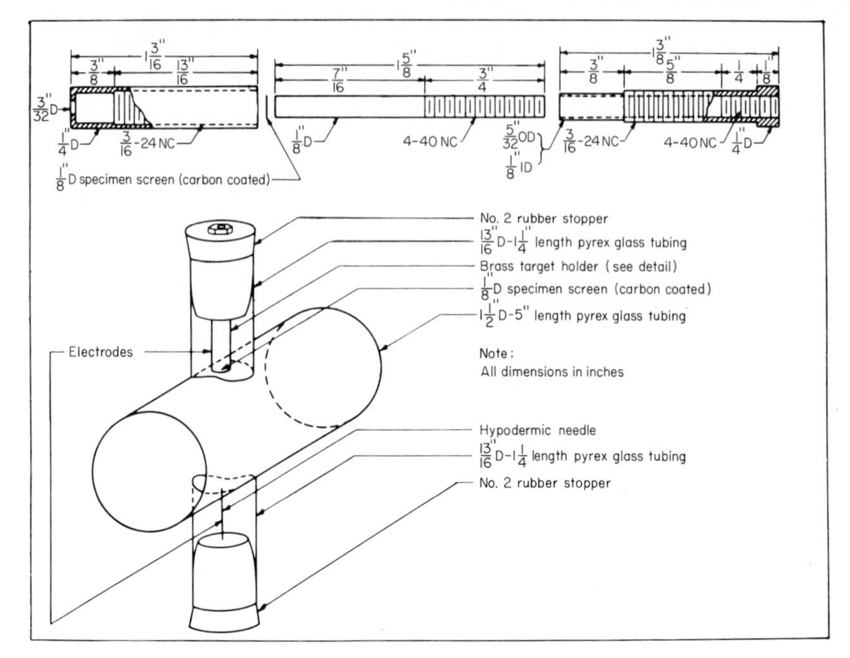

FIG. 3.22. Simple point-to-plane electrostatic precipitator for electron microscopy. (From Billings and Silverman, Ref. 42.)

toxicological investigations and is constructed of Lucite and contains an integral membrane filter holder. For these reasons it is less suitable for stack sampling or use at elevated temperatures.

Use of an electrically conductive carbon coating on the sampling grid in the electrostatic precipitator reduces preferential deposition of particles on the grid wires. The low aerosol velocity past the electrodes provide ample opportunity for deposition of a representative particle sample in the size range of interest for electron microscopy, when the geometry, voltage, and flow rate are adjusted in accordance with discussions presented in Chapter 1 (pp. 5 - 19). After sample collection in the electrostatic precipitator, preparation of electron micrographs is essentially the same as for specimens removed from the thermal precipitator.

In the membrane filter technique described by Kalmus,[45] an air sample of suitable size is drawn through the filter to separate airborne solids. After sampling, the filter is removed from the sampling holder and cut into approximately 5/64-in. (2-mm) squares, which are placed face (sample side) down on Formvar-coated grids. These are placed on a 60-mesh screen held by a perforated brass "bridge" in a petri dish filled with acetone to a level just

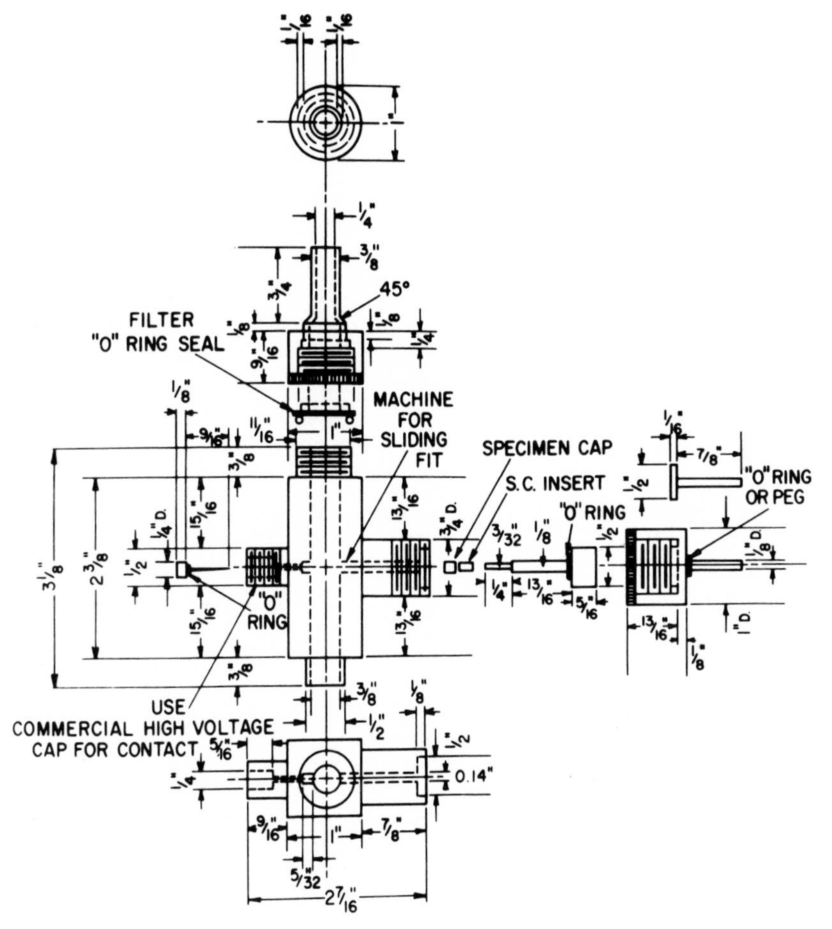

FIG. 3.23. Electrostatic precipitator for direct collection of aerosol particles on electron microscope specimen carrier. (From Morrow and Mercer, Ref. 44.)

touching the bridge. Three changes of acetone, at 30-min intervals, are necessary for complete dissolution of the filter, and the grids are then removed and allowed to dry overnight. The Tracerlab Excited Oxygen Asher, which operates at low temperature, is also a suitable alternative here.[41]

The membrane filter may be used as its own substrate in the procedure suggested by First and Silverman[13] by floating a small piece of the filter, dust side up, on an acetone surface and observing the dissolution of the filter against a strong reflected light. When the filter has partly dissolved, an uncoated grid is placed under the remaining thin film (with tweezers) and slowly lifted to coat the grid. The acetone evaporates in a few minutes,

TABLE 3.4
PHOTOGRAPHIC MATERIALS FOR ELECTRON MICROSCOPY[a]

	Films and Plates	Speed to electrons	Grain	Contrast	Remarks
Kodak:	Fine-grain positive film	fast	very fine	high	SA-I is slightly slower but has finer grain and higher contrast.
	SA-I (film)	fast	very fine	high	
	SA-I (plates)	fast	fine	high	Best Kodak for general electron microscopy.
	Lantern slide contrast	medium	fine	high	Slightly higher contrast than SA-I (plates).
Ilford:	Rapid process experimental	very fast	fine	high	Made for EM, fastest film for general use.
	N-40 Process	fast	very fine	high	Slower than rapid process.
	N-60 Photo-mechanical	medium	extremely fine	very high	Good for resolution and low-contrast subjects.
	Special lantern slide contrast	medium	fine	high	Not suitable in lieu of other Ilford emulsions.
	High resolution	very slow	super fine	very high	Very good for high resolution but too slow for general use.

	Developers	Speed of development	Effect on emulsion speed	Grain size	Keeping qualities	Remarks
Kodak:	Microdol	17–25 min	reduces	very fine	6+ mos	Unfavorable development, too long, brownish image.
	D-76	17–25 min	same	fine	6+ mos	Good—of no advantage because of long development.
	D-72,D-60, D-19,D-11	1–9 min	same	medium	6+ mos	Poor—fast development, but excessive grain.
G.E.:	X-ray	1–3 min	increases some	medium	3–6 mos	

TABLE 3.4 (CONTINUED)

Developers	Speed of development	Effect on emulsion speed	Grain size	Keeping qualities	Remarks
Ilford: Microphene	3–5 min	increases	very fine	12+ mos with replenishment	Very good—short development with fine grain black image.
ID-48	14–24 min	reduces	extremely fine	6+ mos	Good—except for long development.

[a]Courtesy of Philips Electronic Instruments, Mount Vernon, New York.

leaving a substrate-mounted dust sample essentially undisturbed for subsequent examination in the electron microscope.

3-4.3 Exposure and Photography

The specimen of deposited material, prepared as discussed above, is examined in the electron microscope at suitable magnification. Several representative fields are photographed. Since the field of view is very small at high magnifications, it is important to select representative fields at random to avoid bias. It is usually desirable to record the image of an object of known dimensions at the same magnification, as an internal standard.

Objects used for calibration purposes include monodispersed polystyrene latex spheres of known size, electron diffraction grating carbon replicas of known line spacings (28,800 and 54,800 lines per inch), and silver screens with exact wire spacing of 1000 or 1500 mesh. The polystyrene latex particles will shrink on prolonged exposure to the electron beam, so that they must be shadowed, or exposed for very short periods.

The unknown sample and the calibration standard are processed and printed together to reduce errors introduced during handling. In some models two specimen holders can be mounted simultaneously in the microscope, one for the sample and one for the calibration standard. This arrangement reduces errors during changing of sample holders. After representative fields on several grid openings have been photographed at appropriate magnifications, and the calibration standard has been photographed at the same magnifications, the film is removed from the microscope for developing and printing.

Roll film and plates suggested for use in electron microscopy by Philips are included in Table 3.4. On the basis of extensive experience with the Philips 100 B, P-426 roll film (unperforated) has been suggested for making printing negatives and contrast lantern slide plates.*

Much valuable information can be obtained by stereophotomicrography as an aid in three-dimensional visualization of a sample. This is accomplished by rotating the specimen holder a few degrees off horizontal in the beam (the specimen holder is usually provided with sufficient angular freedom for this purpose), focusing this slightly slanted image on the viewing screen, and taking a through-focus series of photographs. The specimen holder is then rotated to its opposite angular extreme (a few degrees off horizontal in the other direction), and another through-focus series is obtained. The best print of each series is mounted in a stereoptican viewer to provide a distinct picture of the sample in three dimensions, which is quite helpful in interpretation in conjunction with specimen shadowing.

Exposure of a sample to the heat and vacuum in an electron microscope column may cause volatilization or evaporation, and this must be considered when interpreting photomicrographs. Stray decomposition products of residual oil and grease vapors, which unavoidably find their way into the column from miscellaneous components and seals, from the vacuum system, and from the specimen itself, gradually coat the sample. These contaminants deposit on the unprotected specimen when it is left in the column for extended periods. Therefore, short exposures and cleanliness are essential.

The transfer of electrostatic charge to the specimen from the beam has been mentioned, and its effects are frequently observed in poorly prepared samples as a rupture and rolling of the substrate. Alternatively, particle migration may be observed from the same cause. As has been indicated previously, the charge effect can be minimized by deposition of a final charge-conducting carbon film on top of the prepared sample.

3-4.4 Measurement

In practice, a print of a calibration standard may be measured with a rule, and the total magnification calculated from known and measured dimensions. Then the electron micrograph of the sample can be sized directly by visual measurement of images of the deposited particles, using a millimeter rule, or an enlarged transparent comparison graticule of the Porton type.

Since both a positive print and a negative are available for analysis, further refinements can be made if many samples are to be sized routinely. The negative can be scanned by an automatic particle counter, as discussed

---

*Ladd Research Industries, Burlington, Vermont; personal communication.

in earlier sections of this chapter, or the print can be evaluated semi-automatically by the variable-aperture light beam device of Zeiss[25]. An optical comparator can be used as a microprojector to enlarge a print or negative for direct measurement with a Porton grid transparency. A standard light microscope microprojector, such as Model 500 of Bausch and Lomb,* which shows an image of a Porton graticule on the translucent viewing screen, can be used with a low-power microscope to evaluate a print or negative. The diameter to be measured on the negative or print can be selected as discussed previously in this chapter (p. 94).

A survey of several competent research groups indicated that differing measures are in common use. Reported results were found to be in fairly good agreement,[46] but there is no general agreement as to which measure of size best represents the health hazard associated with submicroscopic particles.

The statistical interpretation of the size measurements is identical with that described for particle sizing with the light microscope (see Chapter 6). Limitations due to errors caused by the microscope vacuum and electron bombardment, described above, must be considered when evaluating results; an independent check by two different methods of sampling and sample preparation is of distinct importance in this regard, as is the method of calibration.

3-5 Scanning Electron Microscopy

3-5.1 Description

The conventional transmission electron microscope discussed above is used to determine the two-dimensional silhouette or outline of opaque particles deposited on a thin substrate essentially transparent to the electron beam. Many preparative techniques, including shadowing and replication, can be employed to provide additional information about individual opaque particles, such as their general shape, morphology, and surface topographical features, for nonre-entrant geometries.

In the scanning electron microscope (see Fig. 3.24) an electron beam is focused to about 50 to 100 Å (0.005 to 0.1 μm) and deflected (scanned) in a regular pattern across the specimen surface held at an angle to the beam. Secondary low-voltage electrons are emitted from the surface by the primary electron beam in proportion to the primary beam voltage, the nature of the specimen surface at the beam, and the beam incidence angle. Part of

*Rochester, New York.

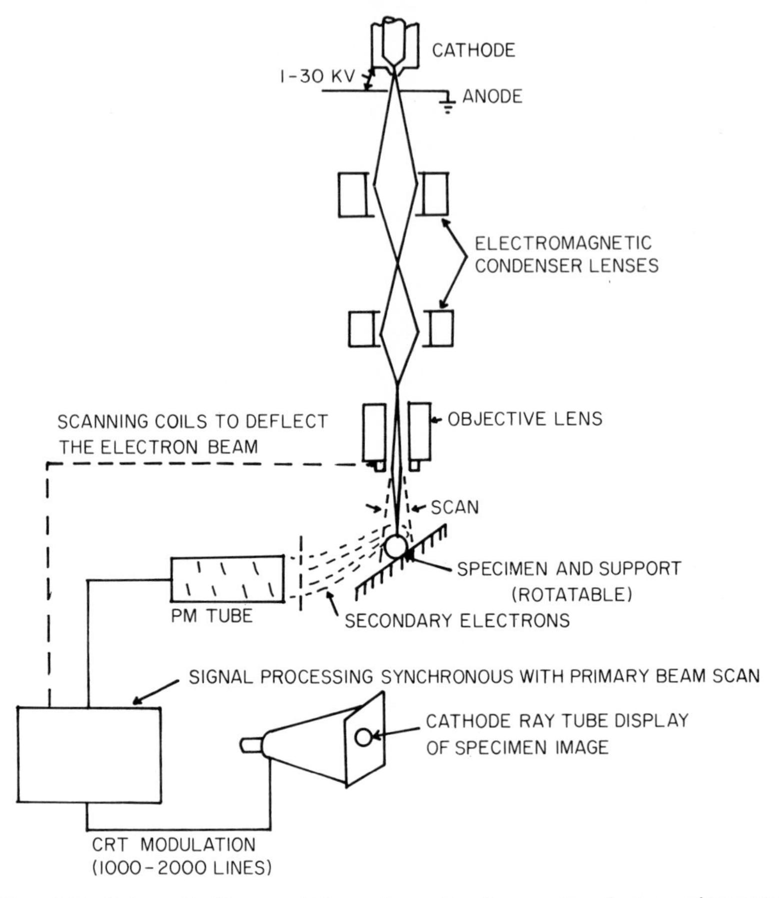

FIG. 3.24. Schematic diagram of image formation in scanning electron microscope.

the emitted electron current is drawn through a curved path by a positive grid to a scintillator–photomultiplier. The output of the photomultiplier is amplified and used to modulate the output brightness of a cathode ray tube whose electron beam is moved in synchronism with the primary electron beam. The tube displays a reconstructed image of the surface of the object, similar to a television picture. Because the secondary electrons are emitted at relatively low voltage, they can be made to follow a curved path, which allows the visualization of curvature associated with re-entrant geometry, holes, valleys, and similar features out of direct line-of-sight of the secondary electron collector. The result is a remarkable representation of particles and surfaces in the beam, as shown in Fig. 3.25. Objects appear with a distinct three-dimensional effect, because surface features closer to the electron collector appear brighter than those further away.

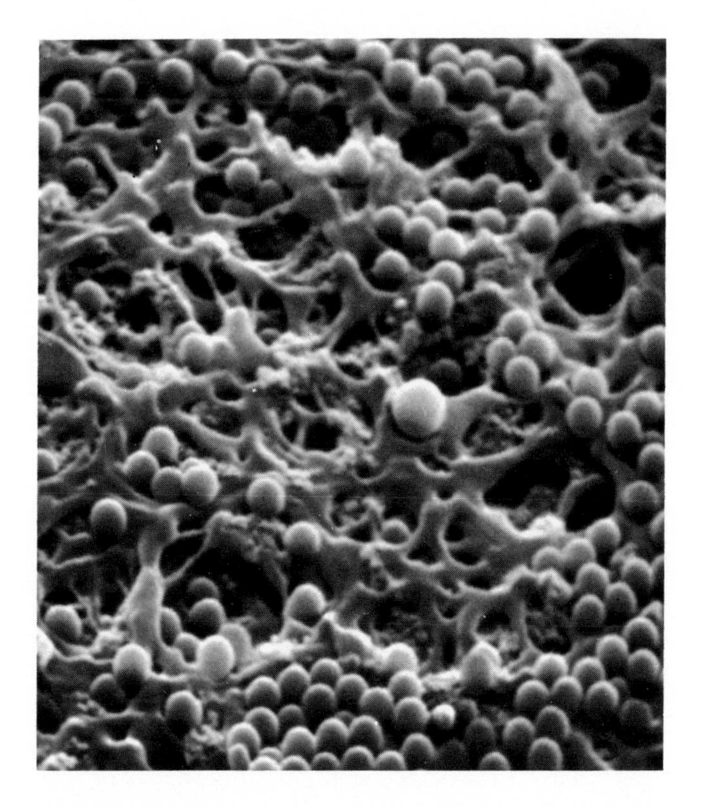

FIG. 3.25. Polystyrene latex spheres on membrane filter by scanning electron microscopy. (From Bulba, Ref. 47.)

The scanning electron microscope (SEM) operates at a large depth of field, some 300 to 500 times that available in a light microscope at the same magnification. Resolving power is presently limited to approximately 150 Å (0.015 μm). Useful magnification is variable from $\times 10$ to about $\times 10^5$. The rotation of the specimen stage facilitates the photographing of stereo pairs, further enhancing the three-dimensional effect, and permitting quantification of surface features.

Primary beam voltage may be varied from 1 to 30 kV. In conjunction with a solid-state x-ray detector, the SEM can provide electron microprobe elemental analysis ($Z > 12$) over small areas, and with an Auger electron emission (electron velocity) analyzer it can detect lower atomic number elements not easily determined by x-ray emission ($Z < 20$). The cost of the SEM is approximately $100,000, and there are presently some eight manufacturers listed.[48]

3-5.2 Specimen Preparation

Particulate materials may be viewed directly on any convenient collector surface, such as the polystyrene latex spheres shown on the membrane filter in Fig. 3.25. In order to prevent electrostatic charging of insulating materials such as these, a thin conducting layer (100 to 400 Å) of gold–palladium or carbon is vacuum-evaporated onto the specimen. Biological material may require more preparation such as fixing, drying, freezing, and sublimation.

3-5.3 Applications

The history and development of applications of the SEM since the first commercial models were introduced in 1965-1966 is given in recent books and symposia.[49–51] General applications of the SEM in fine-particle research both complement and supplement the conventional transmission electron

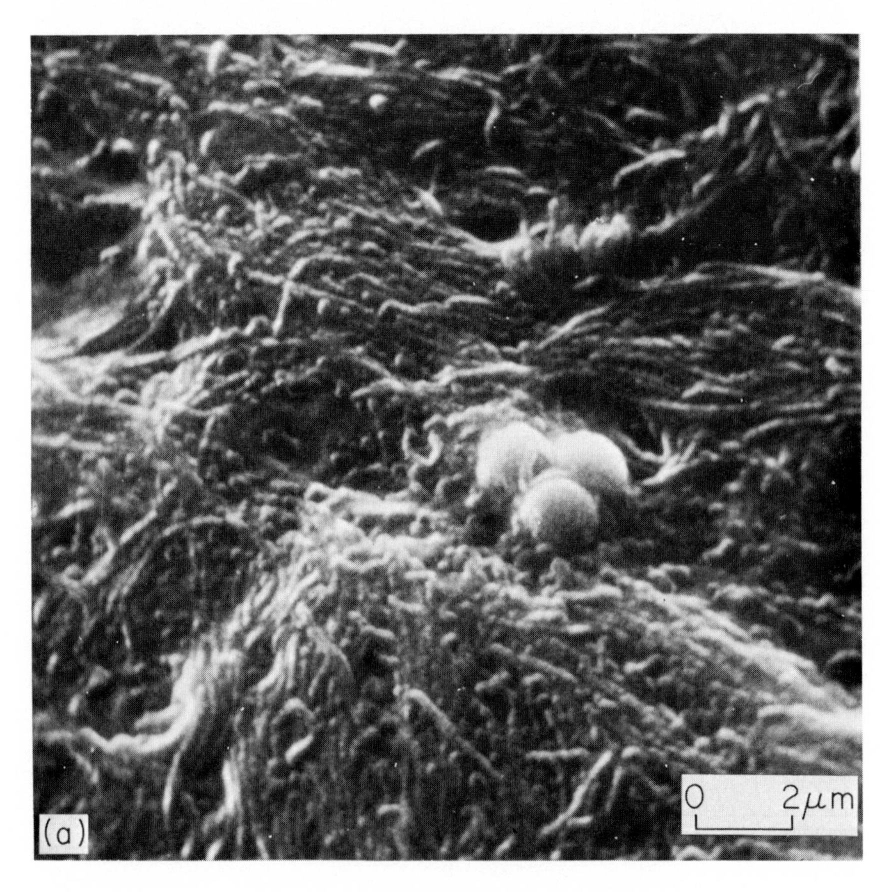

microscope and the light microscope. Details of whole particles as they are naturally deposited can be observed with outstanding clarity, depth of field, and high magnification. A brief résumé of recent SEM studies related to various aspects of particle sizing technology and preparative techniques useful in biological systems is given in Table 3.5. *In situ* investigations of deposited submicron aerosol particulates are providing fresh insights into mechanisms of particle deposition, clearance, and retention in the lung,[61] as shown in Fig. 3.26. The use of the SEM with a digital computer for the development of an automatic particle size analyzer has been recently described.[62]

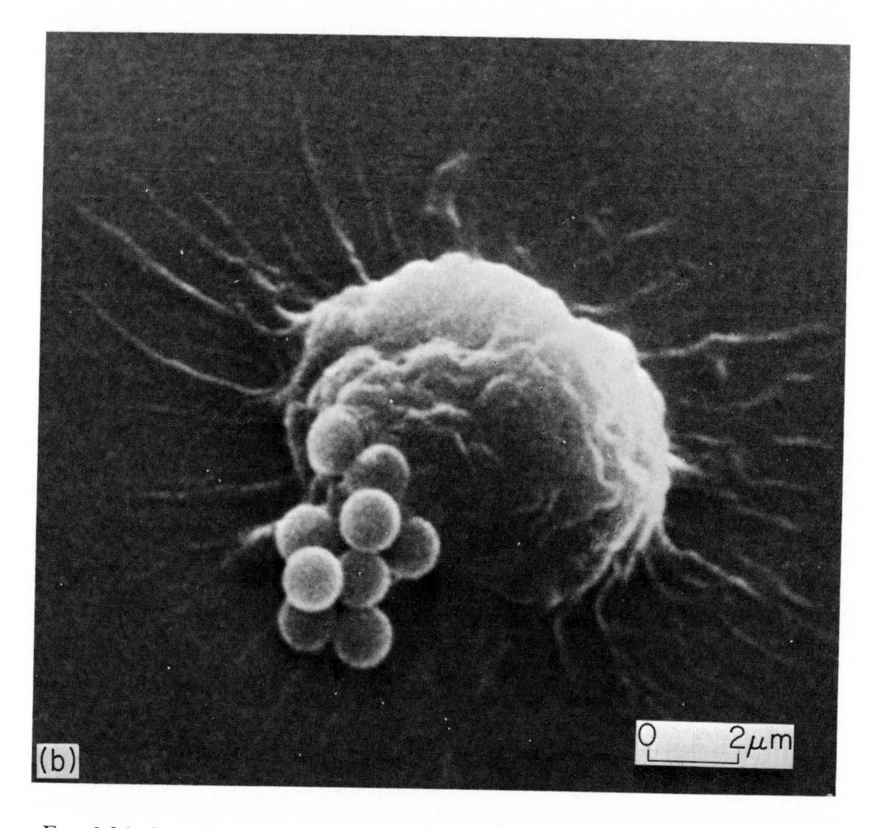

FIG. 3.26. Scanning electron microscopic observation of particles deposited in the lung. (From Holma, Ref. 61.) (*a*) Ciliary bed of mucosa of lower part of main bronchus in left lobe with three deposited 1.3 mμ polystyrene latex particles; individual cilia can be seen. (*b*) Lung macrophage in contact with 1.3 mμ polystyrene particles from study *in vitro* of phagocytosis of these particles by isolated macrophages received by lung washings. Macrophages were fixed with formaldehyde and subsequently dried in air. The microspikes can easily be demonstrated at periphery of macrophage.

TABLE 3.5
RECENT SCANNING ELECTRON MICROSCOPE STUDIES RELATED TO PARTICLE SIZING TECHNOLOGY AND BIOLOGICAL SYSTEMS

Subject	Remarks	References
Biological systems		
Pulmonary deposition	Sodium chloride, rat lung	47
Ciliary motion	Protozoans; includes preparative techniques	52,53
Anatomy, nerve tissue	Submicron cellular processes and preparatory techniques	54
Microorganisms	Planktonic Foraminifera and preparatory techniques	55
Microorganisms	Antibiotic effects of *Staphylococcus aureus, Streptococcus pyogenes,* and preparatory techniques	56
In vivo studies	House fly	55
Particle research		
Filter deposits	Polystyrene latex on membrane (see Fig. 3.25)	47
Filter deposits	Quartz, fly ash, iron oxide, ragweed pollen on membrane.	47
Filter deposits	Serum proteins, membrane pore size distributions	57
Filter deposits	Tungsten on polyester fabric	58
Thermal precipitator deposits	Quart	58
Thermal precipitator deposits	Polystyrene latex	47
Surface nucleation	Chalcopyrite surface defects	59
Powders	Characterization of surface phenomena	60

References

1. P. Drinker and T. Hatch, *Industrial Dust,* 2nd ed., McGraw-Hill Book Company, Inc., New York, 1954.
2. American Society for Testing and Materials, Standard E-11-61, Philadelphia, Pa., 1961.
3. *Testing Sieves and Their Uses,* Handbook 53, W. S. Tyler Company, Cleveland, Ohio, 1969.
4. L. T. Work, Introduction, *Symposium on Particle Size Measurement,* p. 2, American Society for Testing and Materials, Philadelphia, Pa., 1959.
5. R. Schuhmann, Jr., *Powder Metallurgy,* p. 210, American Society of Metals, Cleveland, Ohio, 1942.

6. H. E. Schweyer and L. T. Work, Methods for Determining Particle Size Distribution, in *Technical Publication No. 1,* American Society for Testing and Materials, Philadelphia, Pa., 1941.
7. A. M. Gaudin, *Principles of Mineral Dressing,* p. 51, McGraw-Hill Book Company, Inc., New York, 1939.
8. F. G. Carpenter and J. R. Deitz, Glass Spheres for the Measurement of the Effective Opening of Testing Sieves, *J. Res., Natl. Bur. Std., 47*:139 (1951).
9. H. W. Daeschner, E. E. Seibert, and E. D. Peters, Application of Electroformed Precision Micromesh Sieves to the Determination of Particle Size Distribution, p. 26, in *Symposium on Particle Size Measurement,* American Society for Testing and Materials, Philadelphia, Pa., 1959.
10. H. W. Daeschner, Wet Sieving with Precision Electroformed Sieves, in *International Conference on Powder Technology,* Illinois Institute of Technology Research Institute, Chicago, Ill., May 20-23, 1968.
11. H. O. Suhm, Oscillating Air Column Method for the Dry Separation of Fine and Subsieve Particles Sizes, in *International Conference on Powder Technology,* Illinois Institute of Technology Research Institute, Chicago, Ill., May 20-23, 1968.
12. E. Ioos, Ultrasonic Microscreening, *Staub* (English transl.), *25*:19 (1965).
13. M. W. First and L. Silverman, Air Sampling with Membrane Filters, *A.M.A. Arch. Ind. Hyg. Occupat. Med., 7*:1 (1953).
14. C. P. Shillaber, *Photomicrography,* John Wiley & Sons, Inc., New York, 1944.
15. H. S. Patterson and W. Cawood, The Determination of Size Distribution in Smokes, *Trans. Faraday Soc., 32*:1084 (1936).
16. K. R. May, The Cascade Impactor, an Instrument for Sampling Coarse Aerosols, *J. Sci. Instr., 22*:187 (1945).
17. G. L. Fairs, Developments in the Techniques of Particle Size Analysis by Microscopical Examination, *J. Roy. Microscop. Soc., 71*:209 (1951).
18. C. H. Bosanquet, quoted by Fairs, Ref. 17.
19. M. Corn, The Effect of Dust Particle Orientation on Particle Size Determined by Microscopic Techniques, *Am. Ind. Hyg. Assoc. J., 25*:1 (1964).
20. C. N. Davies, Measurement of Particles, *Nature, 195*:768 (1962).
21. H. H. Watson, Simplified Eye-Piece Graticule for Assessing Thermal Precipitator Dust Samples, *Brit. J. Ind. Med., 9*:80 (1952).
22. H. Heywood, A Comparison of Methods of Measuring Microscopical Particles, *Bull. Inst. Mining Met.,* No. 477 (1946).
23. J. Dyson, The Precise Measurement of Small Objects, *AEI Eng., 1*:1 (1961).
24. R. P. Loveland, Methods of Particle Size Analysis, in *Symposium on Particle Size Measurement,* American Society for Testing and Materials, Philadelphia, Pa., 1959.
25. F. Endter and H. Gebauer, Ein Einfeches Gerät zur Statistichen Auswertung von Mikroskopischen Bzw. Electronenmikroskopischen Aufnahnen, *Optik, 13*:97 (1956).
26. H. Green, A Photographic Method for the Determination of Particle Size of Paint and Rubber Pigments, *J. Franklin Inst., 192*:637 (1921).
27. L. Silverman and W. Franklin, Observations on the Preparation of Slides for Particle-size Determination, *J. Ind. Hyg. Toxicol., 24*:51 (1942).
28. M. W. First, Microscopic Sizing: Notes on Preparation of Dust Samples, *A.M.A. Arch. Ind. Hyg. Occupat. Med., 7*:58 (1953).

29. W. D. Foster and H. H. Schrenk, *Petrographic Identification of Atmospheric Dust Particles,* United States Bureau of Mines, Report of Investigations No. 3368, 1938.
30. S. Laskin, The Selenium-Coating Technic for High Refractive Index Mounting, in C. Voegtlin and H. C. Hodge (Eds.), *Pharmacology and Toxicology of Uranium Compounds,* McGraw-Hill Book Company, Inc., New York, 1949.
31. *Detection and Analysis of Contamination,* Millipore Filter Corporation, Bedford, Mass., ADM-30, 1965.
32. F. J. Viles, Jr., personal communication, 1956.
33. C. E. Hall, *Introduction to Electron Microscopy,* 2nd ed., McGraw-Hill Book Company, Inc., New York, 1966.
34. V. K. Zworykin, G. A. Morton, E. G. Ramberg, J. Hillier, and A. W. Vance, *Electron Optics and the Electron Microscope,* John Wiley & Sons, Inc., New York, 1945.
35. G. L. Clark (Ed.), *The Encyclopedia of Microscopy,* Reinhold Publishing Corporation, New York, 1961.
36. R. D. Cadle, *Particle Size Determination,* Interscience Publishers, New York, 1955.
37. C. Orr and J. M. Dallavalle, *Fine Particle Measurement,* The Macmillan Company, New York, 1959.
38. H. L. Green and W. R. Lane, *Particulate Clouds,* 2nd ed., E. and F. N. Spon, Ltd., London, 1964.
39. *Proposed Tentative Method of Test for Preparation of Aerosol Samples for Electron Microscopy,* American Society for Testing and Materials, Philadelphia, Pa., 1963.
40. S. Posner and E. Miner, A New Method of Preparing Carbon-coated Copper Electron Microscope Grids, *Am. Ind. Hyg. Assoc. J., 24*:188 (1963).
41. C. E. Gleit and W. D. Holland, Use of Electrically Excited Oxygen for the Low Temperature Decomposition of Organic Substances, *Anal. Chem., 34*:1454 (1963).
42. C. E. Billings and L. Silverman, Aerosol Sampling for Electron Microscopy, *J. Air Pollution Control Assoc., 12*:586 (1962).
43. C. E. Billings, W. J. Megaw, and R. D. Whiffen, Sampling of Sub-micron Particles for Electron Microscopy, *Nature, 189*:336 (1961).
44. P. E. Morrow and T. T. Mercer, A Point-to-Plane Electrostatic Precipitator for Particle Size Sampling, *Am. Ind. Hyg. Assoc. J., 25*:8 (1964).
45. E. H. Kalmus, Preparation of Aerosols for Electron Microscopy, *Z. Wiss. Mikroscopie, 64*:7,414 (1960).
46. H. J. Ettinger and S. Posner, Evaluation of Particle Sizing and Aerosol Sampling Techniques, *Am. Ind. Hyg. Assoc. J., 26*:17 (1965).
47. E. Bulba, Scanning Electron Microscopy, *Dust Topics, 6*:4 (1969).
48. E. J. Scherago and B. J. Sheffer (Eds.), Guide to Scientific Instruments, *Science, 165A*:49 (1969).
49. P. R. Thornton, *Scanning Electron Microscopy: Application to Materials and Device Science,* Chapman and Hall, Ltd., London, 1968.
50. O. Johari, The Scanning Electron Microscope — The Instrument and Its Applications, in *Symposium on Scanning Electron Microscopy,* Illinois Institute of Technology Research Institute, Chicago, Ill., Apr. 30-May 1, 1968.

51. *Proceedings of Second Annual Symposium on Scanning Electron Microscopy,* Illinois Institute of Technology Research Institute, Chicago, Ill., 1969.
52. G. A. Horridge and S. L. Tamm, Critical Point Drying for Scanning Electron Microscope Study of Ciliary Motion, *Science, 163*:817 (1969).
53. E. B. Small and D. S. Marszalek, Scanning Electron Microscopy of Fixed, Frozen, and Dried Protozoa, *Science, 163*:1064 (1969).
54. E. R. Lewis, T. E. Everhart, and Y. Y. Zeevi, Studying Neural Organization in Aplysia with the Scanning Electron Microscope, *Science, 165*:1140 (1969).
55. G. A. Bartlett, Scanning Electron Microscope: Potentials in the Morphology of Microorganisms, *Science, 158*:1318 (1967).
56. D. Greenwood and F. O'Grady, Antibiotic-Induced Surface Changes in Microorganisms by Scanning Electron Microscopy, *Science, 163*:1076 (1969).
57. H. G. Fromme and W. Stöber, Establishing the Pore Size Distribution of Classified Membrane Filters by Surface Micrographs, *Staub* (English transl.), *26*:4 (1966).
58. G. Pfefferkorn and R. Blaschke, Dust Analysis with the Aid of the Scanning Electron Microscope "Stereoscan," *Staub* (English transl.), *27*:30 (1967).
59. T. K. Kelly, W. F. Lindquist, and M. D. Muir, Y-Modulation: An Improved Method of Revealing Surface Detail Using the Scanning Electron Microscope, *Science, 165*:283 (1969).
60. O. Johari, Applications of Scanning Electron Microscopy for Characterization of Powders, in *International Conference on Powder Technology,* Illinois Institute of Technology Research Institute, Chicago, Ill., May 20-23, 1968.
61. B. Holma, Scanning Electron Microscopic Observation of Particles Deposited in the Lungs, *Arch. Environ. Health, 18*:330 (1969).
62. P. C. Reist, W. A. Burgess, and D. Yankovich, Development of an Automatic Particle Assaying Instrument Utilizing a Scanning Electron Microscope, annual meeting of Air Pollution Control Association, Paper No. 69-124, Pittsburgh, Pa., 1969.

4

Particle Sizing by Measurement of a Property Related to Grain Size

4-1 Introduction

Numerous methods exist which can be used to measure particle size by evaluating a known characteristic (such as settling velocity) related to grain size, rather than by performing a direct scalar measurement utilizing the techniques described in Chapter 3 (p. 105). A classification of these indirect sizing methods has been devised to distinguish between those that give information on size and size distribution and those that give information on average size alone. The first group is of greater importance to industrial hygienists; hence, most of this chapter is devoted to them.

The preparation of dust fractions and aerosols which contain particles of uniform size is a procedure of considerable importance for industrial hygienists, as well as for those in other branches of the environmental health sciences. For this reason, the sizing methods discussed in this chapter have been further subdivided to indicate those in which uniform size fractions are isolated in a form suitable for recovery and reuse in toxicological and similar studies.

4-2 Instruments Which Give Data on Average Size and Size Distribution

4-2.1 Methods Based on Settling Velocity in a Fluid Medium

Sizing by sedimentation and elutriation enables the classification of particles according to their settling velocities in a fluid. In most commercial applications of sizing by sedimentation, the relation between particle size and settling velocity is empirical or semiempirical. However, if care is taken to perform the sedimentation under certain limited sets of conditions, particle size may be defined in terms of settling velocity by using Stokes' law. If all the particles are spherical, smooth surfaced, and of uniform density, Stokes' law applies exactly. However, in most cases particles differ not only in size but in surface characteristics, packing density, and shape. These differences introduce variables which do not conform to Stokes' law. Further deviations from theory come about because the concentrations of dust commonly used in these methods are not compatible with unhindered settling, and convection currents in the fluid medium tend to reduce the mean rate of fall. Wilson,[1] by studying the settling rate of glass spheres in water, showed that temperature uniformity is the critical factor in the settling of dilute suspensions of fine particles in accordance with theory. In practice, deviations from theory have led to the concept of a diameter of an equivalent sphere of the same specific gravity whose sedimentation rate is identical with the observed settling velocity, or Stokes' equivalent diameter. (See also *aerodynamic diameter* in Glossary.)

Instruments for sizing by sedimentation may be divided into two groups: those which make it possible to separate individual size fractions for further examination, and nonfractionation instruments used for measuring size distribution and estimating total surface.

(a) *Fractionating Instruments Based on Particle Settling Velocity*. Fractionating instruments include the Roller particle size analyzer and the Haultain Infrasizer for the air elutriation of powders. For fractionation of dusts in a liquid medium, the Bureau of Mines Short Column Elutriator and a number of decantation devices are available. In general, elutriation with air requires less time than with water, since settling rates are more rapid and air jets may be used to disperse the dust being analyzed.

Each of the fractionating size separation methods described below is useful for bulk samples (i.e., 5 to 100 g or more) of coarse dust (i.e., 5 to 100 μm), such as may be encountered in investigations of dust-collecting devices, foundry rafter samples, and raw materials. They are of less value for the examination of airborne dusts, which generally are less than 10 μm in size.

These methods, including the Micro-Particle Classifier, may be extremely useful for the preparation of dusts of graded size for use in toxicological studies and of test materials of known size characteristics. All these sizing methods measure the settling velocity of particles, which in practice, is expressed as the diameter of a sphere of identical specific gravity that has the same settling velocity. For all methods, size data are obtained directly in terms of weight fractions.

(1) Air elutriation. The Roller particle size analyzer[2] for air elutriation of dry powders is illustrated in Fig. 4.1. A number of modifications of this

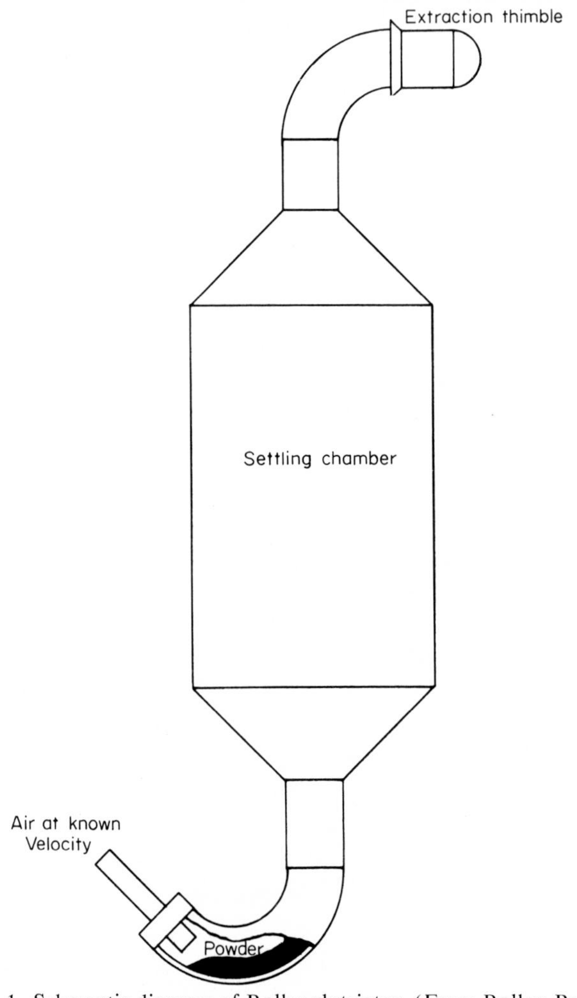

FIG. 4.1. Schematic diagram of Roller elutriator. (From Roller, Ref. 2.)

device are in existence. Essentially it consists of a glass U-tube sample holder which is oscillated by mechanical rapping. When a regulated stream of air is passed through it, particles smaller than the separation size pass upward through a settling chamber (whose diameter can be changed) and are filtered from the air and caught in a weighed paper thimble. Oversized particles and agglomerates return to the U-tube, where the agglomerates are broken up further by the air jet. Ivey[3] found that the Roller method had weaknesses with respect to very poor precision, particularly at the high end of the size scale (i.e., above 40 μm); poor accuracy; attrition caused by the high (246 ft/sec) air jet velocity used to lift dust from the U-tube (it was found also that the degree of attrition was affected by the placement of the jet in the U-tube and by the tension on the hammer spring used to oscillate the sample holder); and electrostatic effects which cause particles to agglomerate and adhere to the walls of the elutriation cylinders. The effects of attrition and charge can be minimized by the following modifications suggested by Matheson:[4] substitution of a porous bronze metal plate in a 1-in. I.D. metal aeration tube for the U-tube, jet, and hammer; and use of 50% to 70% relative humidity (RH) air in place of 29% RH, specified for the original method.

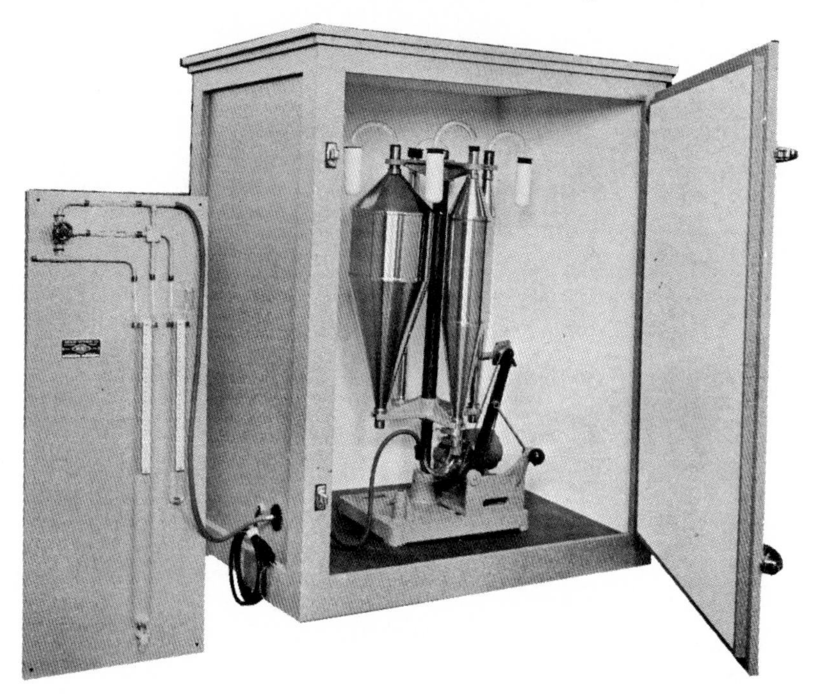

FIG. 4.2. Improved Roller apparatus. (Courtesy American Instrument Co.)

Further modifications of the Roller method of size analysis were suggested by Ivey,[3] who substituted a fritted-glass plate of coarse porosity (40-μm nominal openings) for a bronze plate, substituted a glass aeration tube for a metal tube so that the progress of dust fluidization could be followed visually, and introduced the aeration air into the apparatus at 60% RH by first bubbling it through 38.40% H_2SO_4. Particles having a highly porous structure will adsorb appreciable quantities of water at 60% RH. This increases particle weight without changing particle size and hence increases what Ivey refers to as "elutriation density." Ivey has investigated this phenomenon in connection with fluid cracking catalyst granules.

A commercial model of Roller's particle size analyzer* is shown in Fig. 4.2. Each of the four stainless-steel graduated cylinders may be rotated, in

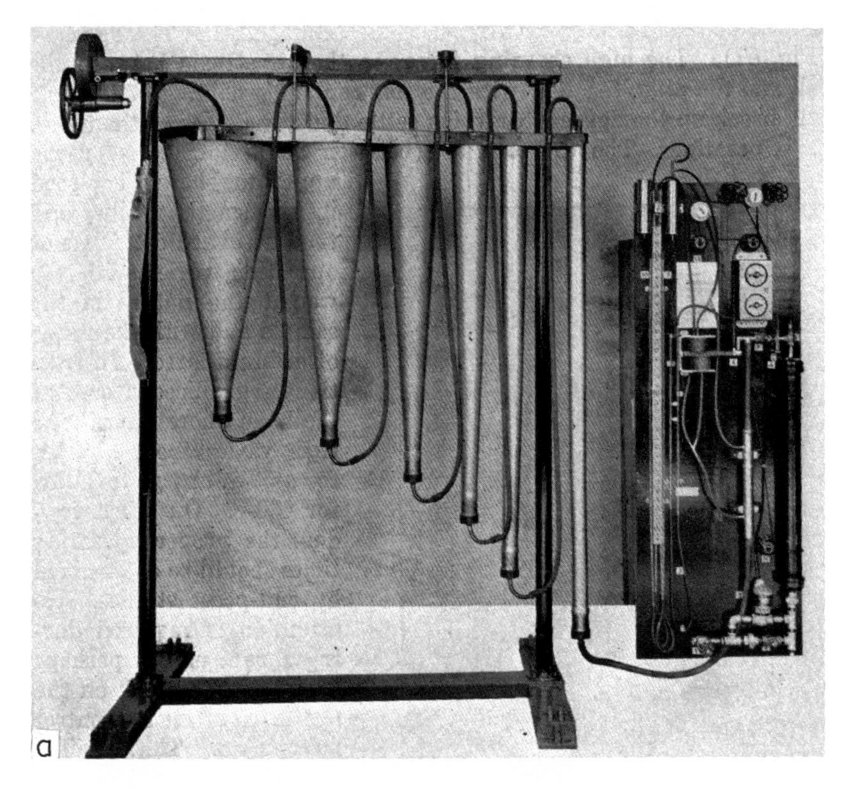

FIG. 4.3. Haultain Infrasizer: (a) complete apparatus; (b) dust disperser at base of each column. (From Haultain, Ref. 6.)

*American Instrument Co., Inc., Silver Spring, Maryland.

turn, to bring the desired chamber into the front operating position. The gooseneck tubes above each settling chamber connect to the paper extraction thimbles in which the separated fractions collect. In operation, a measured volume of compressed air flows through the U-shaped tube, in which a preweighed sample of powder (about 10 g) has been placed. The compressed air is filtered and dried and introduced into the sample chamber through small-diameter nozzles (about 0.02 in.) to stir the dust and break up the clumps. The sample tube is oscillated at the same time to assist in aerating the powder, and the settling chamber is tapped to prevent dust from clinging to the walls. When the sample is composed of material so soft that the air jet produces a grinding action, a special sample tube with a fritted plate is used to introduce the elutriation air more gently.

The lowest air flow rate and largest-diameter chamber are used to carry the smallest size fraction, commonly 5 μm or under, out of the powder sample and into the first thimble. This takes about 1 hr. Coarser size fractions are removed in sequence from the powder remaining in the sample tube, elutriation time decreasing to as little as 10 min as the remaining dust becomes

coarser. Twelve graded-size nozzles plus the flowmeter panel shown in Fig. 4.2 provide a wide range of flow rates. At the conclusion of each run, the weight gain of each sample filter is determined. From this and the known size-separating characteristics of each combination of air flow and chamber diameter it is possible to construct a cumulative particle size (by weight) curve in terms of diameters of spheres having equal settling velocities. Commonly used fractions for size analysis are <5, 5 to 10, 10 to 20, 20 to 40, and 40 to 80 μm. Narrower fractions may be obtained for experimental purposes by careful operation and by rerunning the separated fractions. The percentage weight of separated fractions is reproducible to $\pm 0.5\%$, and the lower cut is usually between 2 and 5 μm. However, when the powder has unusually high density, as for example carbonyl iron spheres, it is possible to separate a fraction containing particles 1 μm and less.

Walton[5] has shown that elutriator chambers fitted with a convergent upper section produce less-sharp size classifications than do those that contain a collection filter across the full diameter of the chamber. In addition, he has shown that particles may be classified by drawing them through a horizontal channel and that smaller equipment is needed for horizontal than for vertical classification.

The Haultain Infrasizer[6] is illustrated in Fig. 4.3*a*. It consists of a graduated series of six vertical tubes with their diameters proportioned by a ratio based on the square root of 2. Since the tubes are connected in series starting with the smallest, the same air flows through all, and the velocity in each tube is half that in the preceding tube. Seven fractions are recovered, including the fines caught in a filter bag following the sixth and largest tube. The frame carrying the elutriation tubes is lifted and dropped about ½ in., twelve or more times per minute, to dislodge dust accumulated on walls and in tubes. The bottom of each tube is fitted with a rubber plug and golf-ball arrangement, as shown in Fig. 4.3*b*, to break up clumps.

This instrument has the same disadvantages as the Roller apparatus; there may be particle attrition, and the fractions are likely to contain considerable amounts of undersized particles. Analysis is lengthy, taking many hours for a sample having a wide size range. However, the apparatus runs virtually unattended for most of this period, and the actual operator's time is small. For routine testing, a minimum amount of skill is required. The optimum size range for the Haultain instrument is 10 to 100 μm, and separate size fractions of dust may be obtained for reuse.

In use, the entire weighed dust sample (usually powder passing a 200-mesh sieve) is deposited in the smallest-diameter tube and the air flow is started. All except the very largest particles are air-lifted out of this cylinder and sent to the next in the series, the process being repeated for each chamber in turn. At the conclusion of the elutriation period, the dust fractions are brushed out of each tube, and, with the final filter, they are weighed on an analytical balance. The air-volume rate through the Infrasizer, the diameter of the chambers, and the true specific gravity of the powder are the only measurements needed to calculate from Stokes' law the diameter of a sphere of equal settling velocity equivalent to each of the chambers in the series. Fig. 4.4 shows the sizes of particles held in suspension by minimum upward flow of air in each cone as a function of true specific gravity.[7] Using only this information and the actual dust weighings, it is possible to prepare a cumulative particle size-by-weight curve on probability paper.

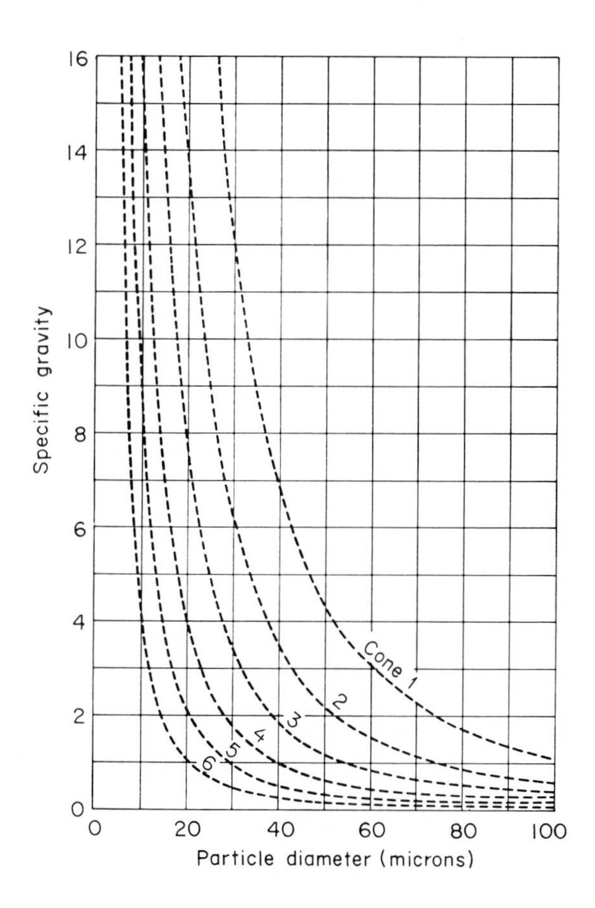

FIG. 4.4. Particles held in suspension by minimum upward flow of air in each cone in Infrasizer. (From Price, Ref. 7.)

(2) Liquid elutriation. Size fractionation by elutriation may be carried out in liquids, such as water or alcohol. Just as in air elutriation, particles of less than a certain size are carried upward through the fluid by a current of controlled velocity, and oversized particles remain behind. By varying the fluid velocity (or keeping fluid volume constant and changing the size of the chambers through which the particles must rise) a size separation into as many fractions as are desired may be accomplished. This method has many of the limitations mentioned above for air elutriation; e.g., size fractions are not sharp, and problems exist when soluble or reactive powders are encountered. However, electrostatic effects which produce clumping and adhesion to the walls of the air apparatus are absent, and particle attrition is less at the lower velocities used in a liquid apparatus.

The Bureau of Mines short-column liquid dust elutriator[8] is shown schematically in Fig. 4.5. The small tank in the diagram has an overflow pipe to maintain a constant liquid head over the capillary spigot. Spigots and cutters are accurately made to provide a series of velocties based on the

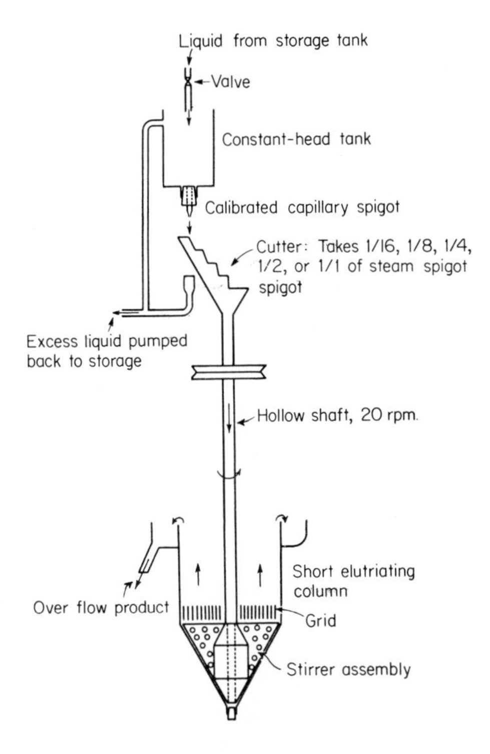

FIG. 4.5. Bureau of Mines short-column elutriator. (From Cooke, Ref. 8.)

ratio of 2 (i.e., to give duct fractionations on the standard square root of 2 scale). The low-speed stirrer maintains the dust in suspension, while the grid eliminates turbulence in the elutriation zone and assists in maintaining a uniform upward liquid velocity over the entire cross section.

Fifty to 100 g of sample is required for an analysis, and a complete fractionation down to a size of approximately 2 μm requires about 60 hr. For much of this time the apparatus will run unattended. Operation of the liquid elutriator is the same as for the Roller apparatus except that each dust fraction removed from the overflow stream must be dried before weighing.

Fractionation of a sample can be achieved by serial sedimentation and decantation with the technique described by Cummings[9] for preparing sized fractions of mineral dusts for animal exposure tests. In this method the powder is stirred into a large volume of water and alcohol and allowed to settle for a predetermined period in a shallow (to minimize convection currents), flat-bottomed glass culture dish. At the end of the first time interval, the supernatant liquid plus the fraction of dust still suspended is decanted into a second vessel of identical size and allowed to settle for twice the initial period of time. This process is continued for eight successive fractions (i.e., from ½ to 64 min duration) and the liquid from the eighth vessel is transferred to another dish and allowed to settle for 4 days, after which time the dust still in suspension is in such violent Brownian motion that it will remain suspended indefinitely. For a clean size separation, each individual dust fraction should be resuspended in fresh liquid and the sedimentation repeated several times. This method will not give absolutely uniform particles, but rather a series of graduated fractions each having a narrow size range. It is a tedious and time-consuming procedure, as illustrated in Fig. 4.6, a modification of Cummings' technique by LaBelle to give successive fractions which do not overlap by more than 10%.[10] It is satisfactory, however, for small quantities of dust (e.g., 5 to 20 g). At the end of the separation, each of the recovered fractions may be dried and weighed to obtain size distributions by weight of the original sample, and each fraction may be examined microscopically to determine its size distribution by count. Cummings found that the theoretical sizes calculated by Stokes' law for the various fractions did not agree with those he had found by direct measurements, and so he prepared an empirical chart (Fig. 4.7) for quartz which gives the observed particle sizes obtained with various settling velocities.[11]

To use the chart, find the diameter of the particle sought in the ordinates; note the number of the column it is in; run horizontally from that point to the index line, and drop vertically from this point to the horizontal column having the same number below to obtain the time of settling, in seconds, of quartz particles.

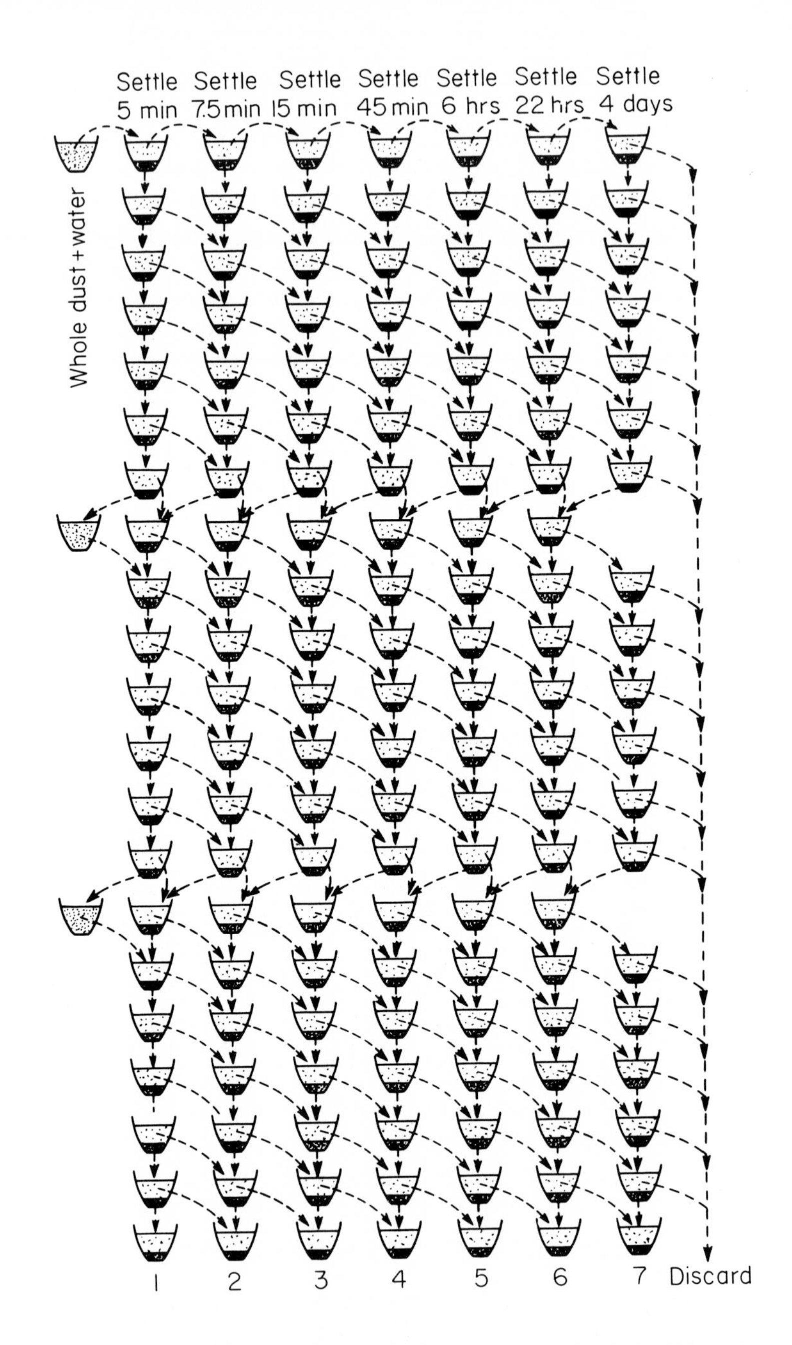

FIG. 4.6. Flow sheet for fractionation of dust. (From LaBelle, Ref. 10.)

(b) *Fractionating Instruments Based on Centrifugal and Inertial Force.* Instruments which utilize centrifugal and inertial forces to separate a dust sample into a number of graded size fractions include small cyclone dust collectors (e.g., the Federal classifier), centrifugal elutriation or winnowing devices (e.g., the Micro-Particle Classifier), cascade impactor, helices, and spinning cone classifiers (Conifuge and Aerosol Spectrometer). The first two require a bulk sample of dust; the remainder, described in Chapter 2, are both sampling and analytical instruments.

(1) Federal classifier. The Federal classifier (Fig. 4.8)[12] utilizes a series of cyclone collectors of graded size to separate dust into a number of

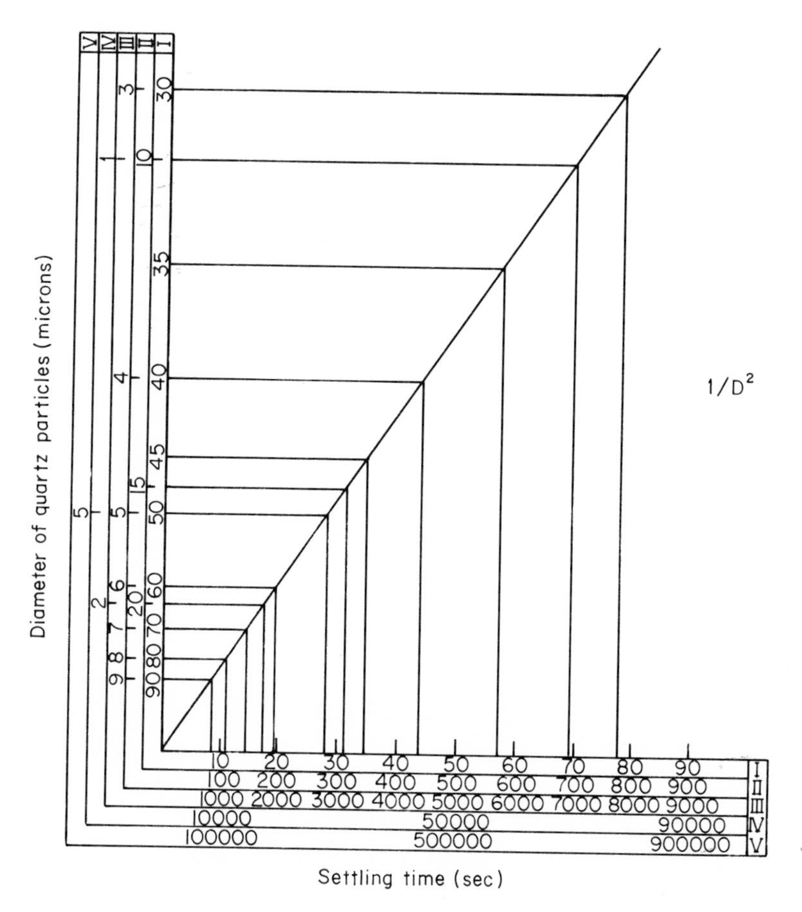

FIG. 4.7. Time of settling of various particles. (From Cummings, Ref. 9.) To use the chart, refer to p. 145.

fractions, including a final sample collected on a filter. Except for the addition of centrifugal force, it is not greatly different in principle from the Infrasizer. The Federal classifier handles large quantities of coarse dusts (i.e., pounds of material in the sieve-size range), but gives a very poor

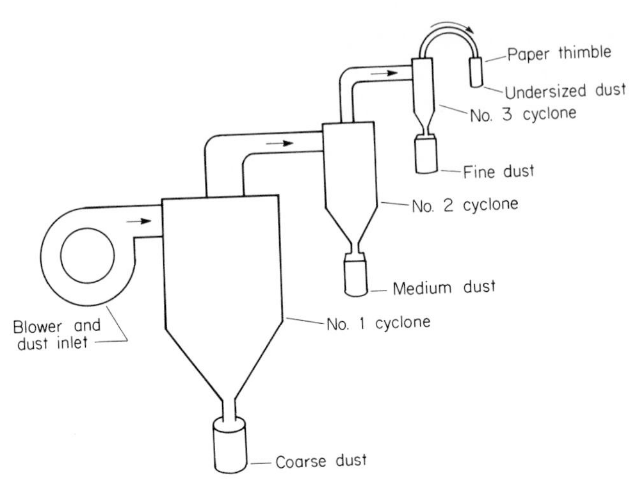

FIG. 4.8. Federal classifier. (From Federal Clasifier Systems, Inc., Ref. 12.)

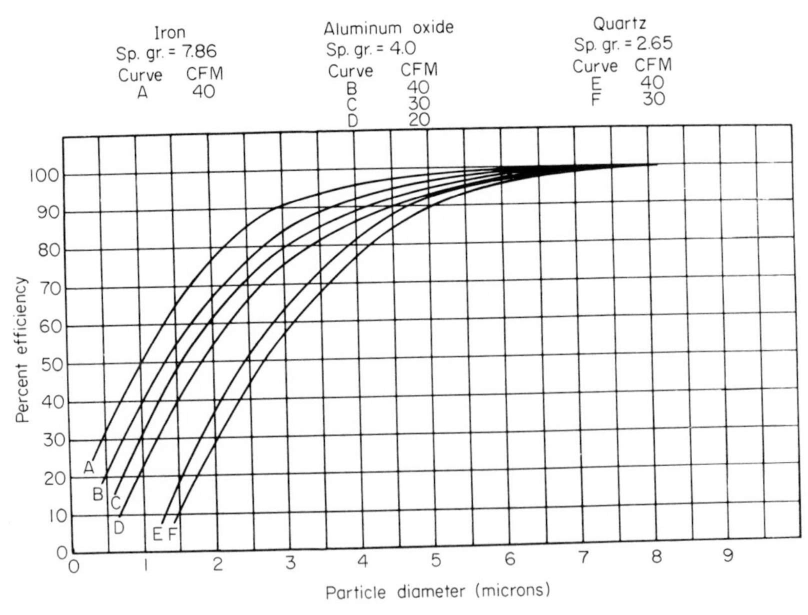

FIG. 4.9. Cyclone particle size efficiency curves. (From Dennis, *et al.*, Ref. 13.)

quality of separation, and is of limited value to industrial hygienists. The limitations of cyclones for sharp size separations are apparent from the typical cyclone size retention curves for mineral dust, shown in Fig. 4.9.[13] This cyclone collector is virtually 100% efficient for the collection of particles greater than 8 μm and has little influence on particles smaller than 1 μm. Between these two size limits, variable fractions of the aerosol particles are retained, depending on particle size.

Cyclones have variable retention efficiency over a wide range of particle sizes because of the manner in which they function. Figure 4.10 shows air and particle flow patterns in a simple tangential rectangular inlet cyclone dust collector.[14] To be retained within the cyclone, a particle must migrate radially outward under the influence of centrifugal force until it contacts the wall of the cyclone body. The cyclone entry has an appreciable width; hence it will be apparent that some particles enter the cyclone body in a more favorable position for separation, because they are closer to the wall. The relationship between the force on a particle, determined by cyclone diameter and entry velocity, and the distance a particle must travel radially in one complete revolution of the cyclone to reach the separating wall is such that there will be a specific particle size that is large enough to reach the outer wall under the operating conditions, even when it enters the cyclone at the most unfavorable location. This and all larger sizes will be completely precipitated in the cyclone. Particles smaller than a few microns experience only small radial velocities in a cyclone collector, and they must be close to the wall at entry to be separated. Even small cyclones are impractical for collecting particles less than 1 μm, although some submicron particles may be collected by diffusional migration through the thin, slowly moving layers in contact with the cyclone wall. Tests with miniature cyclones of 0.5-, 1.0-, and 1.5-in. diameter[15] at 4000-ft/min entry velocity showed efficiencies of 84%, 90%, and 93%, respectively, with resuspended pulverized coal fly ash having a mean diameter by count of 0.6 μm and a geometric standard deviation of 2.7.

A practical application of the size-separating characteristics of small cyclones is their use as the first stage of a dust-sampling train to simulate the aerodynamic classification of suspended particles achieved by the human nose and upper respiratory tract. The cyclone dimensions and air flow rate are selected to permit particles of various sizes to be retained in the same proportions as occur in the nose and upper respiratory tract of the average worker and permit the remainder to pass through to a secondary collector which retains the fractions normally penetrating to the human lung. Cyclone sampling devices of this nature are shown in Fig. 4.11.

The particle size retention characteristics of miniature cyclones must be determined empirically, as there is no adequate theory for predicting performance. If a number of calibrated cyclones of differing diameters are used

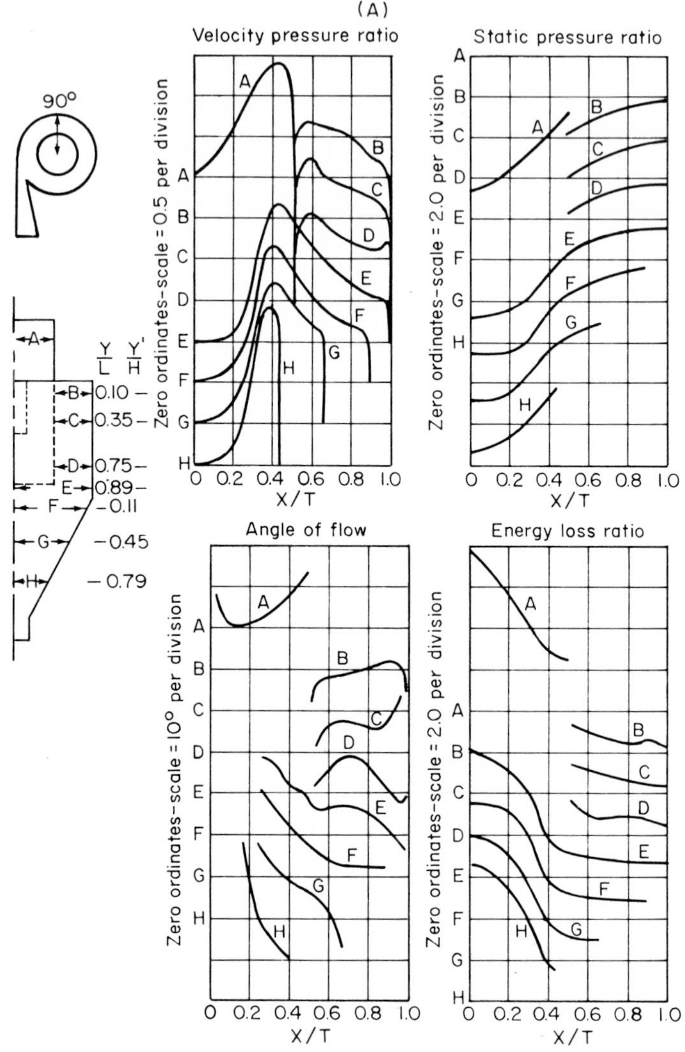

FIG. 4.10. Cyclone flow patterns. (From First, Ref. 14.)

FIG. 4.10. Cyclone flow patterns. (A) Velocity pressure, static pressure, angle of flow, and energy loss ratios at various elevations in the cyclone body and exit duct. (B) Velocity pressure ratios at various distances from the entry duct in the upper part of the cyclone body. (C) Flow pattern in the upper part of the cyclone body. (From First, Ref. 14.)

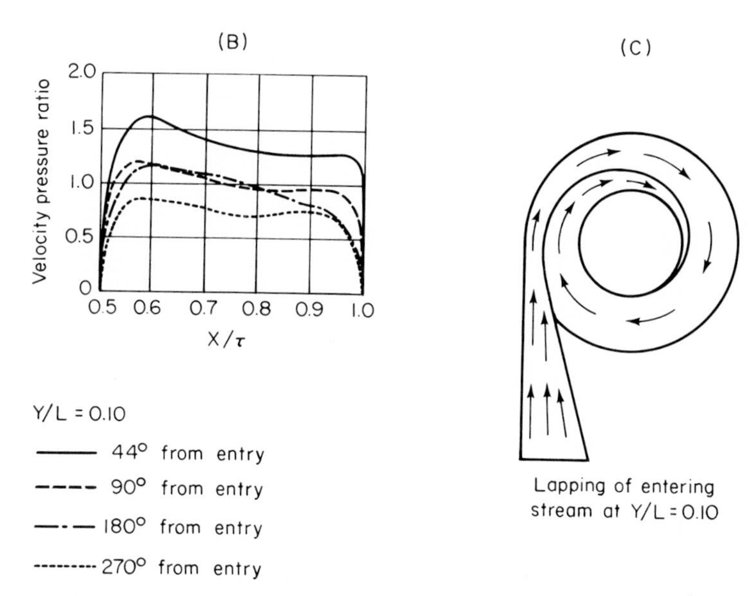

sequentially, it is possible to determine the size distribution of an aerosol by weighing each fraction, including the finest which can be caught on a high-efficiency filter. The gradual slope of cyclone efficiency-velocity curves shown in Fig. 4.9 indicates that larger cyclones do not produce as sharp separations as are possible with the cascade impactor.

*(2) Micro-Particle Classifier.** This classifier utilizes centrifugal force to winnow particles and separate them into graded size fractions. Fig. 4.12a is a schematic diagram of the apparatus. Bulk dust (1 to 20 g) is fed slowly from a feed hopper containing a motor-driven brush and spring plate for

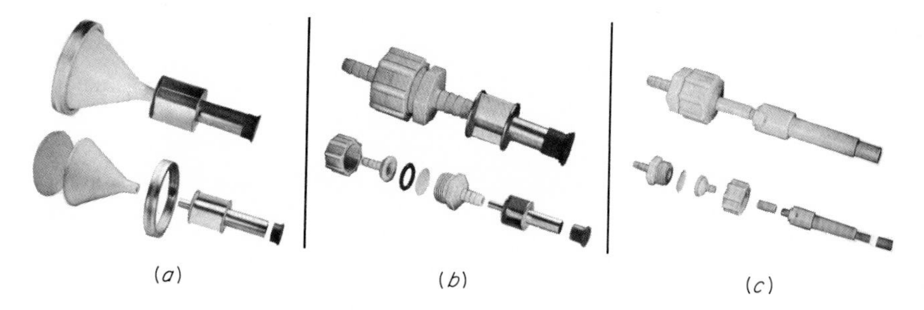

FIG. 4.11. Bendix-Unico respirable dust samplers: (*a*) Model 240; (*b*) Model 18; (*c*) Model 2.8.

*Harry W. Dietert Co., Detroit, Michigan.

dispersion of the sample. The thin spring plate in contact with the brush form a slight barrier, causing the bristles to deflect, then return with a snap in passing the plate. This snap action disperses the sample into the low-velocity air flow moving down the feed nozzle into the feed hole in the rotor. A vibrator is at the base of the hopper to control the rate of feed from the nozzle. The sample leaves the feed nozzle in the form of an aerosol.

The rotor assembly includes fan blades which draw air in a spiral motion through a sifting chamber with a detachable catch basin, and a stack of

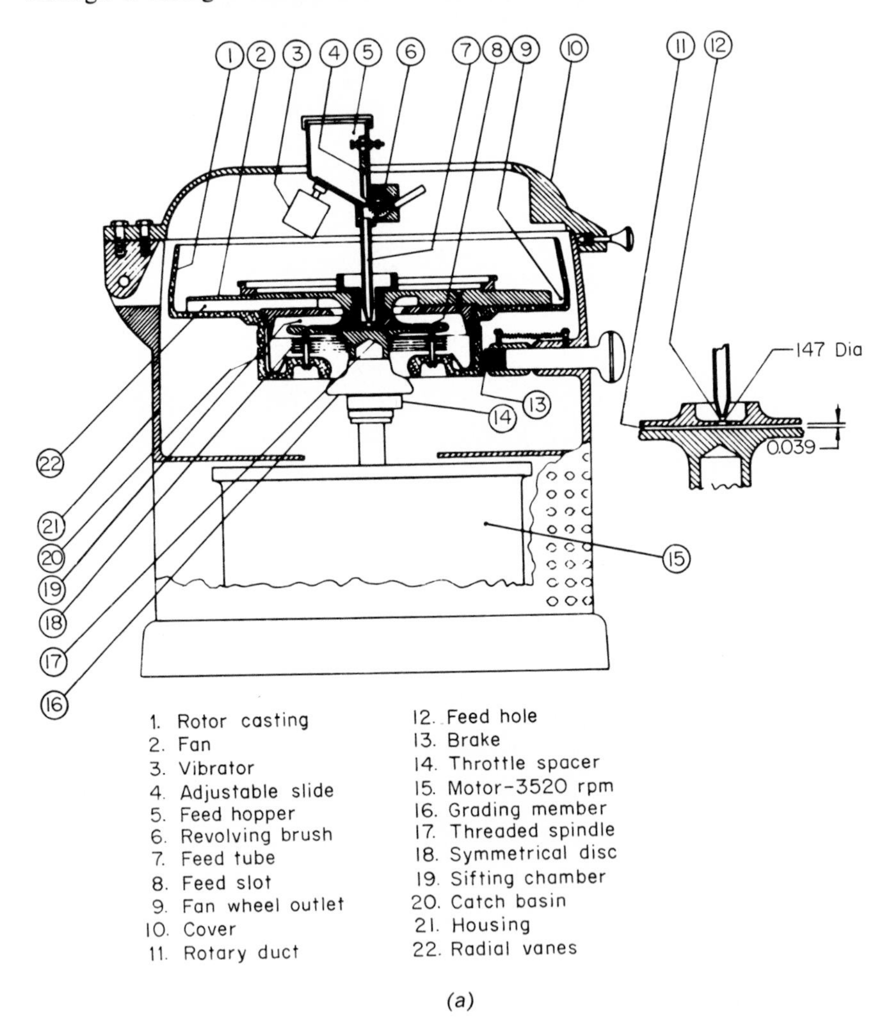

FIG. 4.12. Micro-Particle Classifier: (*a*) schematic; (*b*) photograph.

closely spaced disks to create uniform air velocities in the apparatus. The rotation of the assembly creates a spiral of air in the sifting chamber which is moving toward the center. After aerosolization by the feed mechanism, the sample enters the small rotary duct (No. 11) and is acted on by centrifugal force, which assists in directing it into the sifting chamber with the conveying air. Airborne dust enters the sifting chamber with an outward velocity which is determined by the speed of rotation of the machine and the distance of the dust outlet from the center of rotation (both of which are fixed), plus the size, specific gravity, and geometric and surface properties of the particles. All dust particles in the sifting chamber (No. 19) whose radial outward velocity due to centrifugal force exceeds the entraining velocity of the inward moving air spiral are deposited on the removable container pressed against the rotating wall of the collecting chamber. All dust particles introduced into the sifting chamber whose radial outward velocity due to centrifugal force is less than the inward velocity component of the spiraling air stream are carried through the fan blades and deposited on the inner surfaces of the upper stationary bowl-shaped housing under the influence of the very strong centrifugal field created by the fan blades.

It is customary to carry out the winnowing process by removing the smallest size fractions first. This is accomplished by restricting the annular air inlet port between the "grading member" (No. 16) and the stationary outer

(b)

housing with an adjustable throttling piece so that only a very gentle air flow is created in the sifting chamber. When all the dust has been introduced, the unit is shut off and the lighter fraction deposited in the upper chamber is brushed up and weighed. The heavier, oversized particles in the lower chamber can be recovered and rerun to remove the next smallest fraction, or a new sample identical with the first can be introduced into the unit without disturbing the oversized deposit and run to remove the two smallest fractions together. If the second method is used, the second size fraction will be the difference between the weight of the dust passing through the fan on the second run and the weight found on the first run. By either method progressively larger size fractions are winnowed from the dust sample by stepwise lowering of the grading member by inserting thinner and thinner throttling pieces. This increases the size of the annular air inlet port and permits more air to be drawn through the instrument by the constant-speed fan. By progressively increasing the air opening for each run with graduated throttles, the dust sample can be divided into nine size fractions with the throttling pieces provided by the manufacturer. The material removed after each separation may be retained for further study. It is calibrated by microscopic examination of the fractions removed at each of the graded air settings. Fig. 4.12b is a photograph of the instrument.

Information published by the manufacturer indicates that reasonably clean separations may be obtained with this instrument and that reproducibility on dusts such as fly ash is excellent. It is claimed that attrition during multiple passes of dust through the high-velocity jets, etc., is negligible, but further investigations are required to explore this point more thoroughly. The TA-5 Committee of the Air Pollution Control Association (Dust, Fumes, Mists, and Fog Collectors) has recommended that the Micro-Particle classifier be adopted as a standard instrument for determining particle size when specifying and testing industrial dust-collecting devices, and a standardized procedure has been recommended by the American Society of Mechanical Engineers.[16]

A single separation including recovery and weighing of the residue can be completed within 15 min, and an eight-separation analysis can be completed easily within 2 hr. The upper limit for each fraction can be expressed as the actual size of the particles in microns, as equivalent diameters based on a theoretical specific gravity of unity, or in the form of terminal velocity. This instrument is most useful for particle separations between 50 and 3 to 5 μm, the lower limit depending on the true specific gravity of the powder.

In addition to the applications of this device for particle sizing, sizable quantities of clean fractions of graded dusts for toxicological and analytical studies may be obtained.

(c) *Nonfractionating Instruments Based on Particle Settling Velocity.* For many of the methods included in this category, analysis involves removing serial dust fractions from the main body of settling dust. Since the material removed is in no sense a size fraction but rather a mixture of sizes which differ only in *average diameter,* the process is not analogous to the fractionation procedures described immediately above.

Methods of size analysis based on settling in air, or other gaseous media, are effective for particles from approximately 1 to about 100 μm in size. For sizes below 1 μm, Brownian motion affects the rate of settling and long periods are required for effective separations, whereas the settling rate of particles over 100 μm is too rapid to permit accurate readings. It has been pointed out by Davies[17] that turbulence, or random air movement, in the sedimentation chamber leads to an exponential decay in the number of particles, since only those near the floor settle out, the remainder being continually moved around from one place to another.

(1) Gravity sedimentation in air. Payne[18] has described a gaseous sedimentation method and device called the Micromerograph,* in which a powder cloud dispersed in a gaseous medium is allowed to settle under the action of gravity through an enclosed vertical column onto the pan of a null-seeking servoelectronic balance and the accumulated weight is recorded on a strip chart.

Cumulative weight against particle size is plotted with the use of transparent overlay templates, incorporating a corrected version of Stokes' law. From the strip chart data one can also calculate specific surface and volume surface mean diameter.

The parameter measured by the Micromerograph is Stokes' equivalent diameter—the diameter of a sphere of equal specific gravity which has the same terminal velocity as the particle in question. For a particle whose greatest dimension is less than twice the smallest dimension, the equivalent Stokes' diameter does not vary by more than about 20% from that of a sphere of equal volume, according to Preining.[19]

Deagglomeration of bulk dust samples is achieved in the Micromerograph by viscous shearing action when the dust is projected through the annular space between two conical surfaces by means of a jet of high-pressure compressed nitrogen. The deagglomeration device is shown schematically in Fig. 4.13a. The width of the annular opening is adjusted by moving the setting lever, and the external scale indicates actual width from 10 to 250 μm. After the powder is placed in the sample holder, pressing the charging

*Sharples Corp., Bridgeport, Pennsylvania.

button charges the pressure chamber with a metered volume of nitrogen at a predetermined pressure. Then, pressing the firing button releases the nitrogen, which picks up the powder and carries it through the deagglomerator and into the sedimentation column. By varying the gas pressure and the space between cones, the shear force exerted on the particles can be varied.

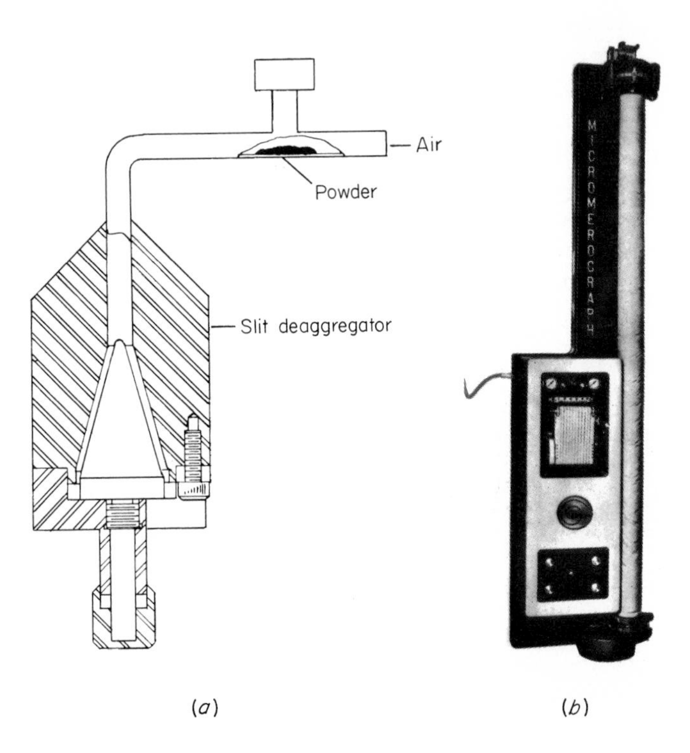

FIG. 4.13. Micromerograph: (*a*) deagglomerator section; (*b*) photograph.

The settling chamber is a vertical tube 3.5 in. i.d. and 85 in. long covered with a glass fiber blanket to provide thermal insulation. Particles rapidly attain their terminal velocity and settle onto the pan at the bottom of the column. The servoelectronic balance is very sensitive, and limit stops are provided to protect the unit from excessive travel. The pan deflection of the null-seeking device is negligible under maximum load. This amount of pan movement creates no air currents to disturb sedimentation. A photograph of the complete instrument is shown in Fig. 4.13b. The time for an analysis varies from a few seconds to several hours, depending on the size and density of the particles.

Sources of error in this technique arise from the fact that the initial powder cloud in the top of the column has finite dimensions and high particle concentration. The size of the cloud is held to a minimum by introducing an amount of gas with the powder which is small compared to the amount of gas already present in the column, and the error due to the finite dimensions of the initial cloud is further reduced by making the settling distance large compared to the dimensions of the cloud.

The error associated with a high particle concentration in the layer at the top of the column is related to the tendency of a dense aerosol to settle as a discrete cloud. This is caused by particle entrainment of the surrounding medium and results in a reduction in the resistance of the medium to the motion of the entire cloud (reference 20, pp. 46-51). This error can be counteracted by reducing the amount of powder in the generated aerosol cloud, but this also reduces the amount of material falling on the balance pan.

The phenomenon of subsidence of a dense cloud through a dust-free fluid below is not confined to the Micromerograph but occurs in all gravity sedimentation devices, whether liquid or gas, in which a thin, concentrated dust layer is placed at the top of the settling column.

For sizing fine particles (i.e., those under 1 to 2 μm in size), settling time becomes excessive and results are unsatisfactory because of the virtual impossibility of eliminating convection currents for extended periods. Approximately 0.1 g of dust is required for an anlysis, and the instrument can operate unattended during most of the time the analysis is taking place. For finer dusts, one or more hours will be required. It has been reported[21] that the dust deagglomeration process induces high electrostatic charges on some dusts which result in their attraction to the walls of the settling chamber and loss from the weighing system. The loss to the walls may be as high as 60% by weight.

Little information is available on the effect of the deagglomeration apparatus on particle attrition, although it seems likely that some materials, at least, may be adversely affected by the high-velocity air jet and narrow passages. In practice, about five trial runs suffice to determine optimum deagglomerator settings for new materials. Starting at a low setting of 50 psi, the pressure is increased for successive runs. The plotted size distribution curves determined by the instrument show progressively finer particles as agglomerates are broken up. Eventually a plateau is reached where additional pressure increase does not cause a finer shift, indicating that minimum particle size has been reached. However, substantial increase in pressure may again cause the distribution curve to move to finer particles, suggesting that breakage of individual particles is occurring.

(2) Gravity sedimentation in liquids. Gravity sedimentation procedures in liquid media have less general application to industrial hygiene than those in air, because the size range over which they give the best results does not reach down to the range of greatest present-day interest (i.e., less than 5 μm), and because bulk dust samples are required. Nevertheless, gravitational liquid sedimentation procedures have been, and remain, of great commercial interest for product quality control, and there is considerable information in the literature relating to sedimentation of fine dusts in liquids.

Two general sedimentation procedures used for determining particle size distribution are those which determine periodically the concentration of solids remaining at a given level in a settling suspension, and methods based on a determination of the total weight that has settled to a given level as a function of time.

The Oden sedimentation balance[22] measures the weight of sediment on a pan placed a fixed distance from the surface of a liquid suspension. Unlike the Micromerograph, in which an effort is made to confine the dust cloud to the top of the settling chamber at zero time, the material on the pan at any given moment may be considered as divisible into two portions. One fraction consists of coarse particles whose settling velocity is sufficiently rapid to make them fall the entire distance from the surface of the settling suspension to the pan in the chosen time interval; the remainder consists of finer particles able to descend to the level of the pan in this same time interval because their original position was lower than the liquid surface. Sedimentation in this fashion tends to indicate a somewhat more disperse size spread than actually exists. It has been found that the original Oden apparatus has a serious defect in that the depression of the balance pan sets up convection currents which introduce errors in the sedimentation data. Undoubtedly this defect could be corrected by means of a null-type automatic recording balance similar to that used in the Micromerograph.

Bostock[23] has improved the sedimentation balance technique by using a special torsion balance which gives only a slight pan displacement as the settling dust accumulates. The pan system of the commercial instrument* (Fig. 4.14) consists of a light stirrup from which a small pan hangs freely immersed in a tank of clear water immediately beneath the sedimentation column so that particles settling to the bottom are caught and weighed. The pan system is suspended from one side of a horizontal torsion wire equipped with an adjustable counterbalance on the other side. Maximum deflection of the pan is about 3 mm, although it is seldom necessary to exceed 2 mm

---

*A. Gallenkamp and Co., Ltd.; Arthur S. La Pine and Co., Chicago, Illinois, United States distributor.

during a determination. The balance is sufficiently sensitive to permit the use of low concentrations of solids, thus allowing sedimentation under practically free falling conditions. There is also the advantage that negligible interference with the settling particles takes place. The instrument has been designed for the determination of particle size distributions within the size range of 5 to 75 μm, but if the top size is not greater than about 20 μm, the lower limit can be extended to about 2 μm, provided the temperature is controlled within narrow limits.

FIG. 4.14. Bostock sedimentation balance. (From Bostock, Ref. 23.)

The sedimentation tube assembly consists of two main parts: a sample container and a sedimentation tube. The sample container is a stoppered cylindrical glass vessel having a conical base to allow the sample to flow freely when introduced into the sedimentation tube, to which it is connected by a short length of rubber tubing fitted with a pinch clamp. The two vessels are connected by a second tube, also fitted with a pinch clamp, which permits air displaced from the sedimentation chamber to enter the sample container.

The sedimentation tube has a 38-mm bore and contains the sample to a depth of 250 mm. The inlet tube at the top has a wide bore to allow the sample to enter quickly and is curved toward one wall to avoid splashing. An engraved scale enables the exact height of the suspension to be measured. The tube is enclosed in a glass thermal stabilizing jacket sealed at the bottom by a gasket. The torsion wire carries a vertical pointer fitted with a graticule, an image of which is projected onto a ground-glass scale by an integral optical system. The graticule has an arbitrary scale of 100 divisions, and the zero can be accurately set by means of a convenient adjusting knob. The counterbalance is used to adjust the sensitivity of the balance, and at its most sensitive setting a full-scale deflection is produced by 0.5 g of powder of specific gravity 1.5 in a liquid of specific gravity 1.0.

Palik[24] has modified the Oden principle further by utilizing an optical arrangement to magnify and record the small deflections produced by particles settling onto the pan of a torsion balance suspended in the settling column. The principle of his sedimentation apparatus is illustrated in Fig. 4.15. The light source (LS) consists of a 6.3-volt tungsten filament lamp operating from a constant-voltage transformer. This light is collimated by the lens (CL) of focal length 15 cm, and is interrupted by an opaque metal shield (S) attached to one end of a torsion balance (TB). A stainless-steel pan (P) immersed in the settling vessel (SV) is attached to the other end of the torsion balance.

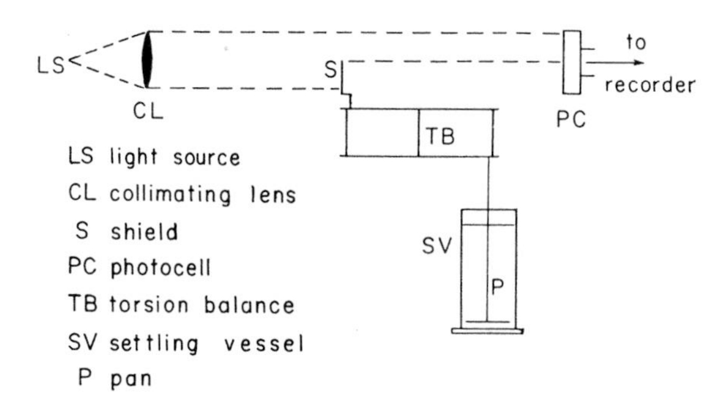

FIG. 4.15. Palik sedimentation balance. (From Palik, Ref. 24.)

As the particles settle on the pan, the right side of the balance descends. The left side of the balance simultaneously ascends, causing the shield to intercept the collimated light beam. The transmitted light falls on a barrier-layer photovoltaic cell (PC). An increase in weight on the balance pan results in a decrease in intensity of the light signal, which is continuously recorded on a strip chart. The settling curve traced by the recorder is analyzed to obtain the particle size distribution in terms of an effective particle diameter calculated from Stokes' equation. The optimum size range of this instrument is 2 to 60 μm.

Another gravity-settling method for size analysis utilizes a special hydrometer to measure (by means of specific gravity determinations) the concentration of solids remaining in suspension after suitable settling intervals. The American Society for Testing and Materials[25] has approved this technique for use with soils, and Schweyer[26] has modified it for use with other materials. Another method, based on measuring the change in density of a settling suspension by means of sensitive manometers, has been proposed by Knapp.[27] Berg[28] has described the use of hollow glass bodies of known specific gravity (called divers) that are sensitive to differences in specific gravity between the suspending liquid and the homogeneous suspensions which exist after recorded time intervals. These methods appear to have limited application to industrial hygiene practice.

Turbidimetry provides another means for following the progress of gravitational settling of particles in a fluid. The concentration of dust required for use of this method may be 1/100th that required for gravimetric sedimentation methods, and this lends itself to the analysis of dusts sampled from the air as well as bulk powders. Turbidimetry involves measurement of the intensity of light transmitted through a dilute suspension of settling dust. As the particles settle, the light intensity increases. This relationship is based on the Lambert-Beer law (equation 1.2), which relates the number of particles in the path of the beam to the fraction of the total light transmitted. When equation 1.2 is rearranged to mass concentrations and particle diameter, it has the following form:

$$\log I_o/I = kcl/d \tag{4.1}$$

where I_o is the intensity of the transmitted light beam when no particles are present.

I is the intensity of the light transmitted by the suspension of particles in the light beam.

c is the mass concentration of particles in the light beam.

l is the width of the suspension traversed by the light beam.
d is the particle diameter.
k is the extinction coefficient of the particles.

Because the extinction coefficient varies with size for small particles and also depends on the optical properties of the particles and the suspending medium, it is usually convenient to substitute the total projected area of the particles (S) for the ratio k/d in equation 4.1.

Let I_1 and I_2 stand for the intensity of a light beam transmitted through a suspension of particles at the start of the settling period and after some time interval. The fraction of the total surface area of the original suspension that has not settled beyond the measuring zone can then be given by the ratio

$$\frac{\log I_0/I_2}{\log I_0/I_1}$$

and the largest particle that remains in suspension in the measuring zone can be calculated with Stokes' equation from the time interval corresponding to I_2 and the total height through which particles must settle to reach the measuring zone. The values of the calculated Stokes' diameters and the corresponding extinction ratios represent the undersize distribution of particles by projected area.

Use of this method is based on the assumption that the general transmission law is valid for all sizes down to zero; therefore, if the sample contains appreciable numbers of particles in the range for which this relation is invalid, large errors may be introduced. In addition, it is assumed that the specific surface value for the portion of the sample finer than the finest size measured is the same as for the larger particles. As a result of these and other factors, the turbidimeter probably does not give size distribution or specific surface with great accuracy. Nevertheless, from a practical viewpoint the turbidimeter is a very useful tool for comparing the fineness of similar materials.

Wagner[29] described an instrument for measuring the specific area (i.e., total surface area per gram) of dust, in which the powder is dispersed in a liquid of known viscosity and density and allowed to settle by gravity past a slit through which light is passed from a source of constant intensity to the photoelectric cell. By measuring the amount of light transmitted (as indicated by the microammeter readings), the settling rate (and hence size) of the suspended dust was calculated. Schweyer and Work[30] made a detailed study of the Wagner method and found that, with the exception of cement and silica, the assumptions regarding the optical relations are not valid, and that results obtained by the Wagner method are in very poor agreement with those from other methods.

Talvitie and Paulus[31] developed a recording photometric particle size analyzer in which the progress of gravitational settling in a liquid medium is followed by means of a collimated beam of light. This instrument differs from those previously described in that provision is made for continuously varying the sedimentation depth at which photometric readings are taken. This permits analysis of an entire range of particles down to approximately 0.5 μm in 1 hr.

Fig. 4.16 is a diagram of the optical system developed by Talvitie and Paulus. Essentially monochromatic conditions are provided by the use of ultraviolet and infrared absorbing filters and a photocell which has a maximal response at 800 millimicrons. The lens and aperture system produces a narrow parallel beam of light having minimum scattered light components. A 20 × 40-mm glass cell with parallel sides is used with a 0.01% dust suspension. In the analyzer, the cell is surrounded by a constant-temperature coil and fastened to a vertical slide which can be raised or lowered by means of a standard micrometer head. The micrometer is coupled to a synchronous clock motor to lower the cell at the rate of 1 in./hr.

To operate the instrument, a weight of dust sufficient to give an absorbance reading near the upper end of the recorder scale is suspended in a volume of fluid which, when transferred to the cell, will provide a 1-in. maximum initial height of fluid above the measuring beam. With a 40-mm cell, a concentration of about 0.1 mg/ml is suitable. The suspension is equilibrated to the temperature of the cell compartment, which is maintained

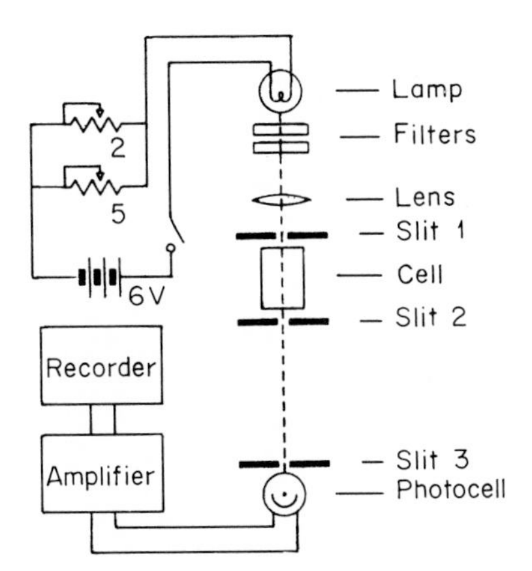

FIG. 4.16. Recording photometric particle size analyzer. (From Talvitie and Paulus, Ref. 31.)

by circulation of water from a constant-temperature bath. Mixing of the suspension is accomplished by quickly inverting and agitating the cell and returning it to the cell compartment without delay. The intensity of the light beam is adjusted and the recorder chart and cell-lowering mechanism are actuated beforehand so that the record will begin the instant the cell is returned to the compartment.

When samples which have a high proportion of large particles are sized, expansion of the scale for the large particles is achieved by operation at a fast chart speed (2 in./min) for a few minutes, after which the speed is reduced to 24 in./hr. Thereafter the instrument operates unattended.

A typical sedimentation curve obtained with the instrument and the grapic means by which the curve is calibrated in terms of Stokes' diameter are illustrated in Fig. 4.17. The initial depth is represented by the distance BC, since the measuring slit was at the lowest point at the beginning of the run (point B) and at the surface of the medium at the conclusion of the run (point C). Any point A on the curve represents the position of the slit at that instant. The distance AC multiplied by the approximate conversion factor gives the height, h, from which a particle originally at the surface has fallen to reach the level of the slit, and the distance BA converted to seconds gives the time, t, for the particle to fall from the height, h, in centimeters. The conversion factors for a chart speed of 24 in./hr and a slit speed of 1 in./hr are $t = 150$ BA and $h = (2.54 \times 150\ \mathrm{AC})/3600$. The velocity of particles which have settled from the surface to the slit is given by:

$$v = h/t = 2.54\ \mathrm{AC}/3600\ \mathrm{BA}, \quad \mathrm{cm/sec} \tag{4.2}$$

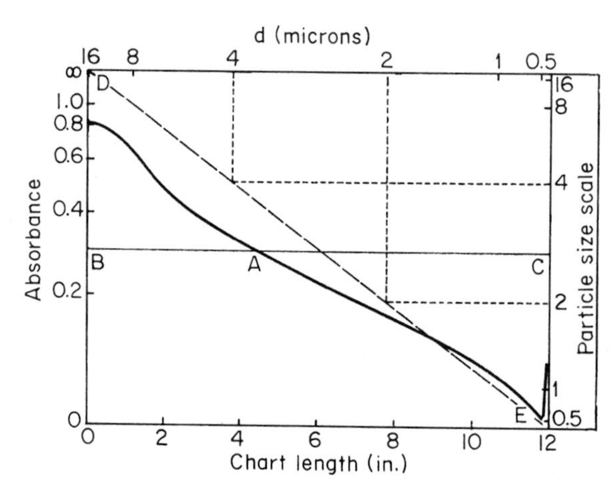

FIG. 4.17. Sedimentation curve of airborne granite dust. (From Talvitie and Paulus, Ref. 31.)

By substituting this expression for terminal settling velocity in Stokes' law, the x-coordinate may be calibrated in terms of Stokes' diameter.

Rose[32] has made extensive investigation of the extinction coefficient–particle size relationship for particles settling under gravity in an aqueous suspension using monochromatic light. The extinction coefficient may be defined as the rato of the light obscured by a particle to the light which would be obscured if the laws of geometric optics held for the particle under consideration; i.e., the ratio of extinguished light to geometrical projected area is not unity. Measurement of the light energy removed by suspensions containing a known number of particles can be translated into average diameter with the aid of an "extinction coefficient-size" curve derived experimentally by Rose. The determination of size distribution by this method is complicated by the necessity of measuring the particle number in the unsettled portion of the suspension each time a transmission measurement is made. As in other gravity settling methods in liquids, the lower practical size limit is about 5 μm, although the curve is useful down to about 2 μm.

The Andreasen pipet[33] is widely used in industry. The standard model (Fig. 4.18) consists of a glass cylinder with about a 6-cm i.d., with a capacity of 550 ml when filled to the upper mark. The ground-glass stopper has a small vent hole and carries a 10-ml pipet with a three-way stopcock.

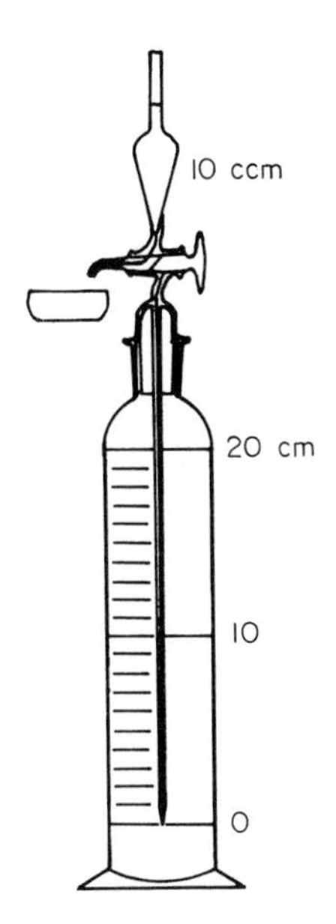

FIG. 4.18. Andreasen pipet. (From Andreasen, Ref. 33.)

Enough powder sample is weighed to form a 1 to 2% suspension with the liquid in the pipet. The apparatus is closed and inverted repeatedly to obtain thorough mixing and is then placed in a thermostatically controlled constant-temperature location. Samples of the suspension are withdrawn periodically through the pipet (i.e., after 1, 2, 4, 8, 16, 32, and 64 min of sedimentation), dried under standard conditions, and weighed to the nearest 0.1 mg. The pipet method is relatively quick and reliable and requires only inexpensive apparatus, but there are disadvantages. When operating with the low dust

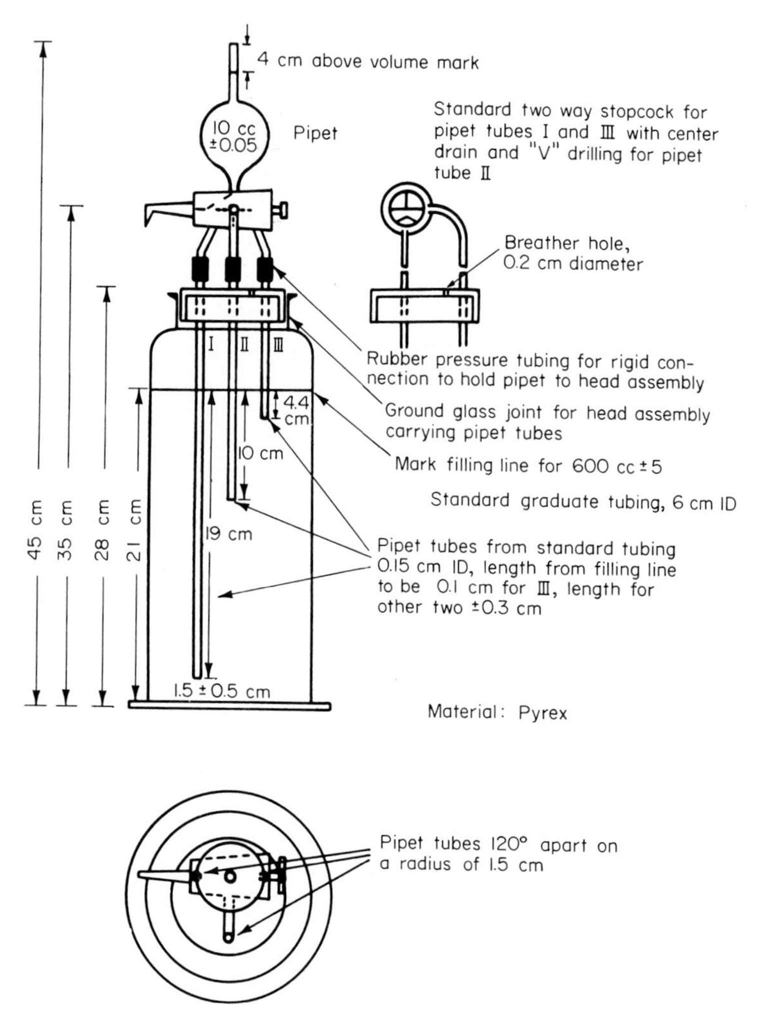

FIG. 4.19. Modified Andreasen pipet. (From Schweyer, Ref. 26.)

concentrations that are necessary to ensure the validity of Stokes' law, a relatively large sample must be withdrawn each time to determine the weight concentration with precision. This means that a numerical correction must be made to account for the change in settling height as sedimentation progresses. More serious is the fact that, when a large fraction of the total column is removed, the thickness of the removed layer is often greater than 1 cm and a considerable disturbance is introduced into the steady settling rate.

Schweyer,[26] Schweyer and Work,[30] and others have suggested improvements to increase the speed and accuracy of the Andreasen pipet method. These consist in introducing two or more pipets into the settling chamber to give additional data and to yield rapid readings, and mechanical improvements in the pipet arrangements for withdrawing samples. Fig. 4.19 shows one such suggested modification of the apparatus, but the method remains essentially unchanged. Schweyer[26] has described its operation as follows:

Forty-five milliliters of a wetting agent solution is added to the sample in a 600-cc beaker and allowed to stand until all the particles are wetted, usually about 5 min. The volume is then made up to about 500 cc with distilled water, and the beaker and contents are kept in a stirring apparatus with a motor speed of 3000 to 4000 rpm for about 15 min, after which the suspension is washed into the pipet cylinder. Distilled water is added through the breather hole to bring the level to a reference mark after the pipet head is inserted. This level (bottom of the meniscus) becomes the starting level mark for every analysis and determines the starting volume of suspension for subsequent calculations. The pipet tubes should be empty and the stopcock closed when the pipet is inserted.

The stopcock permits taking samples in rotation from tubes I, II, and III without admitting the suspension into tubes II or III until desired. The stopcock is turned counterclockwise 90 degrees to fill the pipet from tube I; it is then reversed 90 degrees in order to drain the pipet; it is then returned to the original position to allow the level in tube I to return to the level of the suspension in the cylinder. The stockcock remains in this position until the time for the second sample.

The above procedure is repeated for as many samples as desired. For best results the pipet and suspending medium should be brought to the test temperature before starting.

As it takes about 15 sec to siphon the sample, the midpoint of this interval should be used for calculating particle size. A continuous suction should be applied (by mouth) and the pipet filled, after which the stopcock is closed and the fluid allowed to run immediately into a preweighed crucible held under the drainage spout.

The vessels for collecting the samples may be porcelain crucibles, Gooch crucibles with asbestos mats, or fritted-glass filters. The particle collectors may be dried in an oven at any convenient temperature up to 160°C, but for very rapid results platinum Gooch crucibles may be employed and the asbestos mat dried over a direct flame before and after sampling.

A typical sampling schedule for the special pipet is given in the following tabulation:

Sample No.	Tube No.	Height, h (cm)	Time, t (min)	Diameter,[a] microns
1	1	18.9	1	53
2	1	18.3	3	30
3	1	17.9	7	20
4	2	9.0	15	10
5	2	8.7	45	5
6	3	2.7	60	2.5
7	3	2.2	240	1.25

[a]Actual diameter depends on the particular liquid–solid system used and the calibrated value of h.

Berg[28] has described a pipet apparatus consisting of a sedimentation vessel 90 cm high with an inside diameter of 3 cm and a mark 13 cm from the top, corresponding to a volume of about 500 cc. It has a horizontal sampling tube which is fitted through the side of the vessel, as shown in Fig. 4.20a. Samples are taken with the loose pipet shown in Fig. 4.20b, the upper part of which can be closed by means of a two-way stopcock. The lower part ends in the short tip of the pipet and forms an angle of 90 degrees with the pipet tube. When the tip of the pipet is connected to a rubber tube fitted to the sampling tube, the pressure inside the vessel will make the suspension rise into the pipet beyond the two-way stopcock, which is then closed. Next, the pipet is released from the rubber hose, which is closed by a glass rod serving as a stopper. The suspension above the stopcock is blown out through the side tube of the stopcock, and the pipet is emptied into a weighed vessel which is evaporated and reweighed. The volume of the pipet, including the bore, is 10 cc.

For suspensions in viscous fluids, a wider sampling tube and a pipet having a wider intake aperture are used so that the pipet can be filled in about 3 sec, and the error due to the fact that the pipet is momentarily not filled remains a slight one. By means of this apparatus, measurements can be effected in aqueous suspensions over the whole of the area in which one must use a viscous fluid as a suspending medium when employing Andreasen's

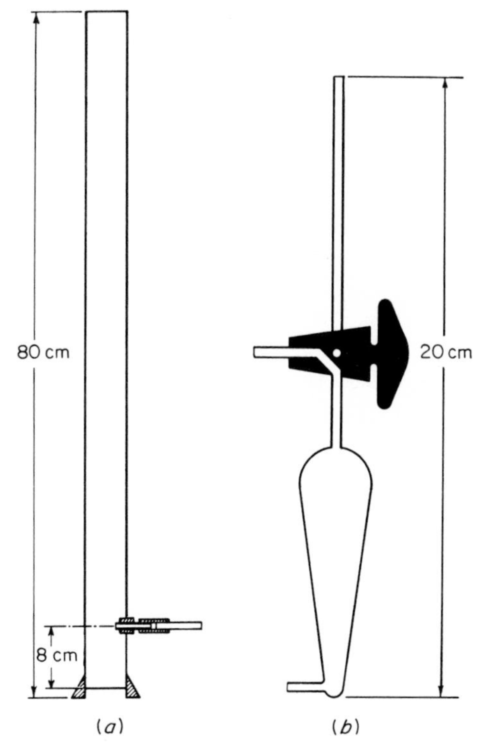

Fig. 4.20. Berg's modification of Andreasen pipet: (*a*) sedimentation vessel with horizontal sampling tube; (*b*) loose pipet with angular tip. (From Berg, Ref. 28.)

apparatus[28] to measure the size of large particles that would not be in Stokes' range when settling in water.

(3) Centrifugal sedimentation in a liquid. Centrifugal methods may be used to extend the useful range of sedimentation techniques below the 2- to 5-μm practical lower limit obtainable with the Andreasen pipet.

Kamack[34] has described a centrifugal pipet which is useful over a size range from about 2 to 0.1 μm, depending to some extent on the particle density. The method requires a 6-in.-long sector-shaped centrifugal flask (Fig. 4.21), because particles in a centrifugal field follow radial paths rather than substantially parallel ones as in a gravity field. In addition, the flask is equipped with means for removing a sample of suspension while the centrifuge is in motion, thereby avoiding the disturbance to the suspension which occurs when the centrifuge is stopped. The plunger closes a length of Tygon tubing between the sector-shaped centrifuge tube and the empty sample cup. When it is time to take a sample, current from a 6-volt battery heats a Nichrome resistance wire, which releases a plunger which presses against the Tygon tube. When the sample line opens, a portion of the suspension at a preset level in the centrifuge tube siphons into the 10-ml sample cup. Calculations for sedimentation in the centrifugal pipet require consideration of the increasing force on a particle as it moves outward in the centrifugal field. Approximate solutions have been presented by Kamack, and the underlying theory is given in standard references on centrifugal devices.[35,36] The Kamack instrument is not available from a commercial source.

Whitby[37] has described a combination gravitational and centrifugal liquid sedimentation technique which is capable of sizing particles from approximately 100 to 0.1 μm in 2 hr. Cartwright and Gregg[38] have summarized the analytical procedures in the following manner:

"A specially shaped tube, shown in Fig. 4.22, is nearly filled with a suitable sedimentation liquid. The sample is dispersed in a second liquid which is miscible with the sedimentation liquid but of slightly lower density. A thin layer of this dispersion is then floated onto the surface of the sedi-

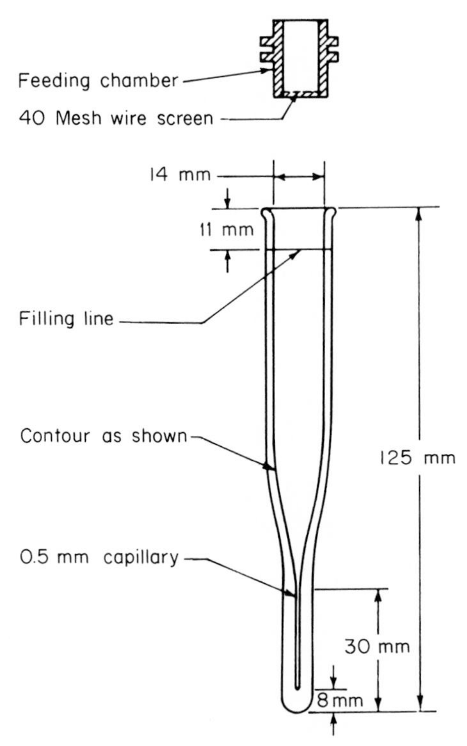

FIG. 4.21. Sector-shaped centrifuge tube. (From Kamack, Ref. 34.)

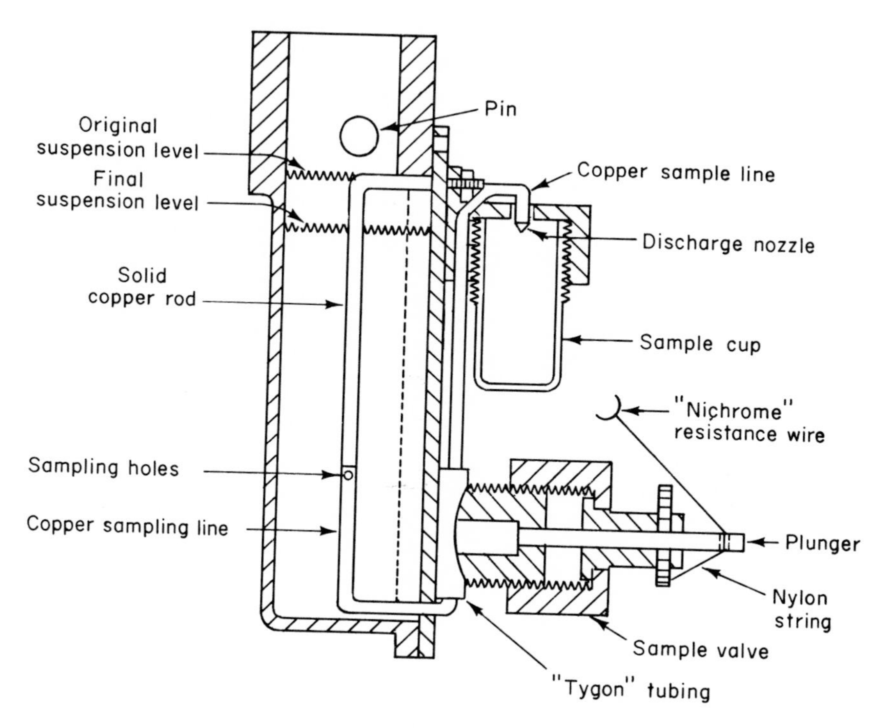

FIG. 4.22. Special sedimentation tube. (From Whitby, Ref. 37.)

mentation liquid, with minimum mixing, so that all of the particles start essentially at the top of the sedimentation tube. Thus, as the sedimentation progresses the particles become classified according to size. After the larger particles have settled to the bottom under gravity, the sedimentation tube is centrifuged to reduce the time required for the smaller particles to reach the bottom. By measuring the volume of particles accumulated as a function of time, the equivalent spherical size distribution of the sample may be computed from formulae based upon Stokes' law."

The major items of the apparatus* used in this method include:

"1. Specially designed sedimentation tubes having specifications as shown in Fig. 4.22.

"2. Two synchronous centrifuges; one 600 and 1200 rpm and one 1800 rpm, specially designed to provide reproducible starting and stopping

*Obtainable from Mine Safety Appliances Co., Pittsburgh, Pennsylvania.

characteristics. Also, the maximum acceleration during starting and stopping has been adjusted to be not greater than 5 radians per sec per sec, which is accomplished through the utilization of an appropriate inertia disk, together with a series resistor, in one of the motor windings. Each centrifuge is equipped with an appropriate electrical timer which can be set to 1 sec for periods up to 1 hr.

"3. A tube holder and tapper assembly stand. This holds the sedimentation tube during the gravity settling period and while taking readings of the sediment height in the tube capillary. The tapper portion of the device applies light blows to the end of the tube about every two seconds to assist the particles in moving down the sides of the tube.

"4. Special chamber, shown in Fig. 4.22. This is used to introduce the sample dispersion onto the surface of the sedimentation liquid.

"The principal steps in this method are as follows:

(1) Select sedimentation and dispersing liquids and determine density of the sample.
(2) Calculate a particle size-settling time table.
(3) Disperse the sample in dispersing liquid.
(4) Prepare the sedimentation tube and initiate the run by introducing the sample dispersion.
(5) Record the readings of the sediment height in the capillary of the sedimentation tube at the times previously calculated for the various particle sizes.
(6) Calculate and plot the cumulative size distribution curve from the data on the size-time table.

"In selecting the liquids to be used, several factors need to be considered. Obviously the liquids must be inert to the material to be examined and must be miscible with each other. Both must also adequately disperse the sample during sedimentation. For convenience, the viscosity and density of the sedimentation liquid should be such that the largest particles present do not settle out sooner than 10 sec after the start of the run, since it is difficult to obtain the first reading before this time. Also, if very large particles are present, Stokes' law does not accurately apply.

"The density of the dispersing liquid must be less than that of the sedimentation liquid so that a stable layer of the sample dispersion may be floated onto the surface of the sedimentation liquid. Under certain conditions, particles may tend to hesitate and accumulate at the interface of the two liquids, resulting in the formation of streamers which upset the proper performance of the sedimentation. It has been found that these effects can be reduced by using a combination of liquids which have a small density difference and for which the viscosity of the dispersing liquid is about twice that of the sedimentation liquid.

"It has been found that benzene or carbon tetrachloride is very suitable as a sedimentation liquid for a wide range of materials. Kerosene is an appropriate dispersing liquid to use with either of these liquids."[38]

The basic steps in particle analysis by the Whitby method are summarized diagrammatically in Fig. 4.23.

4-2.2 MISCELLANEOUS METHODS FOR DETERMINING SIZE AND SIZE DISTRIBUTION

(a) *Optical Methods.* Bailey[39] has developed a method for determining the size and size distribution of particles suspended in liquids from light transmissions measured in the visible and near infrared parts of the spectrum. The method is limited to colorless materials, however. The apparatus requires incandescent and mercury arc light sources to give a wavelength range of 0.4 to 2.0 μm. A series of fourteen transmissions over the full spectral range is measured, and from this information size distribution curves may

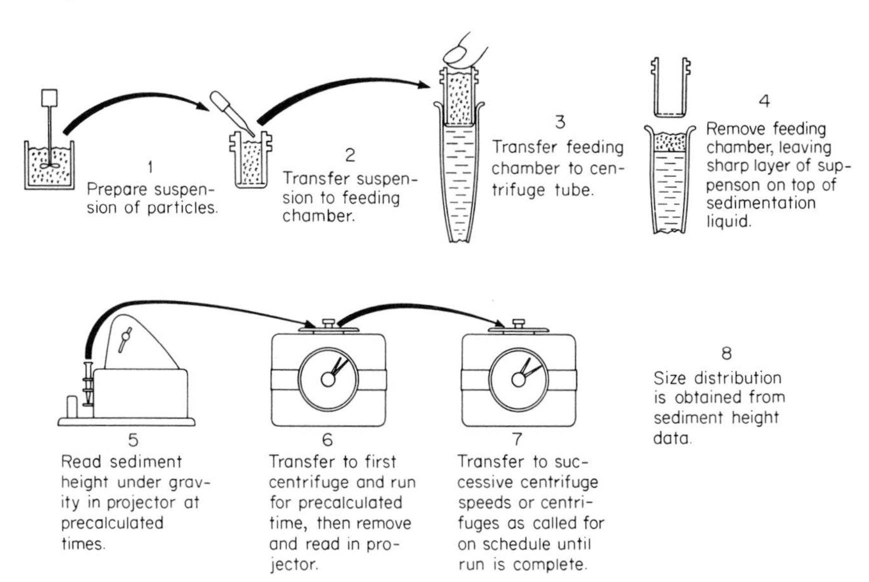

FIG. 4.23. Basic steps in particle size analysis by centrifugal sedimentation. Courtesy of Mine Safety Appliances Co.

be constructed. Particle sizing depends on the fact that each wavelength emphasizes a different part of the particle size spectrum. Bailey describes the method in the following manner:

"Specific extinctions are calculated from the measured transmissions of scattering suspensions according to the Lambert-Beer law as illustrated in equation 4.3:

$$I/I_0 = \exp(-E_M ct) \tag{4.3}$$

where I is the intensity of the transmitted light, I_0 the intensity of the incident light, E_M the average specific extinction for the particle size distribution involved, the subscript M refers to a function determined by the relative refractive index of the particle and the wavelength of the radiation used; c is the volume concentration of the suspension in units of 0.01 cc of suspended material per 100 cc of suspension, and t is the cell thickness in centimeters. Relative refractive index is the refractive index of the particle divided by the refractive index of the suspension medium."

Fig. 4.24 shows the experimental arrangement which has been used for these measurements.

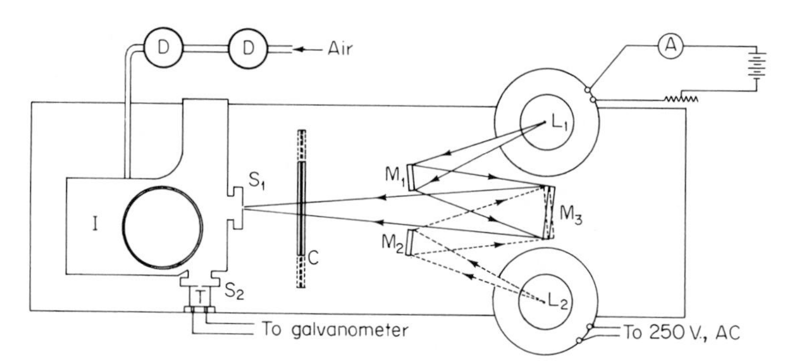

FIG. 4.24. Schematic drawing of apparatus for particle sizing by spectral transmission. (From Bailey, Ref. 39.)

"L_1 is an incandescent lamp and L_2 is an H-3 mercury arc. M_1 and M_2 are plane mirrors which permit interchanging the sources, and M_3 is a concave mirror which produces an image of the source at the monochromator slit S_1. The monochromator, I, is based on a Gaertner Type L235 infrared spectrometer, which was modified by the manufacturer to cover the full spectral range given above. T, the receiver, may be either a vacuum thermopile or a photocell. According to the present operating practice a photocell is used with an amplifier and meter for the visible spectrum and a thermopile-galvanometer combination is used for the spectral region from 0.7 to

2.0 μm. The mercury arc is used in the blue and violet parts of the spectrum. The incandescent lamp is used for wavelengths of 0.5 μm and longer. The cell block containing three cells, two for suspensions and one for the clear suspension medium, is shown at C."

The specific extinctions are calculated from the measured transmissions in the manner indicated in Table 4.1.

TABLE 4.1
SPECTRAL TRANSMISSION DATA[a, b]

Wave length, μm	T^{-1}	$\ln T^{-1}$	$\bar{E}_M$	M	$\bar{E}_M/M$
0.405	3.19	1.16	12.15	1.71	7.11
0.436	3.24	1.18	12.32	1.51	8.19
0.546	2.30	1.16	12.17	1.10	11.16
0.500	3.16	1.16	12.07	1.23	9.81

[a]From Bailey, Ref. 39.
[b]Cell thickness = 0.08 cm; concentration = 1.0×10^{-4} cc/cc; $ct = 0.0955$ ($c_0 t_0 = 10^{-4}$); $\bar{E}_M = (ct)^{-1} \times \ln T^{-1} = 10.47 \times \ln T^{-1}$.

"The factor which converts the logarithm of reciprocal transmission to specific extinction is calculated from the values of concentration and cell thickness as indicated at the top of the table. Column 2 shows the reciprocal of the transmissions measured at the wavelengths of column 1. Columns 3 and 4 show the calculation of specific extinction. The *M* values of column 5 are calculated for anatase in an alkyd varnish and 6 is obtained by dividing 4 by 5. Columns 5 and 6 are plotted as shown" (in Fig. 4.25).

"The method of analysis is rather rapid, the complete operation requiring but 2 hr, and the operations involved are simple enough for a technician to carry out satisfactorily. The results are usually reported as average radius, the *B* value (a measure of uniformity), and the fraction of the material found in the distribution."[39] This method is adaptable to the size range below 2 μm, and size information obtained with this method is reported to be in good agreement with ultracentrifuge data.

(b) *Special Methods Applicable to Liquid Drops.* Volatility, shattering, and coalescence of droplets make it necessary to employ special devices for the collection of artificial and naturally occurring mists and fogs, liquid spray from nozzles, etc. May[40] has suggested a number of methods employing the cascade impactor. For oils and other liquids of low volatility, clean grease-free glass slides may be used and the examination completed before significant evaporation occurs. On clean surfaces, droplets spread uniformly

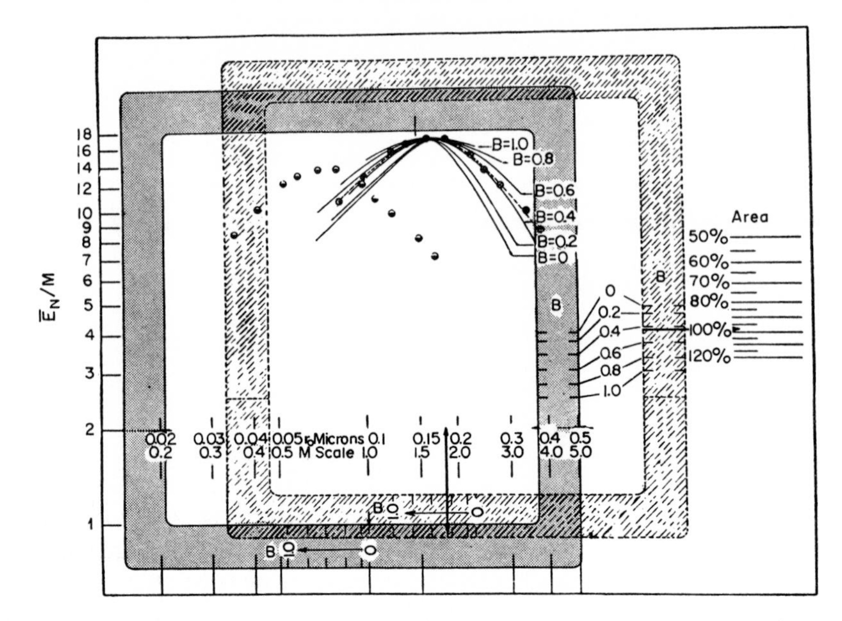

FIG. 4.25. Standard chart for spectral transmission method. (From Bailey, Ref. 39.)

in the form of a planoconvex lens. May has indicated that the size of the original drop may be estimated microscopically from the diameter and focal length of the liquid lens, although Sinclair[41] found this technique to give results about 25% low.

Water droplets may be preserved by impacting them on slides coated with a mixture of mineral oil and Vaseline and then covering them with a similar film of melted oil and Vaseline. When this procedure is carefully performed, the droplets are drawn up into their original spherical form. Liquid droplets impacted on slides coated with a film of freshly formed MgO from burning magnesium ribbon penetrate the surface and leave a permanent round hole which may be seen under the microscope by transmitted or oblique illumination. The lower limit of size detectable by this method is about 3 μm. Jet velocities capable of depositing smaller particles disturb the MgO surface layer. In a similar way, a slide may be coated with a dye soluble in the liquid composing the drops. When a drop strikes the slide, an area of dye will be washed away which will bear a definite relationship to the original drop size. In all the above methods, the lower size limit is approximately 3 to 5 μm and the collected drops are susceptible to microscopic measurement, only.

Houghton[42] and York and Stubbs[43] have developed methods of photographing droplets in the free-floating, airborne state, while Pigford and

Pyle[44] have described a photographic method in which spray droplets are allowed to fall onto greased slides placed inside a moistened chamber. The equipment used in the York and Stubbs method of photographing spray droplets is illustrated in Fig. 4.26. The camera is equipped with a lens having a 50-mm focal length and $f/3.5$ aperture. The distance from lens to film (about 20 in.) provides a magnification of 10×, producing images of convenient size and good resolution. The camera shutter remains open, and the exposure of the high-contrast film is determined by the length and intensity of the illumination. Light from photolamps passes through a water cell (containing nigrosin to control the level of illumination) onto a ground-glass diffusing screen producing a "silhouette picture with the background illuminated and the drops appearing as shadow."[43] When double exposures are taken with a small known interval between exposures, the photograph shows a pair of images for each drop, and drop velocity can be calcula'ed from the distance between images and the interval between exposures. This method is suitable for spray droplets ranging from 15 to 500 μm. Fig. 4.27 shows typical droplet photographs by single and double exposure. Only the drops in sharp focus can be sized satisfactorily.

In another photographic method, described by Pigford and Pyle,[44] droplets are permitted to fall into shallow, transparent pans of mineral oil (or some other suitable nonmiscible solvent of lesser specific gravity), resting on sheets of photosensitive paper. Collection of droplets takes place in darkness, and after a suitable settling period the overhead lights are turned on to expose the sensitive paper. Droplets appear as dark circles after processing the photographic paper. The lower limit of detectability is approximately 2 to 3 μm, and the photographs may be analyzed with the aid of magnifying lenses.

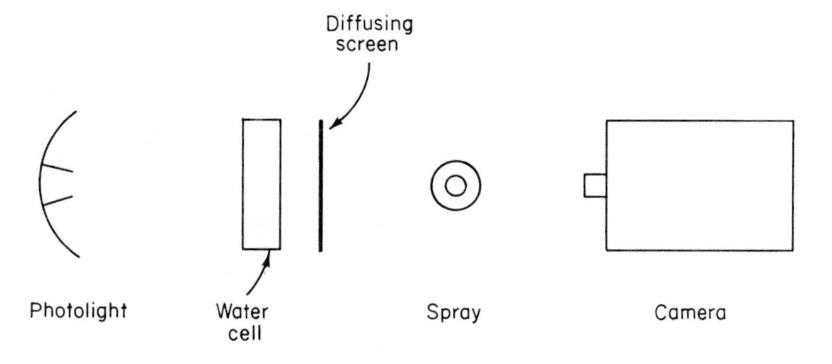

FIG. 4.26. Equipment for photographing spray droplets. (From York and Stubbs, Ref. 43.)

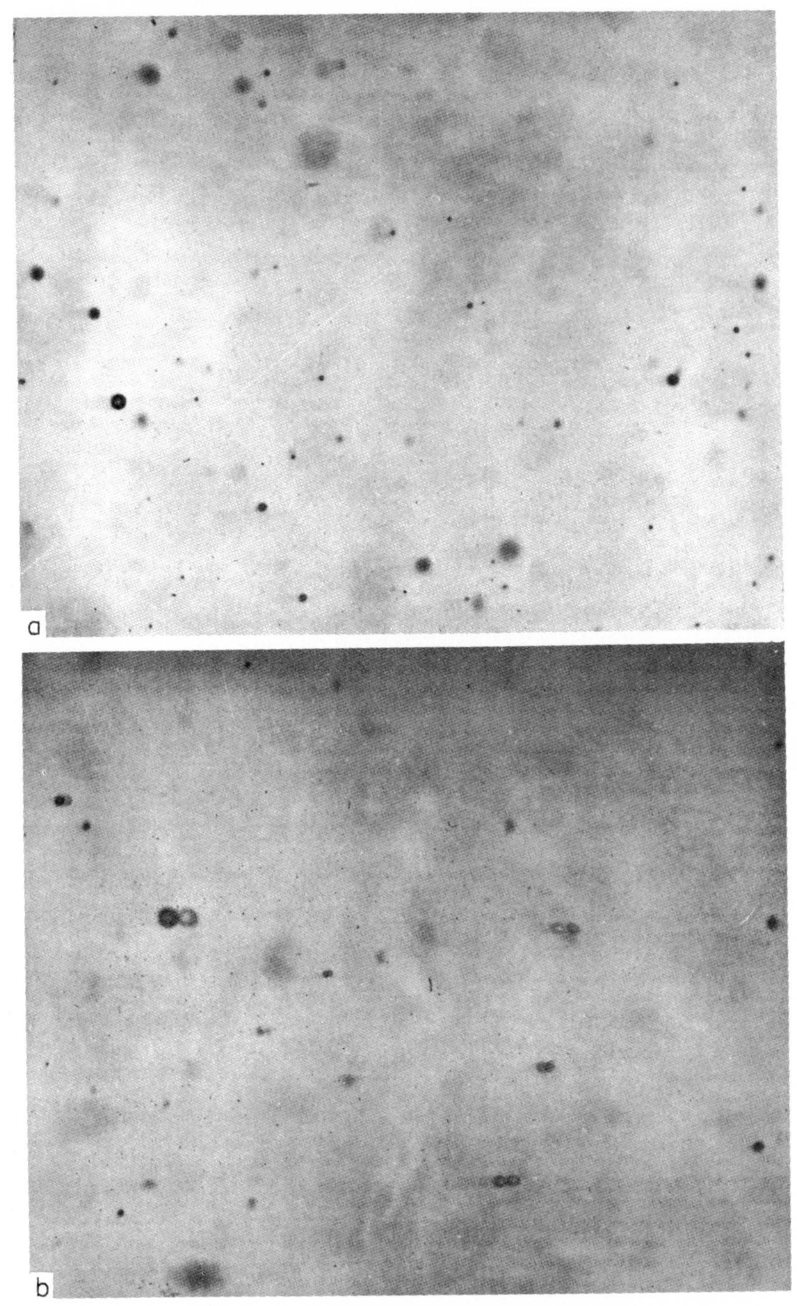

FIG. 4.27. Velocity measurement of spray droplets: (*a*) single exposure; (*b*) double exposure. (From York and Stubbs, Ref. 43.)

Langmuir and Blodgett,[45] Geist *et al.*,[46] and Landahl and Herrmann[47] have proposed devices for sizing liquid sprays based on the selective impaction efficiency of wires (and similar devices) of graded size. An electronic spray analyzer has been developed by Geist *et al.* to be used in conjunction with such a method. These devices are limited to droplets about 5 μm in size.

Under suitable conditions of illumination and quiescence the free-falling velocity of droplets may be measured in the ultramicroscope and their true size calculated from Stokes' equation. The ultramicroscope is a light microscope so arranged that the freely falling particles or droplets are illuminated by a strong pencil of light at right angles to the microscope axis (Fig. 4.28). Visibility is limited by the intensity of the light source, but the method may be used to observe particles in the size range below 0.2 μm. Since size determination is based on observed settling velocity in still air, elaborate precautions are required to eliminate convection currents in the settling chamber due to the high-intensity light source and turbulent air currents introduced with the suspension being measured. In addition, it is difficult to prevent evaporation of volatile substances. The lower size limit is about 0.05 μm with this technique, but measurement of a statistically significant number in a polydispersed cloud is a lengthy procedure.

Particles near the limit of resolution of the ultramicroscope may be detected only in the absence of large particles, since the amount of light scattered by coarse particles obscures that from the fine ones.

Settling of charged particles in the ultramicroscope can be augmented by the use of electric fields of known potential. Increased settling rate minimizes

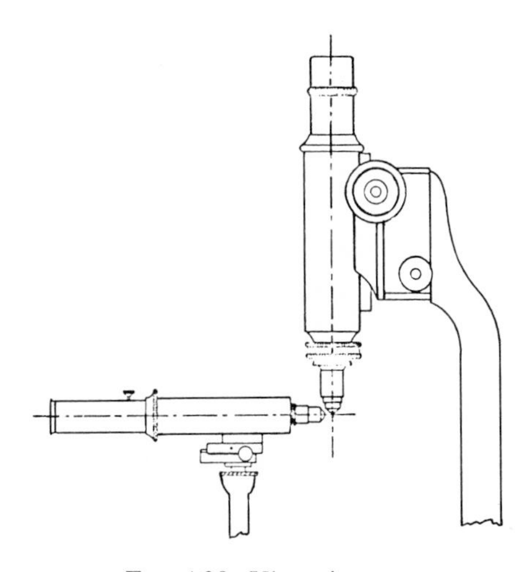

FIG. 4.28. Ultramicroscope.

the effect of random convection and Brownian motion on very small particles. Thomas and Rimberg[48] have described a method for independently determining the number of electrical charges on individual particles in a monodispersed aerosol and their settling velocity, using a parallel-plate electrostatic precipitator and a modified Millikan apparatus. Another method utilizing the ultramicroscope principle is to measure the amplitude of oscillation of airborne particles in an alternating electric[49] or sonic field[50] of low frequency by photographing their paths in the cell of the ultramicroscope. Carey and Stairmand[51] described an apparatus for photographing the paths of fall of particles and computing the size from the exposure time and length of path. The flow ultramicroscope of Derjaguin and Vlasenko[52] consists of an optical microscope, a viewing cell having axial particle flow toward the observer, and a strong transverse light beam. A discussion of general construction details and some modifications to the ultramicroscope for particle photography is presented by Kubie.[53] Techniques of ultramicroscopy are most suitable for laboratory aerosol research on fine particles, such as determination of number concentration, or for the observation of particle motion as a result of an impressed field, but have only limited application to particle sizing in general industrial hygiene practice.

These indirect methods depend for their accuracy on several assumptions that may not be correct. Although they provide valuable assistance in the laboratory, they are too sensitive for routine field use. In all these methods, large numbers of particles must be measured to give statistically sound size distribution data, and skill of a very high order is required to use this type of apparatus. The electron microscope is preferred generally to the ultramicroscope because of greater ease of manipulation and the added information obtainable on particle shape, state of agglomeration, etc.; but for measuring very small volatile liquid drops, the ultramicroscope and its modifications still prove useful.

Taylor and Harmon[54] have described a size analysis method for liquid droplets based on the frozen-drop method of Longwell,[55] in which droplets are sprayed into a collecting liquid maintained well below the freezing temperature of the spray. Taylor and Harmon used hexane cooled to −20°C with dry ice to catch and freeze water droplets in the apparatus shown in Fig. 4.29. The authors describe their apparatus as follows:

"The water is sprayed over the edge of box a, landing in the catching liquid, b. The drops freeze very quickly and fall to shutter c. When all the drops are resting on the shutter, the shutter pull, d, is opened, which allows the drops to descend through the hexane to a scale pan, e. The drops fall approximately according to Stokes' law (neglecting interaction effects) and

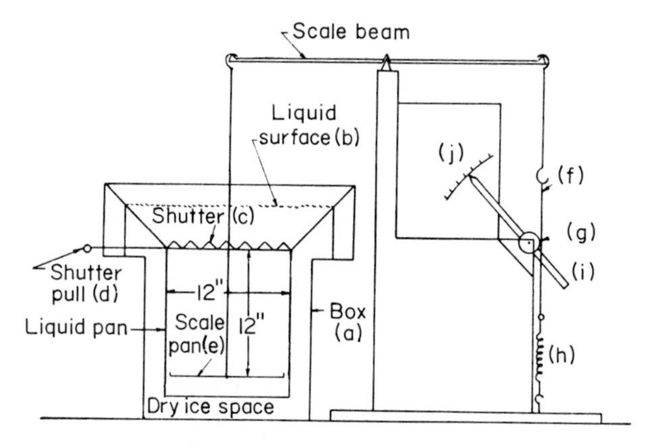

FIG. 4.29. Frozen drop apparatus. (From Taylor and Harmon, Ref. 54.)

the largest drops arrive first at the scale pan. The weight on the scale pan is transferred to a cord, f, which passes around an aluminum cylinder, g, and then to a spring, h, which balances the force on the scale pan. The slight movement of the disk, g, is amplified by the pointer, i, which moves over a measuring scale, j, calibrated to indicate the differential weight on the scale pan. The weight on the pan versus time relationship is determined by the use of a stopwatch, though in a more complex instrument this could easily be done electrically with greater accuracy.

"The weight falling on the pan can be transformed simply into equivalent drop diameters and total number of drops at each diameter by use of a drag coefficient which, for the smallest drops, depends upon the Stokes' law settling velocity. It is then necessary to convert the measured sizes to true drop sizes by allowing for the density difference between the ice and the water which was sprayed.

"The ratio of pointer tip movement to pan movement is 10 to 1. The pan moves approximately ¼ in. The total depth of fall of the drops is 12 in., which gives a possible error of about 3% from pan movement.

"In general, the time required for the frozen drops to fall through the liquid is a function of the density difference between the drop and the supporting liquid, the liquid viscosity, the depth of fall, and the square of the diameter of the drops. For the instrument described, times of fall would range from about 4 min for drops 100 μm in diameter to about 27 hr for drops 5 μm in diameter. These times could be decreased by altering the depth of fall of the drops. For drops of this size range a 6-in. depth might be appropriate. This depth of fall would give times from 2 min to 13.5 hr."

Choudhury and Stevens[56] have made additional modifications in the freeze-out technique for sizing liquid droplets by capturing and freezing them in a bath of liquid nitrogen. A spray nozzle is mounted to spray vertically downward into a liquid nitrogen container 12 in. below the nozzle. After the droplets have been caught and frozen, the entire collection container is transferred to a walk-in cold room kept at a temperature consistent with the freezing point of the material being sprayed. Here the liquid nitrogen is decanted off through a fine screen, and the entire group of collected solid droplets is placed on the top screen of an ordinary set of 8-in. test screens, fitted with a bottom pan and a cover. After the individual fractions have been obtained, each is transferred to a small beaker and allowed to warm up and melt. Then the weight of each fraction is determined on an analytical balance and the data are combined to give a complete drop size distribution.

Frozen sub-sieve droplets can be retained inside the cold room and the size analysis continued by microscopic examination or by one of the indirect methods described previously.

(c) *Particle Sizing of Radioactive Aerosols.* Leary[57] described a particle-sizing technique having special application to radioactive particles. The active material is collected from the air on filter paper, impactor slides, etc., and placed in contact with nuclear track plates for various exposure times. By counting the number of tracks in the emulsion for a given exposure time, the size of each emitting particle can be calculated. Alpha-emitting particles as small as 0.2 μm may be determined by this method when specific activity of the material and exposure time are great enough. If there is a large proportion of coarse particles, however, smaller ones tend to be obscured in the radioautograph. Since the actual particles are not observed on the nuclear track plate, this method does not reveal true particle shape but gives sizes in terms of the effective diameter of a sphere of equal radioactivity. With this method no direct distinction may be made between a discrete particle of pure radioactive material and a larger agglomerate containing the same amount of active material and additional amounts of inert materials, unless the nuclear track plate has been keyed to the filter paper so that the particle responsible for each set of tracks can be located and examined microscopically. Otherwise, the gauging of particle size from the radioactive material alone might underestimate the actual or aerodynamic size.

Details of several methods of size-analyzing radioactive particles are included in Chapter 7.

4-3 Instruments Which Determine Average Diameter Only

Particle size analyses which result in an average diameter have only limited application to industrial hygiene. Frequently they have utility as supplementary methods for determining or checking the specific surface area of a dust sample, to establish shape factor, or as a rapid routine control technique for laboratory dust clouds used in toxicological or other types of testing. Some of these methods are applicable to fine smokes and mists of uniform size.

4-3.1 Adsorption Methods

(a) *Gas Adsorption.* Methods based on a determination of total surface area include low-temperature gas adsorption and adsorption from solution. Lauterback *et al.*[58] described a method for determining the specific surface of uranium dusts by low-temperature adsorption of ethane. Emmett[59] used butane and nitrogen for determining the specific surface of carbon blacks, pigments, etc., and numerous other applications have been reported in recent years. The principle of the method, according to Emmett, depends on the selection from an experimental adsorption isotherm of the volume (and hence the number of molecules of some inert gas such as nitrogen) corresponding to a monomolecular layer. Multiplication by the average cross-sectional area of each molecule determines the absolute surface area of the adsorbent. The accuracy of the method depends on the exactness with which the point corresponding to a monolayer of adsorbed molecules can be selected from the isotherm, and the closeness of the approximation of the average area covered by each adsorbed molecule. From a knowledge of specific surface area, the diameter of a sphere having this same ratio of surface to weight may be easily calculated.

A typical arrangement of apparatus for this determination is shown in Fig. 4.30. It consists of an adsorption bulb which holds the sample, a calibrated gas buret, a manometer, a high-vacuum pump, and a McLeod gauge connected as shown. The adsorption bulb containing the sample is sealed into the rest of the apparatus and heated and evacuated to remove water and adsorbed gases. The sample is immersed in liquid nitrogen or oxygen for cooling, and after the volume of the dead space is determined with pure helium (which is not adsorbed) a measured volume of nitrogen, ethane, or other suitable gas is admitted to the sample bulb. After equilibration the volume of gas adsorbed is measured and the process repeated to obtain a series of values for adsorption as a function of pressure at a given temperature. Particle size determinations by this method are possible for nonporous dusts only, although surface area measurements can be made on porous or nonporous materials. There is no lower size limit for this method, but the information obtained concerning size is limited. Sensitive analytical

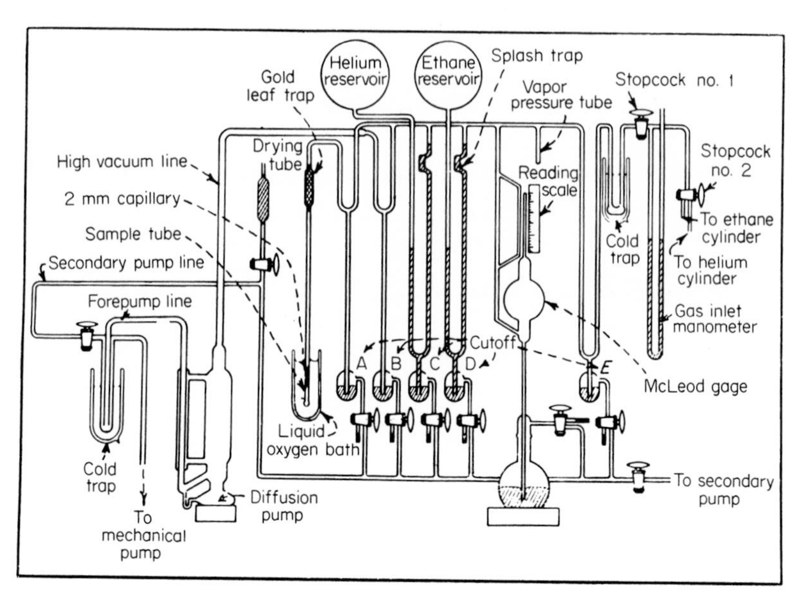

FIG. 4.30. Low-temperature gas adsorption apparatus. (From Lauterback *et al.*, Ref. 58.)

devices for measurement of particle area by the gas adsorption technique are commercially available.*

(b) *Liquid Adsorption.* A liquid adsorption method for fine clays has been described by Orr and Bankston.[60] A monomolecular layer of oriented molecules of surface-active stearic acid is adsorbed on the solid surfaces. "Basically, the technique consists of exposing a dried clay sample to a solution of known fatty acid concentration and determining by the analysis of an aliquot of the solution the fatty acid remaining after a period of agitation. The difference between the acid in solution initially and finally gives the quantity of acid that was adsorbed by the clay. From this quantity, and the size of the fatty acid molecule, the total surface area of the clay can be calculated."

4-3.2 PERMEABILITY METHODS

Permeability of a porous bed as an index of surface area and particle size has been investigated extensively by Carman,[61] Lea and Nurse,[62] and others.

*NUMINCO-ORR Surface-Area Pore-Volume Analyzer, NUMEC, Inc., Monroeville, Pennsylvania.

The method depends on a measurement of the resistance to flow (gas or liquid) of a bed of known dimensions. Fowler and Hertel[63] have adapted the method to the surface measurement of fibers. The apparatus required is simple, but elaborate preparations of the porous bed are required to ensure uniformity of orientation and bulk density. The method is siutable for rigid particles and is not limited to any specific size range.

4-3.3 Low-Angle X-Ray Diffraction

Low-angle x-ray diffraction was employed by West[64] to measure the size of clay particles when dry and in water suspension. The reason the term "low-angle" is applied to this type of x-ray diffraction may be illustrated by an example. If the radius of the particle is 0.1 μm and the wavelength of the x-rays is 1 Å, then the half-width of the diffraction pattern is 56 sec of arc. Observation of such a narrow diffraction pattern requires special instruments. The apparatus used by West is shown in Fig. 4.31. The target of the x-ray tube is molybdenum. Crystal A serves as a monchromator for a wavelength of 0.71 Å. With no sample in place, the diffracted beam from crystal A is reflected again from crystal B when B is parallel to A; but when crystal B is turned, the diffracted intensity quickly drops to zero. The intensity of the reflected beam is measured by a xenon-filled Geiger counter at C. With a sample in place, part of the x-ray beam is diffracted in passing through the sample and, because the wavelength remains unchanged in this diffraction, crystal B can be rotated and the intensity of the refracted beam measured at several scattering angles. When the logarithm of the diffracted intensity is plotted as a function of the square of the scattering angle, the slope of the curve is proportional to the square of the particle radius.

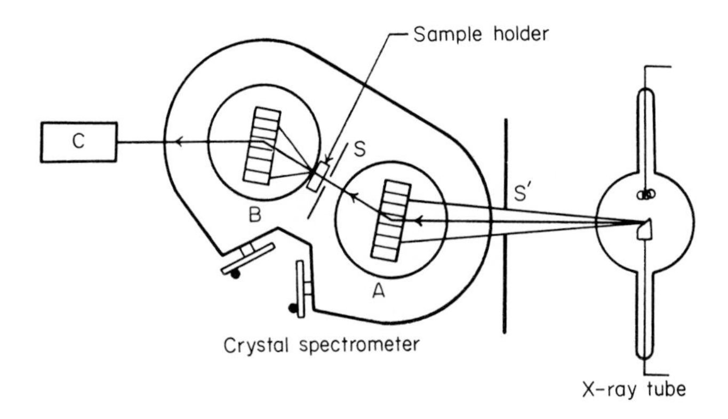

Fig. 4.31. Low-angle x-ray diffraction apparatus. (From West, Ref. 64.)

For particles larger than 0.1 μm, theory ceases to hold exactly because of refraction effects. However, by calibrating with particles of known size the instrument may be used for larger particles.

4-3.4 Cohesion of Surface Films

The cohesion of surface films has been suggested as a measure of particle size by Gray.[65] For this method a monoparticulate surface layer of the dust to be sized is spread on the surface of a liquid with the aid of flotation agents such as fatty acids or castor oil. By film-balance techniques the surface pressure-area relation is determined and a point on the compression curve representing a rigid film (and suitable for determining a reproducible mean film thickness) is selected. It is claimed that this method is suitable for determining mean particle size down to 0.1 μm.

4-3.5 Reaction Velocity

The thermal analysis of reaction velocity has been proposed by Suito and Hirai[66] as a method giving a value for average particle size. It is based on the proportionality between surface area of a powder and rate of solution in a suitable solvent (e.g., limestone dissolved in an acid). The use of heat of wetting and the natural radioactivity of atmosphereic particles are other methods that have been suggested for determining an average size parameter. However, none of these methods are sufficiently well developed to be of practical value for size analysis at this time.

4-3.6 Optical Methods

Optical methods for the determination of average diameter are based on the light absorption, or scattering, of nonsettling suspensions of known weight concentration. The relation between size and light scattering is a complex one, and little accurate size information on small, irregular particles can be obtained from such methods.

Optical measurement of uniformly sized submicron smoke and mist particles (which may be produced in the laboratory for testing of filters, etc.) is of interest to industrial hygienists. An instrument called the Owl utilizes the color and polarization of scattered light to determine the particle size of a homogeneous smoke of spherical particles. As described by Sinclair,[41] it consists of an observation chamber and light source which may be rotated while observing the smoke through a low power microscope. When uniform spherical particles are illuminated with unpolarized light, partially polarized light is scattered in all directions. As the angle at which this polarized light is observed varies from near 0 to 180 degrees, a series of colors may be seen which resembles the spectrum of white light. This series may be repeated several times. The purity and brightness of the colors increase with the uniformity of particle size. The number of times the color sequence is repeated increases with particle size. For example, the color red is repeated five times for an aerosol particle diameter of 1.0 μm.

Fig. 4.32 is a diagrammatic plan view of the Owl. The chamber is rotated by using the light trap as a handle, and the aerosol is blown into the chamber from the bottom and out the top. The polarization photometer is made by mounting a bipartite Polaroid disk in the eyepiece of the microscope with the polarizing axes perpendicular and parallel to the dividing line. A rotatable analyzer and Wratten filter are also placed in the eyepiece for use with aerosol particles less than 0.4 μm. The analyzer and filter are removed for particles between 0.4 and 2.0 μm. The Slope-o-meter (discussed below) may be used for particles above the range of the Owl.

Sinclair[41] described the Slope-o-meter for measuring the particle size of a homogeneous smoke in air by comparing the intensity of transmitted light at two or more wavelengths in a manner similar to that chosen by Bailey for liquid suspensions. It is suitable for particles greater than 1 μm.

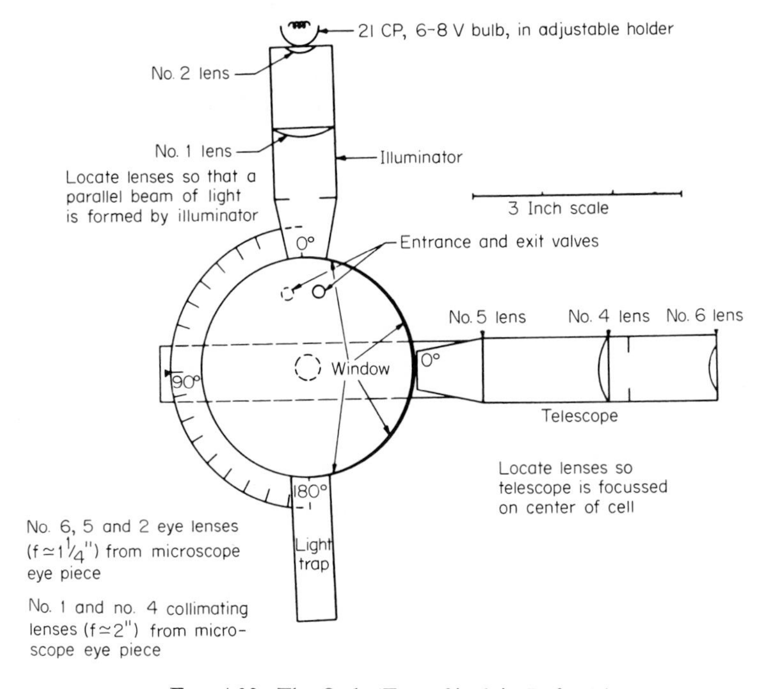

FIG. 4.32. The Owl. (From Sinclair, Ref. 41.)

4-3.7 DIFFUSION BATTERY

The Brownian motion of suspended particles attains a significant amplitude when particles become as small as 1 μm and is the principal phenomenon acting on particles less than 0.2 μm, the lower limit of visibility of the best optical microscopes.

The mean displacement of a particle undergoing Brownian motion increases as particle diameter decreases and approaches the size of gas molecules, about 0.0002 μm. This relationship may be used to determine the size of aerosol particles less than 1 μm with a device called a diffusion battery.[67] The method requires a measurement of the decrease in particle concentration of an aerosol flowing in streamlined motion through a long, narrow channel. Since it is assumed that once a particle reaches the channel wall it will stick, a negative concentration gradient exists from the center of the flow channel to the wall and the random Brownian motion of the particles results in a net migration flux perpendicular to the direction of aerosol flow. For a given apparatus and flow rate, the smaller the particle the greater is the Brownian motion and, as a result, the greater is the percentage of particles that will reach the walls of the diffusion battery.

Circular and rectangular diffusion batteries have been used since 1900 for determining diffusion coeffcients of air ions, condensation nuclei, and smoke particles. Thomas,[68,69] constructed and tested a parallel-plate diffusion battery suitable for measuring the diffusion coefficient and diameter of aerosol particles as large as 1 μm. It was constructed of graphite to minimize distortions due to temperature changes and contained twenty parallel channels each 100 μm wide, 12.70 cm high, and 47.3 cm long. To evaluate the end effects of this instrument, a second parallel-plate diffusion battery of closely identical construction except that the channel length was 5.08 cm instead of 47.3 cm was constructed and operated in parallel with the longer one. This end effect procedure was later shown by Megaw and Whiffen[70] to be in error by about 10%. Fig. 4.33 shows the longer instrument without end pieces (plenums) attached. Heavy stainless-steel plates were placed on both sides of the battery to assure rigidity, and a supporting ridge of graphite was machined through the middle of each plate parallel with the direction of flow.

The theoretical performance curve for a parallel-plate diffusion battery is:[71]

$$F = 0.915e^{-\alpha} + 0.059e^{-11.8\alpha} + 0.026e^{-80\alpha} \tag{4.4}$$

and

$$\alpha^{-\alpha} = 3.77\frac{bDZ}{aQ} = 2.26\frac{nbDZ}{aQ_t} \tag{4.5}$$

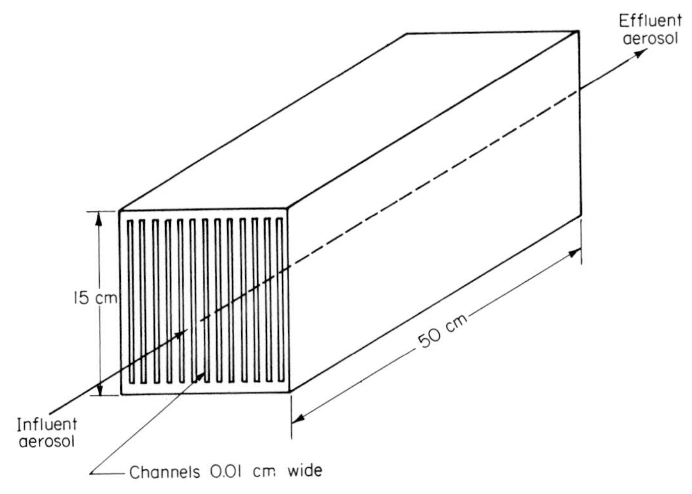

FIG. 4.33. Diffusion battery. (From Thomas, Ref. 68.)

where F = fraction of particles penetrating the instrument.
a, b, Z = channel width, height, and length, respectively, cm.
n = number of identical channels.
Q, Q_t = aerosol flow through a single and all channels, respectively, cm^3/sec.
D = diffusion constant of particles, cm^2/sec.

For the 47.3 cm-long parallel-plate diffusion battery constructed by Thomas, $\alpha = 543{,}000\ D/Q_t$.

Figure 4.34 shows the relation between F, the fraction of influent aerosol penetrating Thomas' 47.3-cm battery, and the particle radius, r, for several values of Q_t. The relationship between diffusion constant, D, in equation 4.4, above, and particle size, as reported by Thomas, is as follows:

$$D = \frac{kT}{6\pi\mu r}\left(1 + A\frac{\lambda}{r}\right) \tag{4.6}$$

where $A = 1.25 + 0.44\ \text{exp.}\ (-1.09\ r/\lambda)$. (4.7)

λ = mean free path of air molecule, cm.
r = particle radius, cm.
T = absolute temperature, °K.
k = Boltzmann's constant (1.38×10^{-16} erg/°C).
μ = air viscosity, poise, g/cm-sec.

This apparatus was tested with 0.3-μm-diameter dioctylphthalate (DOP) smoke using an Owl to check particle diameter and an NRL-E3 smoke

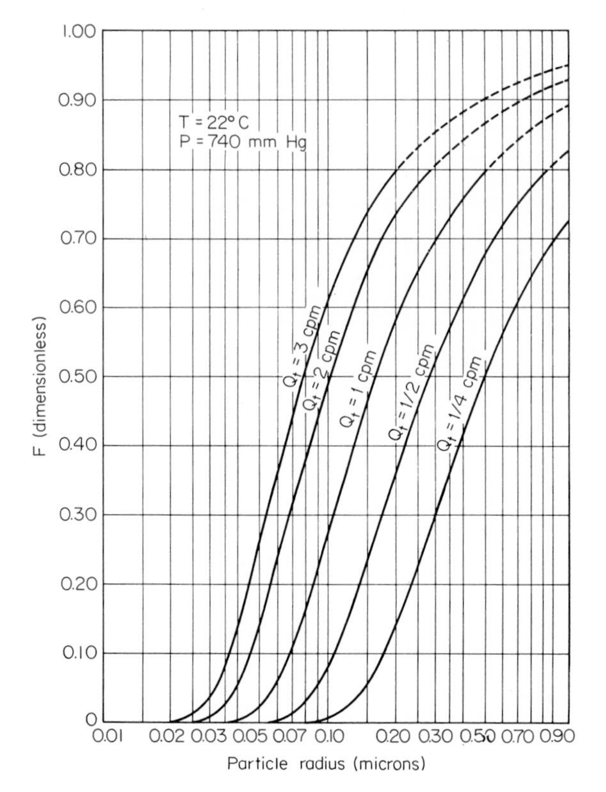

FIG. 4.34. Diffusion battery penetration. (From Thomas, Ref. 68).

penetrometer to measure smoke concentrations, and with 0.8-μm-diameter particles prepared in a LaMer-Sinclair type generator[72] and sized by observing the color of the light scattered by the aerosol at different angles to an incident light beam. It was found that the parallel-plate diffusion battery gave particle sizes close to the sizes determined by light-scattering techniques.

Although calibrating tests were made with smokes of homogeneous size, it was concluded on theoretical grounds that when the battery is used on heterogeneous smoke it will indicate an apparent particle size somewhere between the maximum and minimum sizes present in the aerosol. Thomas concluded that the diffusion battery method of particle sizing is suitable for sizes below 1 μm and that it will check the size determined by light-scattering techniques to within better than 30%. Thomas[73] has applied the diffusion battery method to obtain the percent of radioactivity in different size ranges.

Fuchs and co-workers[74] have presented computations of the penetration of a diffusion battery for an aerosol having a log-normal size distribution in terms of the parameters of the distribution function. Fuchs (reference 20, pp. 204-210) has also critically reviewed the development of diffusion battery research.

4-3.8 Count-Weight Method

When the number of particles in a unit mass can be determined by microscopic count or by indirect methods such as nephelometry, it becomes possible to calculate average grain size. For spherical particles, the average diameter is

$$d = \left(\frac{6w}{\pi n \rho} \right)^{1/3} \tag{4.8}$$

where d = particle diameter, cm.
w = weight of particles in unit volume of sample, g.
n = number of particles in unit volume of sample.
ρ = true specific gravity of particles, g/cc.

For nonspherical particles, the particle shape factors discussed in Chapter 1 must be utilized to determine average size. For example, an aliquot of an impinger sample of quartz dust may be counted in the usual fashion to determine the number of particles per milliliter of suspension. Then the remainder of the sample (of known volume) may be evaporated to dryness and the residue weighed to determine the weight of quartz per milliliter of solution. Using a volume shape factor from Table 1.3 of 0.27 for crushed quartz, it is possible to calculate average diameter from equation 4.8.

References

1. B. W. Wilson, The Sedimentation of Dilute Suspensions of Microscopic Spheres in Water, *Australian J. Appl. Sci., 3*:252 (1952).
2. P. S. Roller, *Separation and Size Distribution of Microscopic Particles,* United States Bureau of Mines, Technical Paper No. 490, 1931.
3. F. E. Ivey, Jr., Particle Size Analysis of Fluid Cracking Catalysts, *Petrol. Refiner, 30*:99, 125, 144 (1951).
4. G. L. Matheson, Modification in Roller Analysis for the Determination of Particle Size Distribution, *Oil Gas J., 26*:307 (Nov. 15, 1947).
5. W. H. Walton, Theory of Size Classification of Airborne Dust Clouds by Elutriation, *Brit. J. Appl. Phys.,* Suppl. 3, p. S29 (1954).
6. H. E. Haultain, Splitting Minus —200 with Superspanner and Infrasizer, *Can. Mining Met. Bull.,* p. 301 (May 1937).
7. E. W. Price, Stokes' Law and the Haultain Infrasizer Unit, *Ind. Eng. Chem., Proc. Design Devel., 1*:79 (1962).
8. S. R. B. Cooke, *Short-Column Hydraulic Elutriator for Sub-sieve Sizes,* United States Bureau of Mines, Report of Investigations No. 3333, 1937.

9. D. E. Cummings, Studies on Experimental Pneumonokoniosis, IV, The Seperation of Particulate Matter Smaller Than Screen Size into Graded Fractions, *J. Ind. Hyg., 11*:245 (1929).
10. C. W. LaBelle, The Relation of Particle Size to Toxicity, in C. Voegtlin and H. C. Hodge (Eds.), *Pharmacology and Toxicology of Uranium Compounds,* p. 508, McGraw-Hill Book Company, Inc., New York, 1949.
11. P. Drinker and T. Hatch, *Industrial Dust,* 2nd ed., p. 183, McGraw-Hill Book Company, Inc., New York, 1954.
12. Bulletin, Federal Classifier Systems, Inc., Chicago, Ill.
13. R. Dennis, R. Coleman, L. Silverman, and M. W. First, *Particle Size Efficiency Studies on a Design 2 Aerotec Tube,* USAEC Report NYO-1583, Harvard University, April 1952.
14. M. W. First, *Fundamental Factors in the Design of Cyclone Dust Collectors,* Sc.D. Thesis, Harvard University, 1950.
15. R. Dennis, L. Silverman, C. E. Billings, D. M. Anderson, W. R. Samples, H. M. Donaldson, Jr., and P. Drinker, *Air Cleaning Studies, Progress Report,* USAEC Report NYO-4611, Harvard University, October 1956.
16. *Determining the Properties of Fine Particulate Matter,* Power Test Code PTC-28, American Society of Mechanical Engineers, New York, 1965.
17. C. N. Davies, The Sedimentation of Small Suspended Particles, in *Symposium on Particle Size Analysis,* Institution of Chemical Engineers and Society of Chemical Industry, London, Feb. 4, 1947.
18. R. E. Payne, The Measurement of the Particle Size of Sub-Sieve Powders, Presented at 3rd National Conference of American Instrument Society, Philadelphia, Pa., Sept. 4, 1948.
19. O. Preining, The Stokes' Diameter of Ellipsoid Particles, *Atmos. Environ., 1*:273 (1967).
20. N. A. Fuchs, *Mechanics of Aerosols,* Pergamon Press, New York, 1964.
21. H. W. Daeschner, Discussion, in *Symposium on Particle Size Measurement,* p. 56, American Society for Testing and Materials, Philadelphia, Pa., 1959.
22. S. Oden, A New Method for Determining the Particle Size Distribution in Suspensions, *Bull. Geol. Inst., Univ. Upsala, 6*:15 (1917).
23. W. A. Bostock, A Sedimentation Balance for Particle Size Analysis in the Sub-Sieve Range, *J. Sci. Instr., 29*:209 (1952).
24. E. S. Palik, Size Distribution Analysis by a Sedimentation Method, *Anal. Chem., 33*:956 (1961).
25. *Standard Method of Mechanical Analysis of Soils* (D422-39), American Society for Testing and Materials, Philadelphia, Pa., 1949.
26. H. E. Schweyer, Sedimentation Procedures for Determining Particle Size Distribution, *Eng. Progr., Univ. Florida, 6,* No. 6 (June 1952).
27. R. T. Knapp, New Apparatus for Determination of Size Distribution of Particles in Fine Powders, *Ind. Eng. Chem., Anal. Ed., 6*:66 (1934).
28. S. Berg, Determination of Particle Size Distribution by Examining Gravitational and Centrifugal Sedimentation According to the Pipet Method and with Divers, in *Symposium on Particle Size Measurement,* American Society for Testing and Materials, Philadelphia, Pa., 1959.
29. L. A. Wagner, discussion of paper by R. N. Trafler and L. A. H. Baum, Measurement of Particle Size Distribution by Optical Methods, in *Proc. Am. Soc. Testing Mater., 35,* Part II; 457 (1935).

30. H. E. Schweyer and L. T. Work, Methods of Determining Particle Size Distribution, in *Symposium on New Methods for Particle Size Determination in the Subsieve Range,* American Society for Testing and Materials, Philadelphia, Pa., Mar. 4, 1951.
31. N. A. Talvitie and H. J. Paulus, Recording, Photometric Particle-Size Analyzer, *Rev. Sci. Instr., 27*:763 (1956).
32. E. E. Rose, Determination of the "Extinction Coefficient" Particle Size Relationship for Spherical Bodies, *J. Appl. Chem., 20*:80 (1952).
33. A. H. M. Andreasen, Über die Feinheits bestimming und ihre Bedeutung für die germanische Industrie, *Ber. Deut. Keram. Ges., 11*:249, 675 (1930).
34. H. J. Kamack, Particle Size Determination by Centrifugal Pipet Sedimentation, *Anal. Chem., 23*:844 (1951).
35. T. Svedberg, *The Ultracentrifuge,* Oxford University Press, London, 1940.
36. K. Rietma and C. G. Verver (Eds.), *Cyclones in Industry,* Elsevier Publishing Company, New York, 1961.
37. K. T. Whitby, A Rapid General Purpose Centrifuge Sedimentation Method for Measurement of Size Distribution of Small Particles, *Heating, Piping, Air Conditioning:* Part I, Apparatus and methods, *27*:231 (January 1955); Part II, Procedures and Applications, *27*:139 (June 1955).
38. L. M. Cartwright and R. Q. Gregg, A Liquid Sedimentation Method for Particle Size Distributions, in *Symposium on Particle Size Measurement,* p. 127, American Society for Testing and Materials, Philadelphia, Pa., 1959.
39. F. D. Bailey, Particle Size by Spectral Transmission, *Ind. Eng. Chem., 8*:365 (1946).
40. K. R. May, The Cascade Impactor: An Instrument for Sampling Coarse Aerosols, *J. Sci. Instr., 22*:187 (1945).
41. D. Sinclair, Measurement of Particle Size and Size Distribution, *Handbook on Aerosols,* Chapter 8, USAEC, United States Government Printing Office, Washington, D.C., 1950; reisued 1963, Report TID-4500.
42. H. G. Houghton, The Size and Size Distribution of Fog Particles, *Physics, 2*:467 (1932).
43. J. L. York and H. E. Stubbs, Photographic Analysis of Sprays, *Trans. Am. Soc. Mech. Engrs., 74*:1157 (1952).
44. R. L. Pigford and C. Pyle, Performance Characteristics of Spray Type Absorption Equipment, *Ind. Eng. Chem., 43*:1649 (1951).
45. I. Langmuir and K. B. Blodgett, *A Mathematical Investigation of Water-droplet Trajectories,* Technical Services Command, Army Air Forces Technical Report No. 5418, Air Matériel Command, Dayton, Ohio, 1946.
46. J. M. Geist, J. L. York, and G. G. Brown, Electronic Spray Analyzer for Electrically Conducting Particles, *Ind. Eng. Chem., 43*:1371 (1951).
47. H. O. Landahl and R. G. Herrmann, Sampling of Liquid Aerosols by Wires, Cylinders, and Slides and the Efficiency of Impaction of Droplets, *J. Colloid Sci., 4*: 103 (1949).
48. J. Thomas and D. Rimberg, A Simple Method for Measuring the Average Charge on a Monodisperse Aerosol, *Staub* (English transl.), *27*:18 (1967).
49. P. V. Wells and R. H. Gerke, An Oscillation Method for Measuring the Size of Ultramicroscopic Particles, *J. Am. Chem. Soc., 41*:312 (1919).
50. H. M. Cassel and H. Schultz, A Sonic Method of Determining Particle Size in Aerosols, in *Proceedings of United States Technical Conference on Air Pollution,*

McGraw-Hill Book Company, Inc., New York, 1952.
51. W. F. Carey and C. J. Stairmand, Size Analysis by Photographic Sedimentation, *Trans. Inst. Chem. Engrs., 16*:57 (1938).
52. B. V. Derjaguin and G. Ya. Vlasenko, Flow-Ultramicroscopic Method of Determining the Number Concentration and Particle Size Analysis of Aerosols and Hydrosols, *J. Colloid Sci., 17*:605 (1962).
53. G. Kubie, On the Ultramicroscopy of Aerosols, in K. Spurny (Ed.), *Aerosols — Their Physical Chemistry and Applications,* Publishing House of Czechoslovak Academy of Sciences, Prague, 1965.
54. E. H. Taylor and D. B. Harmon, Jr., Measuring Drop Sizes in Sprays, *Ind. Eng. Chem., 46*:1455 (1954).
55. J. P. Longwell, *Fuel Oil Atomization,* Sc.D. Thesis, Massachusetts Institute of Technology, 1943.
56. A. P. R. Choudhury and W. F. Stevens, A New Technique for Drop-Size Distribution Determinations, in *Symposium on Mechanics of Bubbles and Drops,* American Institute of Chemical Engineers, Detroit, Mich., November 1955.
57. J. A. Leary, Particle Size Determination in Radioactive Aerosols by Radioautograph, *Anal. Chem., 23*:850 (1951).
58. K. E. Lauterback, S. Laskin, and L. Leach, Specific Determination of Uranium Dusts by Low Temperature Adsorption of Ethane, *J. Franklin Inst., 250*:13 (1950).
59. P. H. Emmett, A New Method for Measuring the Surface Areas of Finely Divided Materials and for Determining the Size of Particles, in *Symposium on New Methods for Particle Determination in the Sub-sieve Range,* American Society for Testing and Materials, Philadelphia, Pa., Mar. 4, 1951.
60. C. Orr, Jr., and B. T. Bankston, A Rapid Liquid-Phase Adsorption Method for the Determination of the Surface Area of Clays, *J. Am. Ceram. Soc., 35*:58 (1952).
61. P. C. Carman, Shape and Surface of Fine Powders by the Permeability Method, in *Symposium on New Method for Particle Size Determination in Sub-sieve Range,* American Society for Testing and Materials, Philadelphia, Pa., Mar. 4, 1941.
62. F. M. Lea and R. W. Nurse, Permeability Methods of Fineness Measurements, in *Symposium on Particle Size Analysis,* Institution of Chemical Engineers and Society of Chemical Industry, London, Feb. 4, 1947.
63. J. L. Fowler and K. L. Hertel, Flow of Gas Through Porous Media, *J. Appl. Phys., 11*:496 (1940).
64. W. J. West, Size Determinations of Clay Particles in Water Suspensions by Use of Low-Angle X-Ray Diffraction, *J. Colloid Sci., 7*:295 (1952).
65. V. R. Gray, Surface Films and Particle-Size Determination of Powders, *Can. J. Res., 28B*:277 (1950).
66. E. Suito and N. Hirai, Measurement of the Particle-Size Distribution of Powder by the Thermal Analysis of Reaction Velocity. I, Theory, *J. Chem. Soc. Japan, 72*:713 (1951).
67. P. G. Gormley and M. Kennedy, Diffusion from a Stream Flowing Through a Cylindrical Tube, *Proc. Roy. Irish Acad., 52A*:163 (1949).
68. J. W. Thomas, *The Diffusion Battery Method for Aerosol Particle Size Determination,* USAEC Report ORNL-1648, Oak Ridge National Laboratory, 1954.
69. J. W. Thomas, The Diffusion Battery Method for Aerosol Particle Size Determination, *J. Colloid Sci., 10*:246 (1955).

70. W. J. Megaw and R. D. Whiffen, Measurement of the Diffusion Coefficient of Homogeneous and Other Nuclei, *J. Rech. Atmos., 1*:114 (1963).
71. W. C. DeMarcus and J. W. Thomas, *Theory of a Diffusion Battery,* USAEC Report ORNL-1413, Oak Ridge National Laboratory, October 1952.
72. V. K. LaMer, E. C. Y. Inn, and I. B. Wilson, The Methods of Forming, Detecting, and Measuring the Size and Concentration of Liquid Aerosols in the Size Range of 0.01 to 0.25 Microns Diameter, *J. Colloid Sci., 5*:471 (1950).
73. J. W. Thomas, Distribution of Radioactivity on a Polydisperse Aerosol by the Diffusion Method, *Health Phys., 12*:765 (1966).
74. N. A. Fuchs, I. B. Stechkina, and V. I. Starosselskii, On the Determination of Particle Size Distribution in Polydisperse Aerosols by the Diffusion Method, *Brit. J. Appl. Phys., 13*:280 (1962).

5

Automatic Particle Counting and Sizing Instruments

5-1 Introduction

The counting and sizing of large numbers of particles with a microscope is a tedious task, liable to human error and quite expensive. This has encouraged the development of automatic instruments for routine particle measurement procedures, such as blood cell counting and sizing, aerosol size spectrum analysis, etc. Automatic devices permit direct size measurement of suspended aerosol particles and give essential information that is difficult to obtain in any other way. This is especially true when analyzing aerosols composed of droplets of volatile liquids.

For the purpose of this discussion, automatic particle counting and sizing devices may be divided into two categories according to whether particles are separated from the gas phase prior to measurement or are drawn into the sizing instrument as an unaltered aerosol. In the first category are devices that scan a microscope slide (or a photomicrograph of the slide) on which a sample of particulate material has been dispersed, as for visual microscopy; and devices that pass a suspension of the particles in a liquid through a continuous particle counter and sizer after the particles have been collected, as with a Greenburg-Smith impinger. The collection of airborne particles in liquids may introduce an additional uncertainty in their subsequent size measurement. Some aggregated materials may separate in liquids because interfacial forces are not sufficient to hold them together.

In the second category are devices which conduct the aerosol particles directly through a sensing zone. The necessary counts and size measurements are inferred from the physical properties of the particles through phenomena such as light scatter or acoustical or electrical effects. The first category of automatic instruments (i.e., those that scan a prepared dust deposit on a microscope slide) substitute an "electronic eye" for the human eye and, unlike the devices that make measurements directly on the unaltered aerosol, add no new capability to the measurement of fine particulate matter.

The design and construction of automatic particle counting and sizing devices are based on established scientific principles. Laboratory studies started during World War II have been continued ever since. Several types of automatic aerosol counters and sizers are now available commercially.[1] Particles as small as 0.3 μm can be detected by some units (approximately the limit of resolution of the best light microscope), although experience indicates that at present field-operated units cannot be relied upon to measure the size of particles much below 1 μm accurately. To attain even this degree of accuracy in the field, frequent calibrations and careful operations are required. The reasons for this will become apparent in the discussion that follows.

5-2 Counting and Sizing Collected Particles

5-2.1 Optical Spot Scanning

Devices for counting and sizing particles mounted on prepared slides consist of a photoelectric scanning device (or detection system) combined with a high-speed pulse counter. Generally speaking, the pulse height analysis and counting equipment has been adopted without change from commercial sources supplying equipment for radiation and similar measurements.

Two scanning methods are in use. In one, a moving spot of light traverses the field, whereas in the other a slit of light is used. In both cases a photocell measures the change in light intensity caused by the presence of a particle in the light beam, but the means used to discriminate whether changes arise from more than one particle or from only a portion of a single particle are different.

Spot scanning has been summarized by Walton[2] as follows:

"A very small light spot, in effect, scans by drawing a line across the field and the signals arising from interceptions of the light beam are 'on' or 'off' type. Measurements of the duration of the signals give the lengths of the intercepts of the scan line across the particles.

"The number of signal intercepts does not, of course, indicate the number of particles in the field of the instrument, because particles which are large in relation to the diameter of the scanning light spot (and a reduction in beam size increases resolving power) are likely to be intercepted by more than one scan line. Neither does the length of the intercepts give the particle size directly, because particles of given size give intercepts of various lengths according to whether they are cut near an edge or across a diameter. However, if a small scanning spot draws a sequence of closely arranged parallel lines across the specimen, it will explore all points of it, and the signals from the photocell will contain all possible information about the specimen, for the signals can be reassembled as in a television set to re-create the complete picture. In reassembling the picture, it is necessary to bring together or associate signals arising at corresponding positions in successive lines. Thus we can recognize the identity of different interceptions of the same particle in successive lines of scan by the fact that they occur at corresponding positions in the line or at the same time interval after the beginning of each scan. A more practical form of memory device is, in effect, to record the signals from a scan line on a moving belt and play back the record synchronously with the scanning of the next line. The recorded signal and the new signal from the same particle will then be received at the same time and are thus identified as belonging to the same particle. In practice, owing to the effects of particle shape, these successive signals may not occur at exactly identical times in the scan. Either a small tolerance must be allowed, or the association can be recognized by the fact that the two signals must overlap. The system of recording scan lines and playing them back in synchronism with the scanning of a new line need not be restricted to one line in arrears. The record can extend over any number of lines and, by the use of a series of pick-up heads, we can obtain simultaneously the signals from corresponding positions in these lines. The effect is that of scanning the specimen with a row of points. A great deal of information can be obtained about a particle from the associated signals obtained in this way. For example, the number of signals from the same particle (i.e., the number of successive lines by which it is intercepted) gives its overall height; the signal of greatest length gives a measure of the particle's width in the direction of scan; the integrated length of all the signals gives the (projected) surface area of the particle."

Dell[3] has described an ingenious two-spot system of scanning which utilizes the two spots to first locate the edge of an individual particle; one spot then explores the particle for size by subsidiary traverse scans, while the second spot holds the particle in view until the exploration is completed and the machine is free for further search.

Development of an automatic particle size analyzer, based on the scanning electron microscope (Chapter 3, p. 127), has been described recently.[4] The electron beam scanning process is used to generate digital information on the presence or absence of particles in the scanned line, and these data are then compared to similar segments stored from the previous line. Line-to-line connections are summed and processed for area and perimeter in a direct-connected digital computer. This paper also contains results of preliminary experiments, discussions of data processing requirements of interface and computer components, and an extensive bibliography on scanning instruments.

5-2.2 Optical Slit Scanning

In contrast to a spot scan, which traces a relatively small area, approximating a line, a slit scan sweeps a wider area, or band. It is possible to count the number of times the slit is partially or completely obscured by a particle and measure the amplitude of the signal (particle size), although only some of these signals will represent individual particles that fall entirely within the swept band. Others will be responses to particles which only partially overlap the scanned band (correct count but incorrect size), and to interceptions of more than one particle simultaneously.

Le Bouffant and Soule[5] have shown that, if the size of the slit and the density of particles on the slide are properly adjusted, errors from edge fractions and coincidences cancel and a reliable measure of the number and size of particles lying within the area traversed by the slit can be obtained. An alternative method is to scan with two slits of different width, the difference in count representing only particles that lie wholly within the smaller slit width. A disadvantage of this system is that the errors of each individual measurement must be added to give an error of the difference figure; but this difficulty can be overcome by using a spot of light at the end of a single illuminated slit to detect particles that overlap the end of the slit. All signals arising from interceptions of the slit are recorded except when there is a simultaneous signal from the spot scan at the end of the slit. This method requires two photocell detectors and two electronic amplifiers. Particle size is determined by the amplitude of the signals given by the photocell, and a complete size distribution curve can be obtained by recording the signal levels of a single scan or by using multiple scans which record only signals greater than a predetermined value to obtain a complete oversize distribution curve.

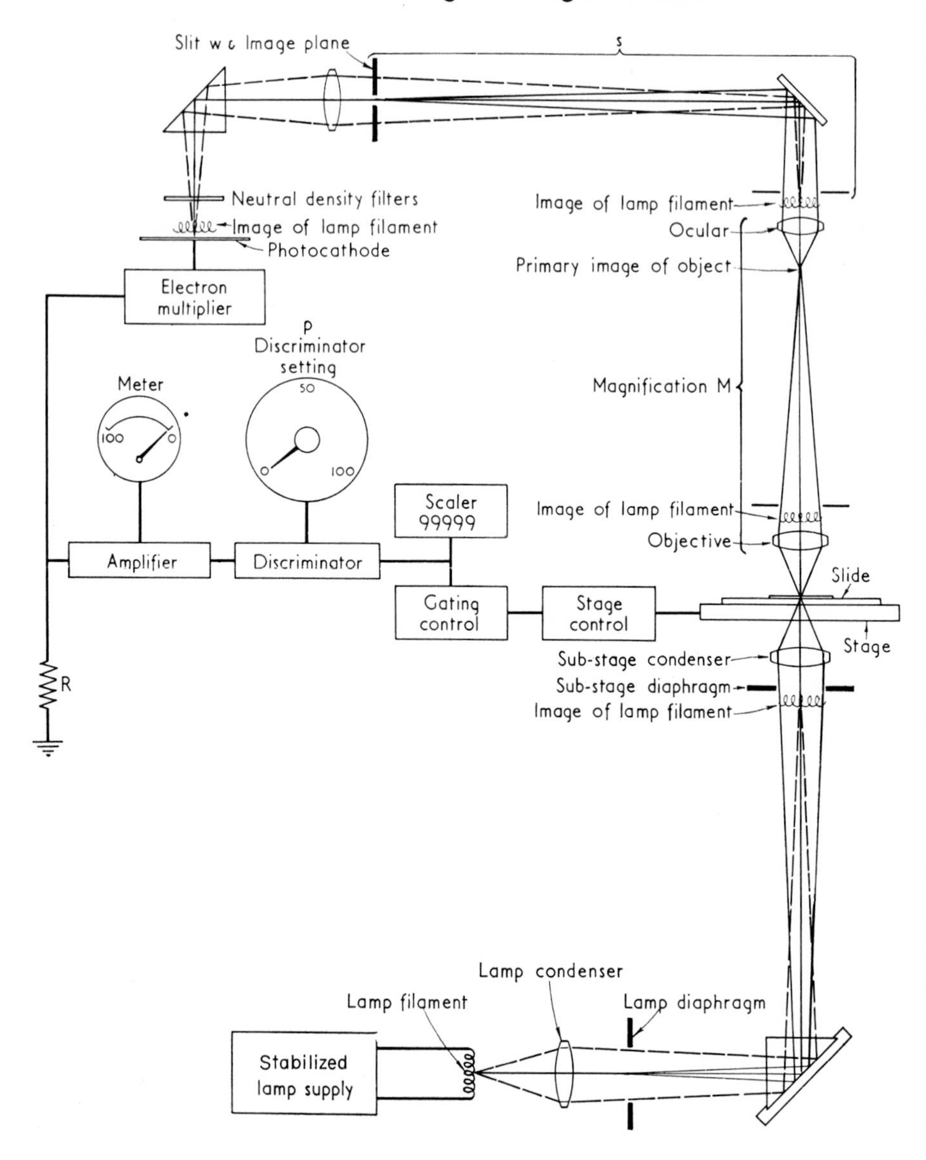

FIG. 5.1. Schematic view of Casella automatic particle counter and size analyzer.

The Casella automatic counter and sizer* is a typical slit scanning instrument. Fig. 5.1 is a schematic diagram of the apparatus. Particles may be mounted directly on a microscope slide by thermal precipitation, or a section

*Formerly available from C. F. Casella and Co., Ltd., London.

of a membrane filter may be analyzed after it has been made optically transparent by infusion with a compatible liquid having the correct index of refraction (described in Chapter 3, p. 107). The slide is placed on the mechanical stage and systematically moved through the stationary scanning slit. The time for one complete scan is 55 sec. Slit width is adjusted by means of the discriminator setting knob, and only particles greater in size than the machine setting are recorded. By successively increasing the discriminator threshold, cumulative counts of all particles greater than the set size will be counted, and these data can be plotted to obtain size-count frequency distribution curves. The instrument is capable of sizing particles in the range of 1 to 200 μm and comes equipped with five discriminator and read-out channels, so that multiple size measurements can be made simultaneously.

5-3 Counting and Sizing Particles in Liquid Suspension

The Coulter counter* for particle counting and sizing, illustrated in Fig. 5.2, determines the number and size of particles suspended in an electrically conductive liquid. When the stopcock is opened, a vacuum pump causes the suspension to flow through a small (10- to 400-μm) aperture with an

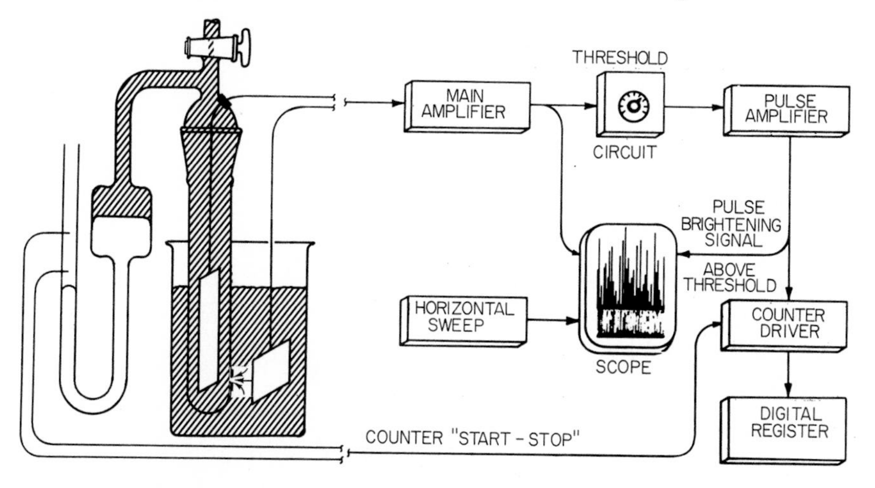

FIG. 5.2. Schematic view of Coulter counter components. Courtesy of Coulter Electronics Industrial Division.

*Coulter Electronics, Hialeah, Florida.

immersed electrode on either side. An electrical current flows from the electrode in the sample beaker to the electrode in the aperture tube. When a particle enters the aperture, it displaces some of the conductive liquid, changing the electrical resistance between the two electrodes. Either the particle or the fluid may be the better conductor; it is only important that the difference between the two conductivities be substantial.

When particle concentration is sufficiently low, particles pass through the aperture one at a time, and the momentary change in the resistance of the path between the electrodes produces a voltage pulse of short duration which has a magnitude proportional to particle volume. When these pulses are amplified, they show on an oscilloscope screen as peaks, with peak height proportional to particle volume. The pulses are sent to a threshold circuit having an adjustable screen-out voltage level, and those which reach or exceed the set level are counted.

An enlarged view of the elements of the electrical sensing zone in Fig. 5.3 shows these parts in more detail. When the stopcock is closed suddenly, the siphoning action of the mercury manometer continues the sample flow. The advancing mercury column activates the counter via start and stop probes, providing a count of the relative number of particles above a given

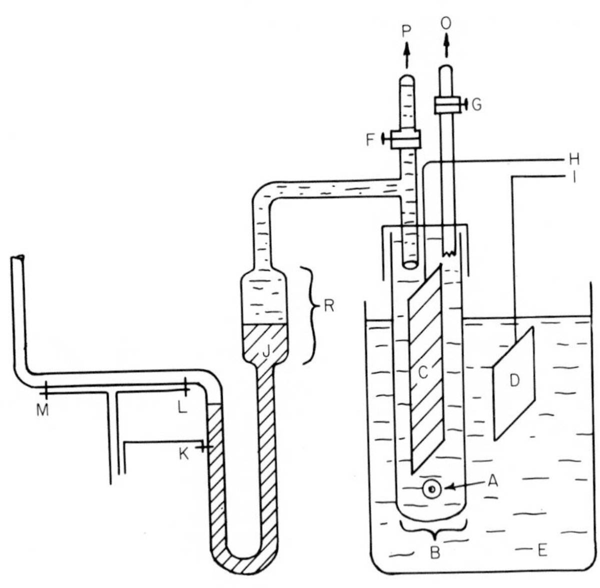

FIG. 5.3. Elements of electrical sensing zone. Courtesy of Coulter Electronics Industrial Division.

size in a fixed volume of suspension (e.g., 0.5 ml). A photograph of the instrument is shown in Fig. 5.4.

Size calibration is accomplished by observing the pulse heights obtained with monodispersed particles of known diameter (plastic, glass, etc.) and adjusting the threshold reading to correspond with this size. For spherical particles, the relationship between the particle volume measured by the instrument and the true particle diameter is a simple one. For nonspherical particles, the dimension that is determined is the diameter of a sphere having equal volume. The size discrimination of the machine is not influenced by particle shape when extreme length-to-diameter radio *(L/D)* is near unity, or by density, composition, etc. However, for fibrous particles which tend to align in a preferred direction when passing through the aperture, an orientation correction may be required. (See below.)

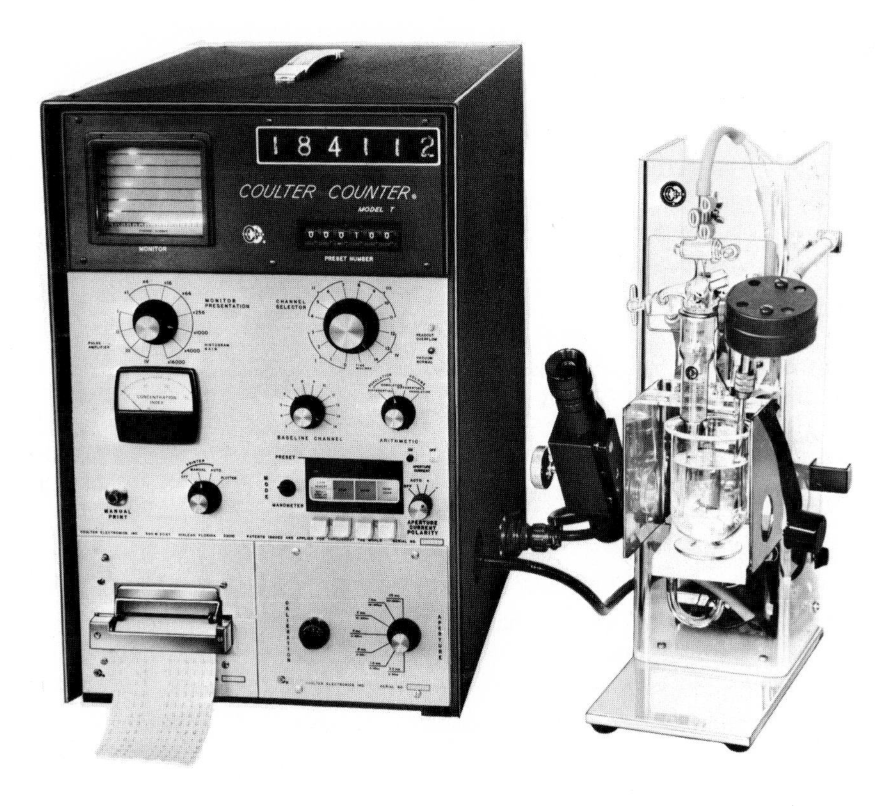

FIG. 5.4. Photograph of the Coulter counter. Courtesy of Coulter Electronics Industrial Division.

When the electrolyte in the aperture forms the principal resistance between the electrodes, the resistance change due to particle passage is

$$\Delta R = \frac{\rho_0 V}{A^2}\left(\frac{1}{1-\rho_0/\rho} - \frac{a}{xA}\right)^{-1} \tag{5.1}$$

where ρ_0 = electrolyte resistivity, ohm-cm.
A = aperture area normal to axis, cm^2.
V = particle volume, cm^3.
ρ = effective particle resistivity, ohm-cm.
a = projected area of particle normal to aperture, as it is oriented when passing through the orifice, cm^2.
x = particle dimension ratio (L/D), dimensionless.

All items on the right side of equation 5.1 except a, x, and V are constant for a given electrolyte and sample composition. When the particle diameter is no greater than 30% of the aperture diameter, and $L/D = 1$, the ratio a/A becomes a negligible quantity and the change in resistance (ΔR) is directly proportional to the particle volume (V). Therefore, different sizes of apertures, ranging from 10 to 400 μm, are used with different particle size ranges. Ullrich[6] has stated that, for irregularly shaped or porous particles, the instrument responds to the enclosed volume as though the particle were wrapped in a tightly fitting membrane. Batch[7] found that particles having resistivities between 1 and 10^{15} ohm-cm (coke, coal, fly ash, aloxite granules, and methylmethacrylate spheres) did not change the instrument performance or initial calibration factor.

Electrolyte solutions for suspending particles for counting and sizing include 0.5 to 4% NaCl in water; 6% Na_3PO_4 and Na_2HPO_4 in water; and 2 to 10% NH_4SCN in methyl, ethyl, or isopropyl alcohol. The use of wetting agents reduces the tendency of particles in suspension to reagglomerate. Duponel ME* (anionic), Naconnal† (anionic), and Triton X-100‡ have been used successfully.[8]

Kinsman[8] states that the particle size range of this instrument is generally between 0.8 and 200 μm but that these limits have been extended recently to 0.3 to 700 μm. This instrument may be adapted for use with impingers and other wet methods of particle collection for counting and size-analyzing particles, but it cannot make direct measurements on aerosols. While this may be no real limitation to its usefulness over a wide range of industrial

---

*E. I. DuPont de Nemours, Wilmington, Delaware.
†Allied Chemical and Dye Corp., New York City.
‡Rohm and Haas Co., Philadelphia, Pennsylvania.

and clinical applications, it is a serious limitation for certain studies in the environmental health sciences.

Anderson *et al.*[9] have used the Coulter counter successfully down to 0.68 μm for the routine analysis of impinger dust samples from bituminous coal mines. They excluded particles larger than about 15 μm, which tend to plug the 30-μm aperture used, by employing a preliminary sedimentation step.

Single-particle optical counters for particles suspended in liquids have been introduced. In the Royco* Model 320, light scattered from a single particle passing through a light beam is converted to an electrical pulse by a photomultiplier. Pulses are classified in an electronic pulse height analyzer. The principles and performance of this instrument are similar to those of optical aerosol analyzers discussed below. An optical instrument manufactured by HIAC† causes particles suspended in a liquid to pass in front of a small light beam. Light attenuation is sensed by a photomultiplier, and the resulting pulses are sorted with respect to size in a pulse height analyzer.[10]

Gayle *et al.*[11] described the performance of the Sperry Ultrasonic Particle Counter‡ for solid particles (10 to 100 μm) in hydraulic fluid. A 5.0-MHz pulse is generated by a Li_2SO_4 ultrasonic transducer crystal, and the beam is directed across a short space containing particles in a liquid suspension. The energy in the pulse reflected from a single particle in the beam is converted to an electrical signal by the same transducer and is then amplified and gated for pulse height discrimination. Unfortunately, the calibration data used to evaluate the feasibility of this device were obtained with heterodisperse test material. The device requires modification and additional testing before its utility can be determined.

At present, the lower size limit for optical and ultrasonic liquid suspension particle analyzers appears to be of the order of 5 to 10 μm,[12] which limits their application for industrial hygiene studies.

5-4 Counting and Sizing Airborne Particles

Several types of automatic particle counting and sizing machines have been proposed to measure particle size spectra in flowing aerosols. Only a few have been commercially developed.[1] Phenomena that have been used in these systems include detection of light scattered from individual particles passing through a light beam, measurement of electrical properties or motion of charged particles, and other physical phenomena (acoustic, momentum

---

*Royco Instruments, Inc., Menlo Park, California.

†HIAC, Claremont, California.

‡Sperry Microwave Electronics Co., Clearwater, Florida.

and heat transfer, etc.). Light-scattering instruments operate in the size range of the optical microscope (>0.3 to 100 μm). An electrical particle mobility analyzer has been developed for the range 0.015 to 1 μm. Most of this section contains discussions of light-scattering and particle mobility instrumentation. The other methods are reviewed only briefly.

5-4.1 Light-Scattering Methods

(a) *Description.* The most common method of constructing automatic aerosol particle sizing instruments utilizes light scattered from individual particles. The theory of the light-scattering process is discussed by van de Hulst.[13] Applications of light-scattering theory to design of specific counting instruments is considered by Hodkinson.[14] Recent advances in the theory and application of light scattering have been summarized in reviews by Kratohvil.[15,16] When white light strikes an aerosol particle, a certain amount of the light is scattered irregularly in all directions. The amount of light scattered at any given angle is a function of the particle size, shape, composition, and surface configuration. A system for optical aerosol particle counting and sizing contains (1) an optical component to furnish the light beam and collect the scattered light, (2) an aerosol flow component to cause the particles singly to flow through the light beam, and (3) an electronic counter and pulse height analyzer to measure and record the pulsed light.

In operation, a small sensing volume of the order of 1 mm^3 is illuminated from a light source, and aerosol particles contained in a flowing air stream are individually passed through this volume. Scattered light from each particle is collected by a lens at a fixed angular orientation and directed to an electron photomultiplier tube. The resulting photocurrent is amplified, and may be counted or discriminated by a conventional pulse height analyzer.

Optical (light-scattering) aerosol particle size analyzers are called by their respective manufacturers aerosol photometers, particle counters, particle analyzers, and so on. In what follows we use these names interchangeably.

O'Konski *et al.*[17] have listed the following advantages for the scattered light detection systems: (1) particles can be measured with a minimum of disturbance of the aerosol; (2) the limit of size detection by light scattering is 0.3 μm in diameter, suitable for many aerosol studies; (3) light scattering can be calculated, at least for isotropic spherical particles, by application of the Mie theory; and (4) highly uniform illumination and light collection can be achieved by use of a coaxial flow system in conjunction with appropriate optics.

The electronic circuitry associated with these devices is complex. Reference should be made to the original sources cited in this section for details of photometers and pulse height analyzers which incorporate independently adjustable discriminators, anticoincidence circuits, and pulse counters. This is also true to some degree of the optical and aerosol flow and metering systems, but since these parts have a direct bearing on the sensitivity and reliability of the machines, a description of these elements is in order.

The problem of keeping the optical system of the instrument free of dust deposits from the aerosol under investigation has been solved in an ingenious manner by introducing the sample stream into the illumination zone encased in a sheath of dust-free air. Fig. 5.5 illustrates a coaxial aerosol flow system which incorporates means for conducting the aerosol stream through the illuminated chamber without the use of a solid transparent confining tube, which would refract, reflect, and scatter the illuminating rays. In addition, particles near the wall of a solid tube would move more slowly than those near the center and result in pulses of varying length for particles of equal size. The flow is essentially parallel and constant across the section of the

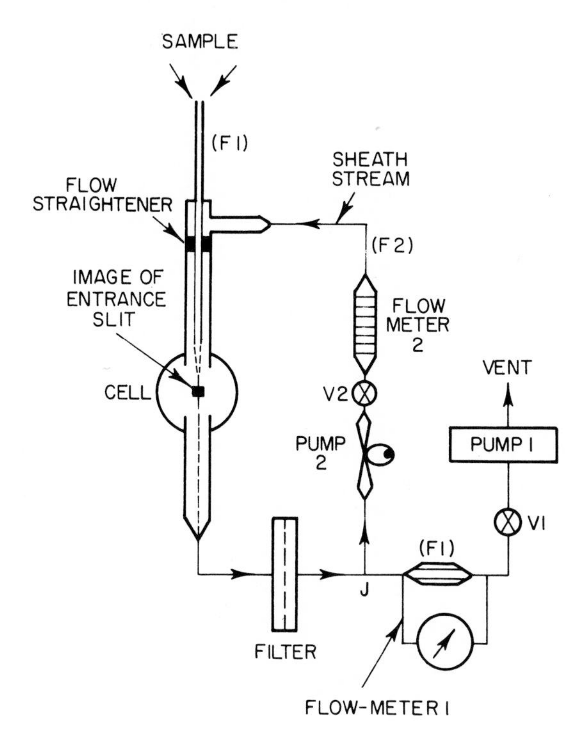

FIG. 5.5. Coaxial aerosol flow system. (From O'Konski *et al.*, Ref. 17.)

aerosol stream in the optical zone. The entire exhaust stream from the cell is filtered and sent to the junction (J). At that point a small and variable fraction is withdrawn through flowmeter 1 by means of pump 1. The remainder of the stream is recirculated as sheath air by means of pump 2. Since the only flow entry to the system is the aerosol tube connected to the sample space, and the only flow exit is through flowmeter 1, it follows that the sampling rate is equal to the pump 1 exhaust rate, F1, and the total flow rate through the cell is F1 and F2. The filter, flowmeter 2, and pump 2 are dry units. Thus the aerosol particles are removed but the vapor in the stream is recirculated through the sheath air tube. After an equilibration period, determined partly by the volume within the recirculation loop and partly by the ratio F2:F1, the sheath air stream approaches equilibrium with the aerosol vapor. This minimizes particle size changes due to evaporation of volatile constituents. The aerosol flow rate is determined accurately because it is equal to the flow rate measured directly by flowmeter 1. This makes it possible to reduce flow rate F1 to a very small fraction of F2. When this is done, the sensing volume decreases because the aerosol stream is drawn out to a very small diameter within the sheath stream. Accordingly the concentration range of the instrument is moved upward. The system performs satisfactorily at flow rates (F1) from 100 down to 3 cm^3/min.

Optically, there are several light-scattering systems in use. In the one illustrated in Fig. 5.6, small-angle forward scattering with dark-field illumination is used to take advantage of the smooth response with respect to size which occurs between 0.15 and 0.50 μm according to Mie's light scattering theory. In Fig. 5.6, the circular dark-field stop (DS) is placed between a brightly illuminated slit (S_1) and an achromatic condensing pair (L_3L_4); this forms a sharp image of the dark stop beyond the bright image of the slit.

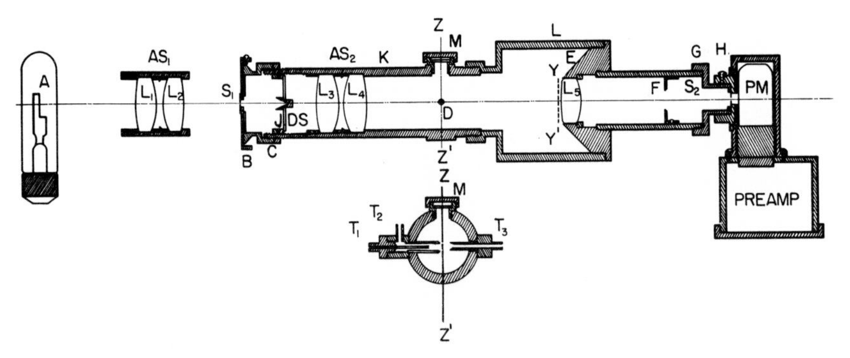

FIG. 5.6. Small-angle forward-scatter optical system. (From Gucker and Rose, Ref. 18.)

As the aerosol stream flows through the bright slit image (at point D) each particle is illuminated for an instant. The light pulses scattered in a forward direction between 1 and 20 degrees are collected by a lens and strike the cathode of a photomultiplier (PM). Uniform illumination of the image space at point D is necessary to allow correlation of pulse height and particle size. With a source slit up to 0.5 mm wide and a dust tube diameter up to 2.3 mm, the variation in the illumination of the particles will not exceed ±4%.

In the flow system shown in Fig. 5.6, the aerosol and the surrounding sheath of filtered air enter, respectively, through T_1 and T_2 at pressures slightly above atmospheric, and are exhausted through T_3 at a pressure slightly below atmospheric. With a central sample tube 2.26 mm in diameter and a flow rate of 1 lpm, the aerosol flow is 416 cm/sec, corresponding to a Reynolds' number of 602 (well below the turbulent region) and giving, with a slit 0.5 mm wide, a pulse length of 120 μsec. The outer tube diameter of 4.7 mm gives the sheath air the same linear flow rate at 3 lpm.

At a flow rate of 1 lpm the machine has a maximum counting rate of 3000 particles per liter for 1% or less coincidence. Fig. 5.7 shows the voltage output of this instrument plotted against particle size for DOP (dioctyl phthalate) smoke and indicates that a smooth curve is obtained below a radius of 0.5 μm. The size range of this forward-scattering counter increases

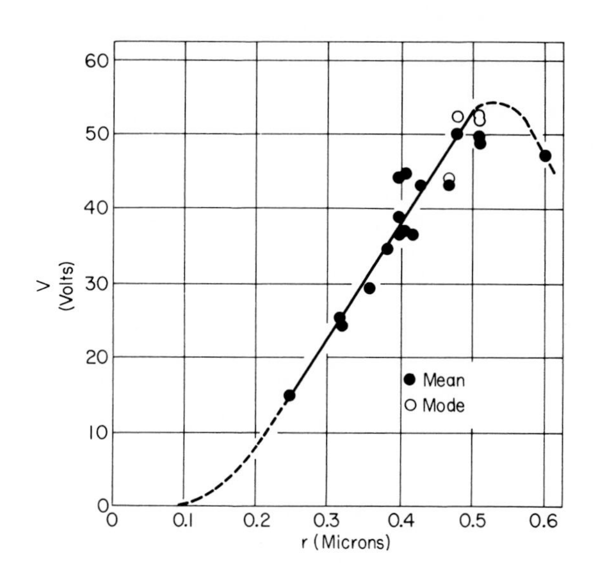

FIG. 5.7. Calibration of small-angle forward scatter aerosol analyzer with dioctyl phthalate particles. (From Gucker and Rose, Ref. 18.)

with decreasing refractive index of the particles and would extend up to about 0.8-μm radius for water fogs of refractive index 1.33 with green or white light.

A second light-scattering particle counting and sizing system utilizes right-angle rather than forward-angle scattering, in spite of the fact that the response curves of right-angle scattering counters increase with particle size through a series of plateaus between 0.3 and 1.0 μm in diameter, as shown in Fig. 5.8 (see also reference 27, Fig. 6). Advantages of the right-angle configuration include[17,18] the ability to use the full aperture of the illuminating and collecting lenses; the fact that the stray light scatter from lenses, edges, and contaminants in the optical path is smaller at right angles; and the fact that the optical alignment is less critical and less subject to shock and vibration effects in the right-angle design.

The light-scattering power of this instrument is proportional to the square of particle diameter between 0.365 and approximately 1.0 μm, but is about

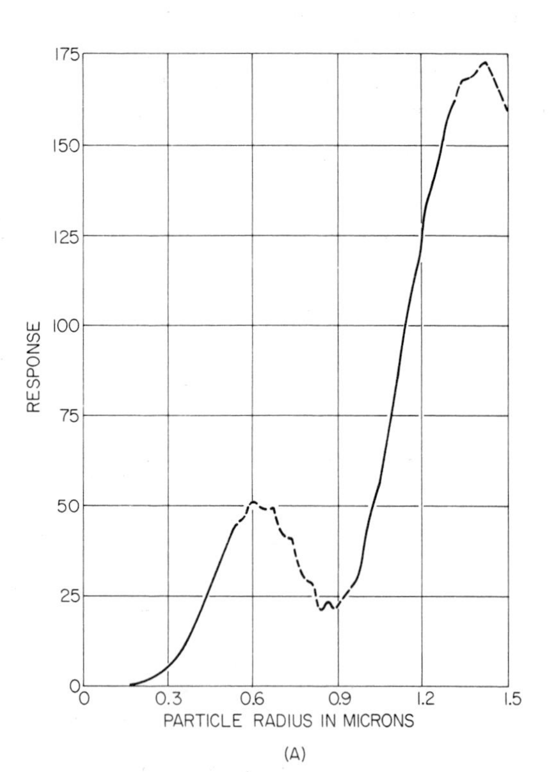

FIG. 5.8. Response curve from water drops: (*a*) forward scatter; (*b*) right-angle scatter. (From Gucker and Rose, Ref. 19.)

10% less than a square-law relationship for 1.17-μm-diameter particles. Therefore, it is necessary to determine experimentally the response curve for particle diameters which cover the range of the size distributions to be examined. In the regions of the response curves above 1.2 μm, where there are plateaus, a reasonable degree of accuracy can be obtained by a judicious selection of size intervals to avoid ambiguities. For example, Gucker and Rose[19] have shown that ambiguity of response could be avoided for sizes greater than 1.2 μm in diameter by choosing the following size intervals: 1.2 to 1.5 μm, 1.5 to 1.8 μm, and 1.8 to 2.4 μm. This is illustrated in Fig. 5.8, where the response curves of water droplets for forward scattering and right-angle scattering are shown. From these curves it can be seen that the right-angle counter gives a response about one-fiftieth that of the forward scattering instrument, and that for particles above 1.4 μm in diameter the right-angle response curve oscillates.

In the field of commercially available automatic particle counting and

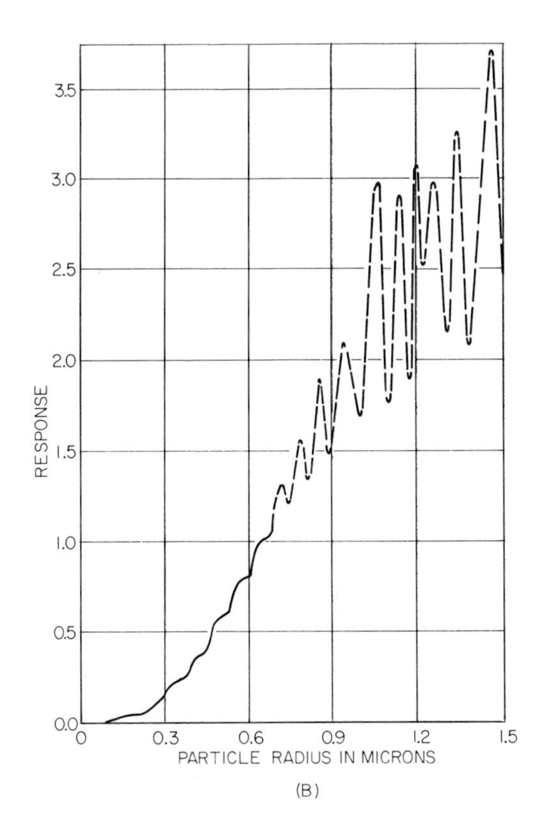

(B)

sizing devices, the Royco instrument shown in Fig. 5.9 uses a right-angle light-scattering optical system similar to that described above. However, no use is made of the clean air sheath system of introducing the aerosol sample into the scanning light beam. Whitby[20] has suggested a design for a sheath air inlet tube for the Royco PC-200.

Zinky[21] has described the Royco PC-200 particle counter and analyzer shown in Figs. 5.9 and 5.10. In operation, an aerosol sample is drawn through the optical volume and illuminated by a projecting lens system, a slit, and a lamp. The sample air stream is viewed by a collecting lens, a slit, and a phototube positioned at 90° to the main projection axis. Thus, a small volume located in the center of the sensor head is optically defined by a lamp and viewed by the phototube. When no particle is passing through this volume, the phototube sees a dark field. If a particle is present, light is scattered into the phototube during the time that the particle travels through the illuminated volume. An electronic pulse height analyzer is used to determine the number of particles that should be classified in each size range.

The instrument is calibrated in two ways. Because the projection lamp decreases in intensity with service and the photomultiplier output varies with temperature, the instrument comes equipped with a built-in light pulse generator that simulates pulses of light from particles. This is done with the mechanical chopper mechanism shown in Fig. 5.9. Light pulses pass through the light pipe and the optical cell to the phototube, and amplification is adjusted with a front panel control until the calibration voltmeter indicates

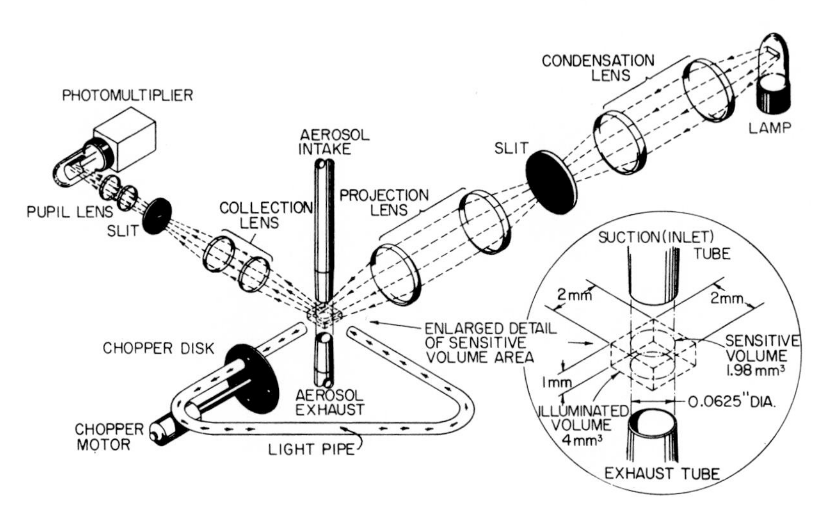

FIG. 5.9. Optical system of the Royco particle counter PC-200A. (From Zinky, Ref. 21.)

the proper value. In addition, air may be made dust-free by passage through a membrane filter before entering the machine to provide a zero check of the instrument for internal contamination and light leaks. These internal calibrations are rapid and give reasonable assurance that the instrument is functioning.

A second calibration, with particles of known characteristics, is required to interpret the light pulse heights in terms of the sizes of actual particles undergoing analysis. Royco instruments are factory-calibrated before ship-

FIG. 5.10. Photograph of the Royco particle counter Model 202. Courtesy Royco Instruments, Inc.

ment with monodispersed polystyrene latex spheres by aerosolizing very dilute water suspensions of selected size with the aid of a compressed air-operated atomizer. Although it is recognized that the quantity of light scattered from particles having the same diameter but different composition, shape, color, surface texture, etc., may not be identical with that from latex spheres, the suggested calibration procedure is justified on the assumption that no single substance could be found that would represent adequately the diverse matter normally found in indoor and outdoor air. Nevertheless, it is not possible to agree with the instrument manufacturer that the particle counter operates in accordance with physical law such that response is proportional to particle area, and that therefore two-point calibration (with latex spheres) is sufficient to provide proper response in all ranges.[22]

The counting rate of one Royco model is 5000 particles per minute at a flow rate of 100 cm^3/min for a coincidence loss of less than 10%. Concentrations of particles up to 10^8 per cubic foot can be handled with a 50-to-1 dust-free air dilution system (standard in the Model 202). This instrument has up to 23 particle size channels covering a size range from 0.3 to 100 μm.

Other commercial optical aerosol particle counters operate essentially by the principles discussed above. Characteristics of some of the present commercially available automatic particle counter systems are given in Table

TABLE 5.1
CHARACTERISTICS OF COMMERCIALLY AVAILABLE AUTOMATIC AEROSOL PARTICLE COUNTERS

Counter	Optical specification[a]				Size range (μm)	Sampling rate (cm^3/min)
	ψ	γ	β	η		
Royco PC 200	90	23.75	23.75	—	0.3–10	300
Southern RI	45	13.75	13.75	—	0.5–8	20
B and L	—	13.75	53.75	23.75	0.3–10	78
IITRI Aerosoloscope	90	—	—	—	1–64	1800
Sinclair-Phoenix	3–20	—	—	—	0.3–17	2800
Climet	—	—	—	—	0.3–10	1500
Block—NASA	—	—	—	—	0.5–10	472

[a] ψ = inclination of illuminating and collecting cone axis.
γ = semi-angle of illuminating cone.
β = semi-angle of collecting cone.
η = light trap semi-angle.

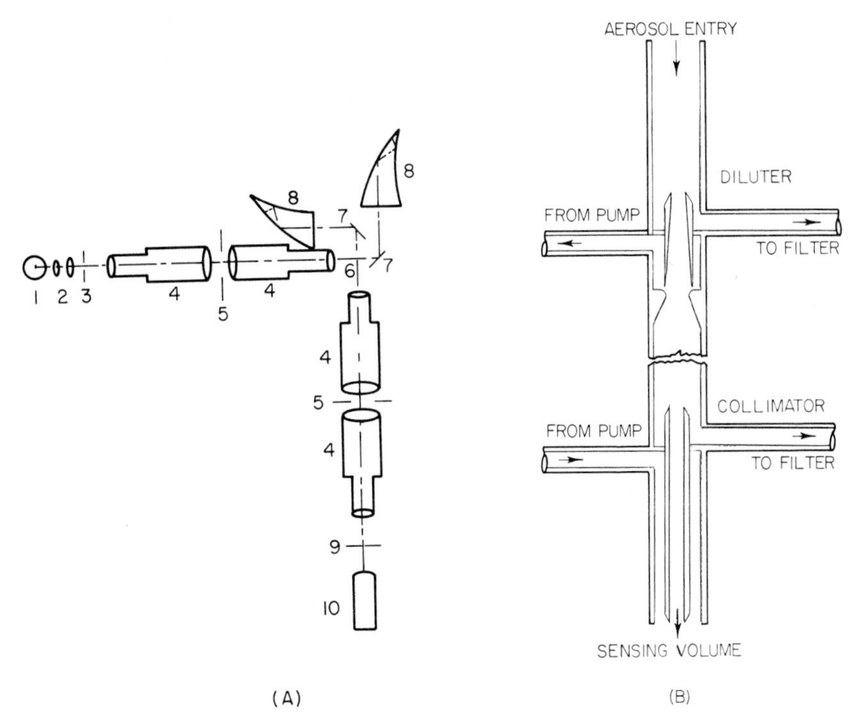

FIG. 5.11. Details of the Aerosoloscope (IITRI) aerosol particle counting and sizing instrument: (*a*) schematic; (*b*) diluter and collimator. (From Fisher *et al.*, Ref. 23.) 23.)

5.1. The Aerosoloscope system developed by the IIT Research Institute has been described by Fisher *et al.*[23] It uses 90° scatter together with an internal diluter and sheath flow collimator, as shown in Fig. 5.11.

The counter produced by Southern Research Institute* utilizes the light scattered at 45° in an optical sensor, as shown in Fig. 5.12. It also contains an internal diluter and flow collimator. This unit has been miniaturized and is available commercially in a portable design as discussed by Thomas *et al.*[24] Another microminiaturized portable multichannel aerosol particle analyzer is currently produced by Block Engineering† for the National Aeronautics and Space Administration in its program of evaluation of particle contamination in space flight life-support systems and space capsules. A description of the instrument and characteristics of its performance have been given by Lavery and Leavitt.[25]

---

*Southern Research Institute, Birmingham, Alabama.

†Block Engineering, Inc., Cambridge, Massachusetts.

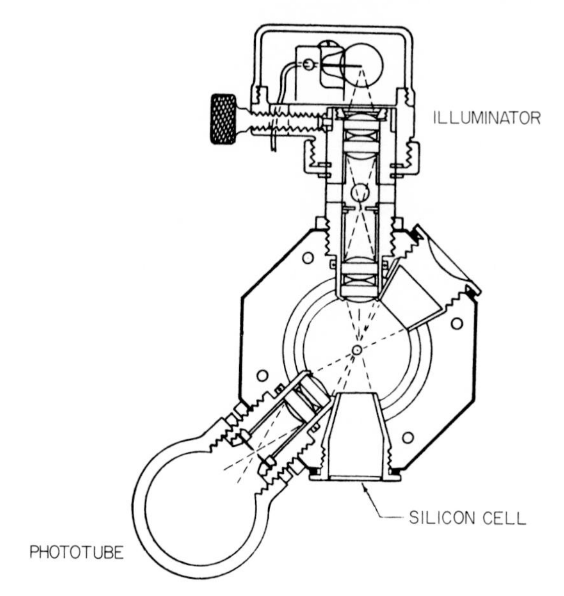

FIG. 5.12. Optical system of Southern Research Institute aerosol particle size analyzer. (From Thomas *et al.*, Ref. 24.)

In the dust counter developed by Bausch and Lomb,‡ light scattered forward by the particles is collected by a parabolic mirror over the solid angle of 30°, as shown in Fig. 5.13. This instrument has been described by Randall and Keller.[26]

The Sinclair-Phoenix* aerosol particle size analyzer has been discussed by Sinclair.[27] A schematic diagram of the optical and aerosol system is shown in Fig. 5.14. The aerosol flow system contains a sheath air collimator together with excess air provisions for additional cell-purging air to reduce particle deposition on the optical parts. The small-angle forward scattering optical system design is similar to that described above by Gucker and Rose.[18]

Climet Instruments† has introduced a particle analyzer (Model C1-200, optical and flow systems) and a compatible particle counter (Model C1-205-1-2-3). The optical system collects light scattered over a 200° solid angle in a system design which minimizes effects of particle physical characteristics. The flow system utilizes a clean air sheath design.

(b) *Performance Calibrations.* Monochromatic (or coherent) light scattered from a single optically homogenous isotropic sphere can be calculated

‡Bausch and Lomb, Rochester, New York.

*Phoenix Instrument Co., Philadelphia, Pennsylvania.

†Climet Instruments, Inc., Sunnyvale, California.

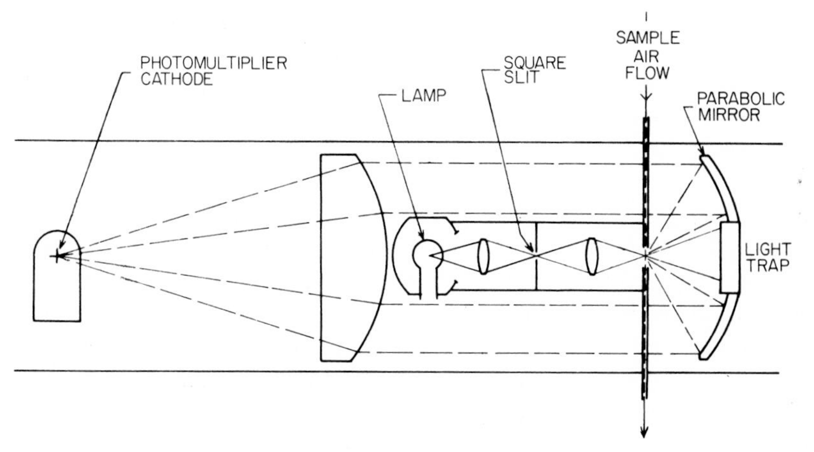

FIG. 5.13. Optical system of the Bausch and Lomb aerosol particle size analyzer. Courtesy of Bausch and Lomb.

from theory.[28] The scattered light intensity and its angular distribution vary with the particle size, the refractive index, and the wavelength of the incident light. The amount of light scattered by the particle and received by the electron photomultiplier tube is a function of the internal design, construction, and maintenance of the optical and aerosol flow components of a specific instrument configuration. The electrical pulse produced by the photomultiplier is subsequently amplified, and can be counted, or measured

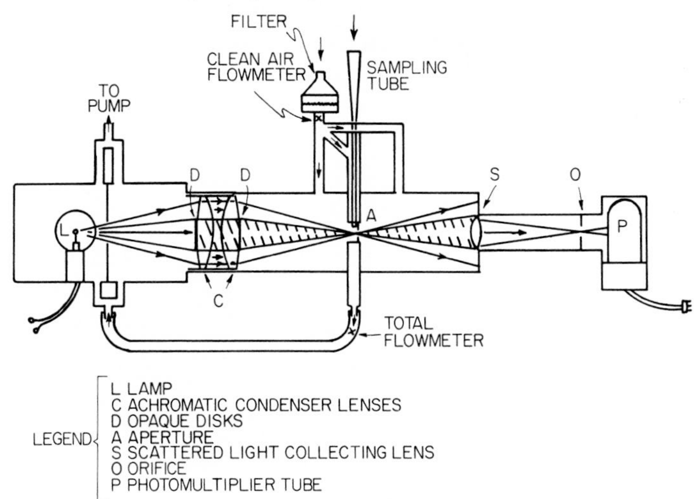

FIG. 5.14. Optical and aerosol systems of Sinclair-Phoenix aerosol particle size analyzer. (From Sinclair, Ref. 27.)

electronically, introducing further factors in the interpretation of an indicated pulse output. Calibration of the proper functioning of the photomultiplier tube and associated electronic components with a standard light pulse is provided by an internal calibration scheme (e.g., the chopped pulse provision on the Royco instrument shown in Fig. 5.9). Output of the instrument must be calibrated with particles of known size. This is usually done with an aerosol of monodisperse polystyrene latex* spheres about 1.0 μm in diameter. This calibration procedure is discussed in ASTM Tentative Standard F 50-65T.[29] Monodisperse PSL particles are furnished as a stabilized liquid suspension and must be diluted 10^4 to 10^5 times with distilled, demineralized, membrane filtered water (and stored in polyethylene). The diluted liquid suspension is aerosolized with a compressed-air nebulizer (e.g., a DeVilbiss† No. 40 nebulizer at 6 lpm) and mixed with dry air. The final aerosol concentration is such that particles proceed through the optical volume individually and far apart to avoid pulse coincidence.

Calibration with monodisperse polystyrene latex particles is specific for materials with similar optical properties (i.e., refractive index $m = 1.6$, transparent, spherical). Calibration data can be related to other particles whose light-scattering characteristics are known. Aerosol particles of practical interest usually consist of many sizes and shapes with variable composition (particularly related to refractive index), and it is usual to calibrate a particle size analyzer with an aerosol of interest or with a simulant having known characteristics. There have been several studies of the comparative calibration of various optical analyzers on different kinds of particle materials. Methods used are summarized in Table 5.2.[30] In general, these studies have been obtained on a single instrument and may be considered as representative performance only. Individual instruments will have their own specific calibrations.

The study of Hosey *et al.*[32] was conducted on an analyzer built by Southern Research Institute. Aerosols of the materials shown in Table 5.2 were formed by air-jet dispersal into a chamber. Simultaneous samples were obtained with an analyzer set to count above a given particle size level, a midget impinger, and a membrane filter. Results of coal dust particle counts indicated by the analyzer (AP) and obtained from counts of midget impinger (MI) samples are shown in Table 5.3. The ratio of dust counts (AP/MI) ranged from 0.4 to 1.9, but, based on an average of all 18 samples, the ratio was near unity. About half of the samples indicated a count ratio from

---

*Bioproducts Center, Dow Chemical Co., Midland, Michigan.
†DeVilbiss Co., Somerset, Pennsylvania.

TABLE 5.2
CALIBRATION METHODS FOR OPTICAL PARTICLE SIZE ANALYZERS USED BY VARIOUS INVESTIGATORS[a]

Reference	Material	Shape	Surface	Refractive index	Aerosol distribution: Mean sizes	Aerosol distribution: Spread	Preparation
O'Konski and Doyle[31]	Polystyrene latex	Sphere	Smooth	1.6	0.132, 0.333, 0.514	$2.1\% < \sigma < 6.8\%$	Atomizer, no stabilizer removal
	Polyvinyltoluene latex	Sphere	Smooth	—	0.144, 0.470, 0.986	$1.1\% < \sigma < 2.8\%$	—
Fisher *et al.*[23]	Dioctyl phthalate	Sphere	Smooth	1.49	0.8–2	—	LaMer
	Spores	Sphere	Rough	—	2.5–25, 30–35	—	Air jet
	Glass beads	Sphere	Smooth	—	5–25	—	Air jet
	Carbonyl iron	Sphere	Smooth	—	0–7	—	Air jet
	Dibutyl phthalate	Sphere	Smooth	—	0–6, 20–35, 0–50	—	LaMer, spinning disk, air jet
Hosey *et al.*[32]	Diatomaceous earth (natural)	Irregular	Rough	—	0.7	2.5	Air dispersed
	Diatomaceous earth (calcined)	Irregular	Rough	—	1.0	2.8	Air dispersed
	Portland cement	Irregular	Rough	—	0.5	2.6	Air dispersed
	Silica flour	Irregular	Rough	—	0.5	1.8	Air dispersed
	Coal	Irregular	Rough	—	0.5	2.5	Air dispersed
	Iron dust	Irregular	Rough	—	0.4	2.0	Air dispersed
Thomas *et al.*[24]	Polystyrene latex	Sphere	Smooth	1.6	0.27–3.2	Monodisperse	Atomizer
Eldridge[33]	Glass	Sphere	Smooth	1.53	0–65	Polydisperse	Air classifier
				1.9	0–40	Polydisperse	Air classifier
				2.25	0–15	Polydisperse	Air classifier
				2.5	0–15	Polydisperse	Air classifier

TABLE 5.2 (CONTINUED)

Reference	Material	Shape	Surface	Refractive index	Aerosol distribution: Mean sizes	Aerosol distribution: Spread	Preparation
Royco instructions[22]	Polystyrene latex	Sphere	Smooth	1.6	0.32–2.85	Mondisperse	Atomizer
	Special dust	?	?	—	1.05	$\sigma = 2.26\ \mu m$	Air dispersed
	Iron oxide	Sphere	Smooth	—	2.8	$\sigma = 0.07\ \mu m$	Air dispersed
	Puff ball spores	Sphere	Smooth	—	4.1	$\sigma = 0.035\ \mu m$	Air dispersed
	Carbonyl iron	Sphere	Smooth	—	7.9	$\sigma = 8.8\ \mu m$	Air dispersed
Zinky[21]	Polystyrene latex	Sphere	Smooth	1.6	—	Monodisperse	Royco atomizer
Channell and Hanna[34]	Polystyrene latex	Sphere	Smooth	1.6	0.156–3.04	—	Atomizer
	Silica flour	Irregular	Rough	—	1.6	Polydisperse	Air dispersed
	Cement	Irregular	Rough	—	—	Polydisperse	Air dispersed
	Diatomaceous earth	Irregular	Rough	—	—	Polydisperse	Air dispersed
	Iron dust	Irregular	Rough	—	—	Polydisperse	Air dispersed
	Coal dust	Irregular	Rough	—	—	Polydisperse	Air dispersed
ASTM[29]	Polystyrene latex	Sphere	Smooth	1.6	Available sizes	Monodisperse	Atomizer
Whitby and Vomela[30]	Polystyrene latex	Sphere	Smooth	1.6	0.36–1.3	Monodisperse	Atomizer
	Dioctyl phthalate	Sphere	Smooth	1.5	1.90–5.6	1.04–1.18	Spinning disk
	Polystyrene	Sphere	—	1.6	1.7–5.5	1.07–1.1	Spinning disk
		Irregular	Dented				
	India ink	Sphere	Rough	—	1.2–10.2	1.09–1.2	Spinning disk
		Irregular					
	Methylene blue	Sphere	Dented	—	0.9–5	1.08	Spinning disk

[a]From Whitby and Vomela, Ref. 30.

0.8 to 1.2 (within about 20% of the microscopic count). Summary data from all tests are shown in Table 5.4.

Particle size information was also presented for coal dust. The count median diameters determined from the aerosol instrument and from membrane filter samples were 0.35 and 0.31 μm, respectively. The associated geometric standard deviations were 2.9 (AP) and 2.3 (MI). The particle size of the coal dust before dispersal was reported as 0.5 μm ($\sigma_g = 2.5$). The particle size distribution obtained with the machine (as compared to membrane filter determinations) indicated a slightly larger median size and a slightly broader distribution.

The aerosol analyzer was also tried in a coal mine where wet drilling

TABLE 5.3
COMPARATIVE RESULTS ON COAL DUST SAMPLING WITH THE AEROSOL PHOTOMETER AND MIDGET IMPINGER[a]

AP average concentration mppcf[b]	MI concentration mppcf[c]		MI average concentration mppcf	Ratio AP/MI
36	29	32.2	30.6	1.2
7.5	8.9	12.7	10.8	0.7
15.5	13	17.4	15.2	1.0
9	11	19	15	0.6
12.5	19.5	24	21	0.6
44	104	125	114.5	0.4
18	13.4	12.5	12.9	1.4
14	9.6	—	9.6	1.5
15.5	17.5	11.0	14.3	1.1
14	8.8	6.3	7.5	1.9
14.3	11.2	11.8	11.5	1.2
86	185[P]	156	170	0.5
94	167[P]	162	164	0.6
83	67[P]	76	72	1.1
26	32.2[P]	24.2	28	0.9
34	34[P]	37.7	35.8	0.9
40	29[P]	30.6	30	1.3
44	24[P]	28	26	1.7
	Range of ratios (18 tests)	0.4–1.9		
	Average	1.03		
	Median	1.0		

[a]From Hosey *et al.*, Ref. 32.
[b]Discriminator set to count particles $\geqq 0.75$ μm.
[c]Duplicate Dunn cells from same sample counted by two individuals by light field microscopy. Those items marked with superior P were counted by phase-contrast microscopy.

TABLE 5.4
COMPARATIVE RESULTS ON VARIOUS DUSTS:
SUMMARY OF ALL TESTS[a]

Material	AP/MI at 1 μm[b]	AP/MI at 0.75 μm[b]	AP/MI at 0.6 μm[b]
Natural diatomaceous earth	1.3	2.1	—
Flux-calcined diatomaceous earth	0.3	—	0.8
Silica flour	0.5	0.9	—
Cement	1.1	—	4.8
Coal	0.6	1.0	—
Sintered iron	1.0	—	3.3

[a]From Hosey *et al.*, Ref. 32.
[b]Discriminator set to count particles of stated size and larger.
Note: Median values are based on 7 to 18 samples for each dust.

methods were employed. Light-scatter instrument counts were found to be much higher than simultaneous impinger samples, indicating that the analyzer was probably recording associated water mist. This finding demonstrates the need for careful interpretation of field data obtained with aerosol analysis instruments using light scatter. The authors[32] concluded that the average overall performance data obtained with the particle counter were acceptable for industrial hygiene purposes with dry dusts. They emphasized the need for careful calibration on every dust.

The study of Channell and Hanna[34] cited in Table 5.2 was an extension of the above work of Hosey *et al.* to the Royco PC-200A aerosol particle size analyzer in the laboratory, using the same test dusts. In addition they presented the results of a study of the Southern Research Institute instrument in an underground metal mine, with dry drilling methods but containing some associated oil mist. Comparative performance of the Royco and SRI units was obtained with laboratory simulants.

The Royco unit was first calibrated with four sizes of monodisperse polystyrene latex spheres from 0.557 to 3.04 μm in diameter. The instrument response was found to be satisfactory except in the lowest size class interval (0.32 to 0.40 μm), where spurious background counts of 100 to 1000 per minute were observed. The authors attributed this to effects of decrease in light intensity with lamp aging and the need for increased photomultiplier gain with its associated thermal noise. They also obtained high counts in the lowest channels when calibrating with polystyrene latex aerosols. This effect has also been shown to arise from the production of high concentrations of small particles (< 0.3 μm) of the stabilizer from the hydrosol during nebulization.[30,35]

Simultaneous particle counting and sizing samples were obtained[34] on industrial dust simulants with the Royco and with membrane filters. It was found that the Royco instrument indicated substantially higher counts in the channels less than about 2 μm, as shown in Table 5.5. Size distributions determined with the Royco instrument were smaller in median diameter and of greater geometric standard deviation than comparable analyses obtained by microscopic analysis of membrane filters. The authors concluded that the Royco instrument gave data acceptable for industrial hygiene purposes with proper calibration on the dust to be monitored, and with careful interpretation of the results.

TABLE 5.5
RATIO OF OBSERVED PC-200A COUNTS WITH MICROSCOPE COUNTS[a]

Channel	Ratio: $\frac{\text{Counts, PC-200A}}{\text{Counts, microscope}}$
>0.4 μm	2.0
>0.5 μm	1.7
>1.0 μm	1.4
>2.0 μm	≃1

[a]From Channell and Hanna, Ref. 34.
Note: These ratios are derived from determinations with flux-calcined diatomaceous earth, natural diatomaceous earth, and iron dust, and they should be considered approximations only.

careful calibration of light-scattering aerosol particle size analyzers on the
Similar studies were made by Channell and Hanna[34] with the Southern Research Institute instrument, and its performance was compared to that of the Royco instrument. As shown in Table 5.6, the Royo machine produced substantially higher counts. These studies serve to emphasize the need for

TABLE 5.6
COMPARISON BETWEEN PC-200A AND AEROSOL PHOTOMETER COUNT[a]

Size (μm)	Ratio: $\frac{\text{Counts, PC-200A}}{\text{Counts, AP}}$
>0.32	3.3
>0.4	2.6
>0.5	2.6
>1	1.8
>2	1.3

[a]From Channell and Hanna, Ref. 34.
Note: These are average values determined from sintered iron, Portland cement, silica flour, and coal dust samples.

dust to be monitored. The instruments can indicate the correct order of magnitude of dust counts in a selected size range based on an average of several determinations, but can be in error by a factor of 2 or greater.

Whitby *et al.*[30,35] have presented the results of laboratory studies of the performance of aerosol particle size analyzers manufactured by Royco (PC-200A), the Southern Research Institute, and Bausch and Lomb. Aerosol materials investigated included spray-dried monodisperse polystyrene latex (PSL), and spinning disk-generated dioctyl phthalate (DOP), polystyrene (PS), methylene blue (MB), and India ink (II). Particles were dispersed into an aerosol tunnel, and samples were obtained on membrane filters or impactor slides for particle size analysis. Each of the automatic particle sizing instruments was operated on one or more of the test aerosols, and comparable size distribution data were obtained. Average results for polystyrene latex, dioctyl phthalate, and India ink are presented in Table 5.7. Their specific conclusions from this study were:

1. Optical counters calibrated with polystyrene latex aerosols do underestimate the size of absorbing aerosols, such as India ink, by factors of 2 to 5. The size underestimation is of the order of magnitude predicted by theory. There are significant differences between the Southern Research Institute, Bausch and Lomb, and Royco counters, the Bausch and Lomb counter showing the least underestimation, and the Royco the most, for India ink particles in the 1 to 10-μm range.

2. The median indicated size seems to be relatively independent of particle shape for both transparent and absorbing particles for sizes less than 5 μm. For particles larger than 5 μm the effect of particle shape and surface characteristics seem to become greater, as evidenced by increased scatter of the data for both transparent and opaque particles.

3. The ability of the counters to resolve a narrow size distribution (as measured by σ_g) is a function of particle size and type of aerosol, and is different for the three counters. For both opaque and transparent irregular particles larger than 2 μm the resolution of the Bausch and Lomb counter is near the theoretical value, whereas the Royco and Southern Research Institute counters respond to particle irregularity by spreading the indicated size. All three counters show near theoretical resolution of the dioctyl phthalate aerosols larger than 1 μm. There are indications that the resolving power of all three counters on polystyrene latex aerosols decreases below 1 μm. However, this must be studied further because the atomizer-dispersed polystyrene latex aerosol is not completely satisfactory for evaluating resolving power, since it is a solid and because of its inevitable background of solution residue particles.

TABLE 5.7

Calibration of Three Optical Particle Size Analyzers with Three Different Aerosols

Microscopic sizing		Royco			Southern Research Institute			Bausch and Lomb			Run numbers from Ref. 30
nmd	σ_g[a]	nmd	σ_g[a]	c/m[b]	nmd	σ_g[a]	c/m[b]	nmd	σ_g[a]	c/m[b]	
Polystyrene latex (transparent solid spheres)											
0.365	—	0.37	1.52	1.01	—	—	—	1.15	1.48	3.2	3, 40, 68
0.814	—	0.86	1.25	1.04	0.93	1.30	1.14	1.30	1.29	1.6	2, 33, 41, 65
1.17	—	1.19	1.26	1.02	—	—	—	1.33	1.19	1.14	1, 42, 67
1.30	—	1.21	1.53	0.93	1.20	1.15	0.92	1.37	1.20	1.05	32, 43
Dioctyl phthalate (transparent liquid spheres)											
1.90	1.18	1.84	1.25	0.97	—	—	—	3.1	1.15	1.63	46
2.70	1.04	2.00	1.23	0.74	—	—	—	—	—	—	38
—	—	2.50	1.39	—	2.80	1.17	—	—	—	—	30
5.15	1.07	5.50	1.08	1.07	—	—	—	7.50	1.1	1.45	47
5.60	1.04	5.45	1.09	0.97	—	—	—	—	—	—	37
—	—	5.90	1.19	—	7.5	1.08	—	—	—	—	27
India ink (opaque irregular spherical aggregates)											
1.23	1.16	0.51	1.31	0.41	1.08	1.53	0.83	0.64	1.83	0.55	23, 49
2.40	1.11	0.46	1.32	0.20	0.74	1.46	0.29	1.20	1.33	0.51	24, 50
4.60	1.12	0.92	1.35	0.20	1.31	1.36	0.28	1.60	1.15	0.36	25, 52
7.60	1.09	3.50	1.17	0.46	—	—	—	3.30	1.21	0.43	54
10.2	1.19	1.93	1.92	0.19	1.87	1.72	0.18	—	—	—	26

[a]Determined from the square root of the ratio of the 84% size to the 16% size obtained from cumulative frequency–log size (log-prob) plot.

[b]Ratio of the number median diameter (nmd) as measured by an optical particle size analyzer to that obtained by microscopic sizing.

4. The absolute calibration of the counters for transparent aerosols was found to be good between 1 and 1.5 μm for all counters; good below 1 μm for the Royco and the Southern Research Institute; fair for the Royco and Bausch and Lomb above 1.5 μm; and poor for the Southern Research Institute above 1.5 μm. This should not be considered as a serious deficiency, since it is always possible to calibrate a given counter on a given aerosol so that it will indicate the correct size. However, the data reported here do show the necessity of calibrating optical counters over the complete indicated size range.

Whitby and Liu[35] have investigated the performance of the Royco instrument on high concentrations of particles smaller than the countable range (<0.3 μm). Light-scatter analyzers respond to all light received by the phototube. In principal, individual countable particles are carried through the optical sensing volume in an air stream. Each of the gas molecules in the optical path will scatter a small amount of light. Since molecular concentrations are high (10^{16} molecules/mm^3), random fluctuations in their numbers present a small and relatively constant background of scattered light to the phototube. If other small sub-countable (< 0.3 μm) particles are present in the optical sensing volume, they will also scatter some light. The ability of the optical analyzer to respond to these additional particles depends on their concentration and optical scattering characteristics. For example, if a number of 0.1-μm particles were present simultaneously in the optical sensing volume, the phototube and pulse height analyzer might respond with a pulse classified as a single 0.3-μm particle. In practice, one is often dealing with an aerosol containing background atmospheric dust in addition to the particulate material of interest. The aerosol analyzer will therefore respond to the total concentration of all particles in the sensing volume at any instant, even though many particles may be substantially below the minimum detectable threshold. For example, atmospheric dust contains a continuous spectrum of particles down to about 0.002 μm.[36,37] The concentration of all particles in the atmosphere usually ranges from 10^2 to 10^6 particles per cubic centimeter, depending on the source, history, and age of the particulate material, and it is possible to have a large and variable background concentration of sub-countable particles present in the optical volume.

In addition, calibration of optical analyzers with spray-dried polystyrene latex spheres may also introduce an error caused by a sub-countable particulate background. Polystyrene latex spheres are stabilized in the original hydrosol with less than 1% of a surface-active agent which is contained in all spray droplets. Upon drying, droplets not containing a polystyrene sphere give rise to another aerosol particle species whose size is a function of the concentration of the stabilizer in the diluted hydrosol and of the droplet size produced in the atomizer, but always less than 0.3 μm.[38] Although the atomization of monodisperse polystyrene latex yields a monodisperse polystyrene latex aerosol, the atomization of the stabilizer gives rise to a distribution of smaller sizes. Whitby and Liu[35] concluded from their study that "high concentrations of aerosols below the normal counting range of these counters may be indicated as a relatively monodisperse aerosol having a median size which is a function of concentration."

To reduce the background produced by the coincident presence of subcountable particles, washing the particles with distilled water[38] or the use of some form of dilution is required.[14] One can dilute the hydrosol by incremental factors of 10 and check the counter background calibration on a filtered sample of stabilizer without particles after aerosolization.

Compressed air supplied to the atomizer must be filtered to remove particulate contaminants and passed through a granular charcoal adsorbent and molecular sieves to remove condensable hydrocarbon vapors. Particles in the generated aerosol can be dried and reduced in concentration by adding measured amounts of dry filtered air. Manufacturers of aerosol particle analyzers furnish equipment and information on methods for calibration and aerosol dilution for their particular instruments.

5-4.2 Electric Charge

An electric charge instrument described by Guyton[39] (not commercially available) employs initial electrostatic charging of all particles, followed by measurement of the number and magnitude of the electrical impulses transmitted to conductive particle collecting surfaces (such as an array of filaments of graduated diameter) when the charged particles impinge upon them. The amount of charge which a particle will accept and subsequently give up to the collecting electrode is in great part dependent on the conductivity of the particle and its surface characteristics, so that particles of the same size but of different materials may give widely varying responses (see Chapter 1, p. 7). Gucker and O'Konski[40] concluded on the basis of theoretical considerations that this principle of charge transfer on impact would not be effective for particles 1 μm or smaller.

An instrument based on the measurement of the electrical mobility of charged aerosol particles has been described by Whitby and Clark.[36] A schematic diagram of the apparatus is shown in Fig. 5.15. Aerosol particles (0.015 to 1.0 μm) are intimately mixed with unipolar (−) gas ions produced in a jet ionizer. The combined aerosol–ion cloud is stored in a chamber

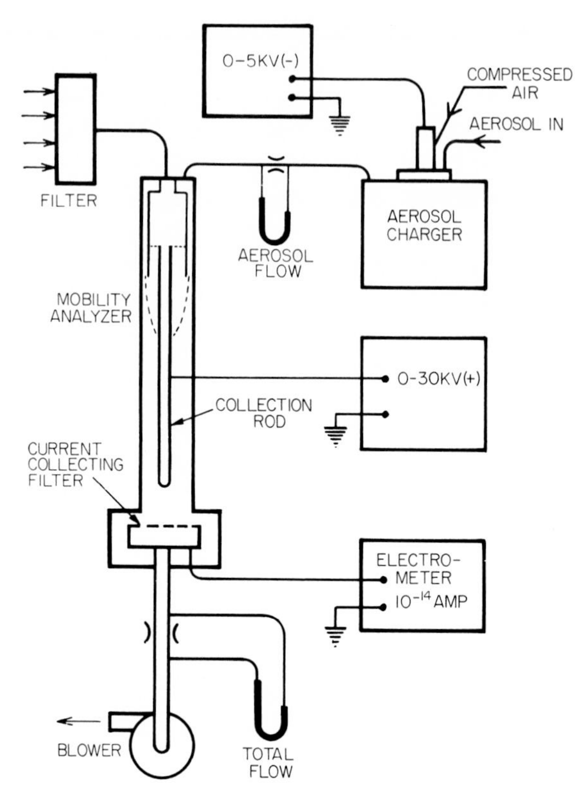

FIG. 5.15. Schematic of the electrical particle counter system. Courtesy Thermo-Systems, Inc.

for a short period (about 30 sec) until an equilibrium charge distribution is established on the particles due to ion thermal motion. A theory for the charging process has been proposed by Liu *et al.*[41]

Charged particles are led to a thin annual circumferential slot of the cylindrical mobility analyzer and carefully combined with a filtered core of laminar air. The center of the cylindrical mobility analyzer contains a charged (+) inner collector electrode provided with a variable power supply (0 to 30 kVd.c.). All particles having a mobility greater than a certain specific value established by the collector rod voltage are collected. Particles of lesser mobility pass the rod and are deposited on a current-collecting filter, where their charge is measured with an electrometer. By successively changing the collector rod voltage, a cumulative distribution of electrical mobilities (0.0002 to 0.01 cm/sec volt/cm) is obtained from the filter currents. From a knowledge of the flow geometry, the cumulative particle size distribution is calculated. The system operates at 1 to 3 cfm aerosol flow rate and can be used as a classifier to collect all particles above or below a certain size (from deposits on the rod or from the collecting filter). It can also be used to measure the natural particle charge on ambient aerosols. This unit is available commercially.* Performance data and calibrations have been reported by Whitby and co-workers.[36,42]

5-4.3 Thermal Effects at a Heated Filament

An automatic aerosol counting and sizing device described by Vonnegut and Neubauer[43] responds to a change in temperature when a volatile or combustible particle impinges upon and is vaporized or oxidized at a hot metallic surface. However, the magnitude of the signal from a heated wire depends on the temperature of the detecting element, the vaporization or combustion properties of the particles, and the velocity of the air stream, as well as the particle size. In addition, the limit of resolution of this system is no better than 2 μm. Similar system developments have been described by Goldschmidt[44] and Reist and Burgess.[45] These are not commercially available.

5-4.4 Other Methods of Aerosol Sizing

A particle counting and sizing instrument has been proposed by Langer[46] which uses acoustic phenomena associated with aerosol particle flow out of a capillary. Particles flow at 6 lpm through a converging nozzle and capillary system shown in Fig. 5.16.[45] As they leave the lower end of the capillary at 100 m/sec, an audible click is produced which can be detected as a pulse with a microphone placed at the inlet to the nozzle as shown. This system is presently under development and is not commercially available.

In addition to the instruments considered above, Lieberman[47] cites the following automatic aerosol particle size analysis systems in various states of development:

"Particles are forced to pass through a microwave-excited plasma or through a hydrogen flame. The emission spectra are analyzed for intensity at a single wavelength if a single material is present, or for composition if there is more than one material.

"Aerosol particles forced to cross an electric beam capture electrons from it. Measurement of the beam's intensity determines its electron loss.

"Diffraction patterns produced by suspended particles (which are illuminated by a beam of coherent light) are compared with previously well-

---

*Thermo-Systems, Inc., St. Paul, Minnesota.

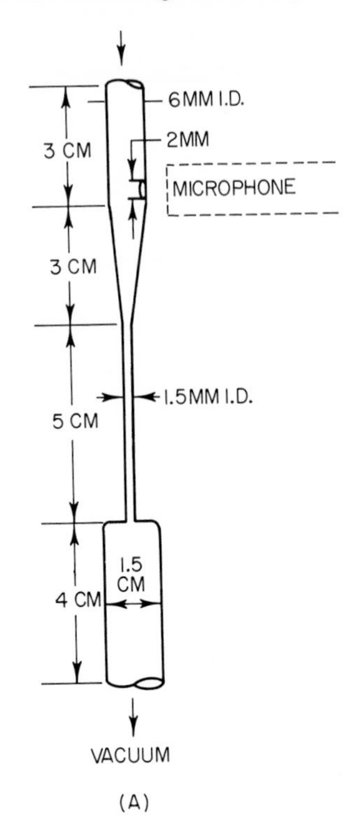

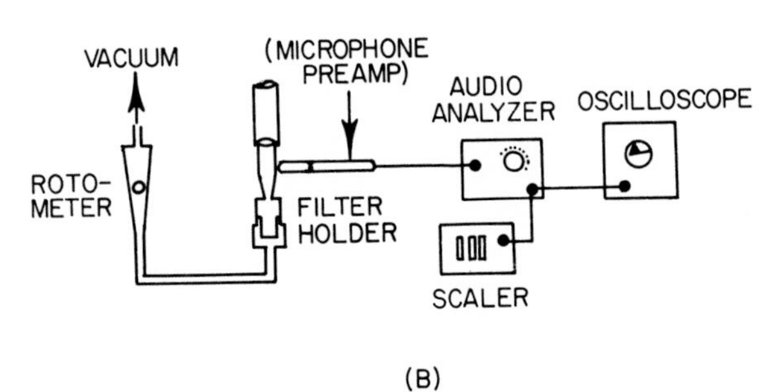

FIG. 5.16. Schematic of the acoustic particle counter: *(a)* counting element; *(b)* experimental apparatus. (From Reist and Burgess, Ref. 45.)

established patterns. Characteristic dimensions in the patterns then permit accurate determination of particle diameters (laser holography).

"Particle-size distributions are determined by observing the polarization ratio of scattered light, i.e., the ratio of the intensities of scattered components —two mutually perpendicular-plane polarized components—of a beam of light."

These instrument developments indicate that a rapid growth may be anticipated in commercial automatic particle size analyzers.

The instruments described in this chapter can be used together for analysis of specific parts of the aerosol spectrum to obtain an analysis of the total particle size distribution when it covers a wide range. Whitby and McFarland[42] have discussed the application of the General Electric Condensation Nuclei Counter (total count of all particles greater than 0.002 μm) in conjunction with the electrical particle (mobility) analyzer (0.015 to 1 μm) (Fig. 5.15) and the Royco instrument (0.3 to 10 μm) for measurement of the count and mass distribution of atmospheric aerosols.

Automatic instruments intended for use on polydisperse aerosol particles of unknown or variable physical and chemical properties must be calibrated for proper interpretation of the instrument reading. At this stage of development, careful instrument calibrations by methods described in Chapter 3 are necessary with identical kinds of particles. The comparative performance of a few individual machines on a limited number of aerosols has been evaluated, but further calibration studies are required to make more extensive applications of these machines.

While we believe that light-scattering aerosol counting and sizing machines can perform a needed and important service in many dust monitoring situations, we deplore the tendency of uncritical users to accept the output readings as exact representations of particle size, in spite of evidence to the contrary by more exact and reliable sizing methods. It may be concluded that the development of automatic particle sizing instruments, as applied to the analysis of aerosols, has made excellent progress but that additional improvements may be anticipated. The popularity of these instruments is evident from the fact that the American Society for Testing and Materials has established a tentative method of Test for Continuous Automatic Counting and Sizing Airborne Particles in Dust Controlled Areas (i.e., clean and white rooms).

References

1. E. J. Scherago and B. J. Sheffer (Eds.), Guide to Scientific Instruments, *Science, 165A*:95 (1969).
2. W. H. Walton, Survey of the Automatic Counting and Sizing of Particles, *Brit. J. Appl. Phys.*, Suppl. 3, S121 (1954).
3. H. A. Dell, Stages in the Development of an Arrested Scan Type Microscopic Particle Counter, *Brit. J. Appl. Phys.*, Suppl. 3, S156 (1954).
4. P. C. Reist, W. A. Burgess, and D. Yankovich, Development of an Automatic Particle Assaying Instrument Utilizing a Scanning Electron Microscope, annual meeting of Air Pollution Control Association, Paper No. 69-124, Pittsburgh, Pa., 1969.

5. L. Le Bouffant and J. L. Soule, The Automatic Size Analysis of Dust Deposits by Means of an Illuminated Slit, *Brit. J. Appl. Phys.,* Suppl. 3, S143 (1954).
6. O. A. Ullrich, Size Analysis of Fine Particles and Results Obtained with an Electrical Sensing-Zone Particle Analyzer, annual meeting of Instrument Society of America, Paper No. 40-NY 60, New York, 1960.
7. B. A. Batch, The Application of an Electronic Particle Counter to Size Analysis of Pulverized Coal and Fly Ash, *J. Inst. Fuel, 37*:455 (1964).
8. S. Kinsman, Particle Size Distribution Measurement, *Ceramic Age, 78*:53 (August 1962).
9. F. G. Anderson, T. F. Tomb, and M. Jacobson, *Analyzing Midget Impinger Dust Samples with an Electronic Counter,* United States Bureau of Mines, Report of Investigations No. 7105, April 1968.
10. J. O. Romine and J. B. Gayle, Evaluating the HIAC PC-101 Automatic Particle Counter, *J. Am. Assoc. Contam. Control, III*:10 (January 1965).
11. J. B. Gayle, W. A. Riehl, R. H. Hollinger, and T. I. Haigh, An Evaluation of the Sperry Ultrasonic Particle Counter, *J. Am. Assoc. Contam. Control, II*:16 (October 1963).
12. K. T. Whitby, Developments in Laboratory and On-Stream Size Distribution Analysis, in *Proceedings of 27th Annual Mining Symposium,* p. 113 (1966).
13. H. C. van de Hulst, *Light Scattering by Small Particles,* John Wiley & Sons, New York, 1957.
14. J. R. Hodkinson, The Optical Measurement of Aerosols, in C. N. Davies (Ed.), *Aerosol Science,* Chapter X, Academic Press, New York, 1966.
15. J. P. Kratohvil, Light Scattering, *Anal. Chem., 36*:458R (1964).
16. J. P. Kratohvil, Light Scattering, *Anal. Chem., 38*:517R (1966).
17. C. T. O'Konski, M. D. Bitron, and W. I. Higushi, *Light Scattering Instrumentation for Particle Size Distribution Measurements,* Special Technical Publication No. 234, pp. 180-206, American Society for Testing and Materials, Philadelphia, Pa., 1959.
18. F. T. Gucker, Jr., and D. G. Rose, A Photoelectric Instrument for Counting and Sizing Aerosol Particles, *Brit. J. Appl. Phys.,* Suppl. 3, S138 (1954).
19. F. T. Gucker, Jr., and D. G. Rose, The Response Curve of Aerosol Particle Counters, in *Proceedings of 3rd National Symposium on Air Pollution,* Pasadena, Calif., Apr. 20, 1955.
20. K. T. Whitby, Sheath Air Aerosol Inlet Tube for the Royco PC-200, in K. T. Whitby, B. Y. H. Liu, and R. A. Vomela, *Evaluation of Optical Particle Counters,* Particle Laboratory Publication No. 110, Mechanical Engineering Department, University of Minnesota, June 1967.
21. W. R. Zinky, A New Tool for Air Pollution Control: The Royco Aerosol Particle Counter, *J. Air Pollution Control Assoc., 12*:578 (1962).
22. *Operating and Service Manual, PC-200A Particle Counter,* p. 3-1, Royco Instruments, Inc., Palo Alto, Calif. (1961).
23. M. A. Fisher, S. Katz, A. Lieberman, and N. E. Alexander, The Aerosoloscope: An Instrument for the Automatic Counting and Sizing of Aerosol Particles, in *Proceedings of 3rd National Air Pollution Symposium,* Pasadena, Calif., Apr. 20, 1955.
24. A. L. Thomas, Jr., A. N. Bird, Jr., R. H. Collins, III, and P. C. Rice, A Portable Photoelectric Aerosol Counter and Particle Size Analyzer, Instrument Society of America, Preprint No. 23-SF60, 1960.

25. A. L. Lavery and W. Z. Leavitt, A Portable Multi-channel Aerosol Particle Analyzer, in *Proceedings of 13th National Conference,* Aerospace Instrumentation Symposium, San Diego, Calif., June 1967.
26. J. M. Randall and J. D. Keller, Electro-Optical Aerosol Counter Instrumentation, presented at American Industrial Hygiene Conference, Houston, Tex., May 1965.
27. D. Sinclair, A New Photometer for Aerosol Particle Size Analysis, *J. Air Pollution Control Assoc., 17*:105 (1967).
28. A. B. Block, *Scattering of Laser Light by Uniform Spherical Particles,* Ph.D., Dissertation, Rutgers, The State University of New Jersey, *Diss. Abstr., 68*:4527 (1967).
29. *Tentative Method of Test for Continuous Counting and Sizing of Airborne Particles in Dust-Controlled Areas by the Light-Scattering Principle (for Electronic and Similar Applications),* ASTM Designation F-50-65T, American Society for Testing and Materials, Philadelphia, Pa., 1965.
30. K. T. Whitby and R. A. Vomela, The Response of Single Particle Optical Counters to Non-Ideal Particles, *Environ. Sci. Technol., 1*:801 (1967).
31. C. T. O'Konski and G. J. Doyle, Light Scattering Studies in Aerosols with a New Counter–Photometer, *Anal. Chem., 27*:694 (1955).
32. A. D. Hosey, H. H. Jones, and H. E. Ayer, Evaluation of an Aerosol Photometer for Dust Counting and Sizing, *Am. Ind. Hyg. Assoc. J., 21*:491 (1960).
33. R. G. Eldridge, A Few Fog Drop-Size Distributions, *J. Meteorol., 18*:671 (1961).
34. J. K. Channell and R. J. Hanna, Experience with Light Scattering Particle Counters, *A.M.A. Arch. Environ. Health, 6*:386 (1963).
35. K. T. Whitby and B. Y. H. Liu, Generation of Countable Pulses by High Concentrations of Sub-Countable Sized Particles in the Sensing Volume of Optical Counters, *J. Colloid Interface Sci., 25*:537 (1967).
36. K. T. Whitby and W. E. Clark, Electric Aerosol Particle Counting and Size Distribution Measuring System for the 0.015 to 1 μ Size Range, *Tellus, XVIII*:573 (1966).
37. R. E. Pasceri and S. K. Friedlander, Measurements of the Particle Size Distribution of the Atmospheric Aerosol: II, Experimental Results and Discussion, *J. Atmos. Sci., 22*:577 (1965).
38. P. C. Reist and W. A. Burgess, Atomization of Aqueous Suspensions of Polystyrene Latex Particles, *J. Colloid Interface Sci., 24*:271 (1967).
39. A. C. Guyton, Electric Counting and Size Determination of Particles in Aerosols, *J. Ind. Hyg. Toxicol., 28*:133 (1946).
40. F. T. Gucker, Jr., and C. T. O'Konski, Electronic Methods of Counting Aerosol Particles, *Chem. Rev., 44*:373 (1949).
41. B. Y. H. Liu, K. T. Whitby, and H. H. S. Yu, On the Theory of Charging of Aerosol Particles by Unipolar Ions in the Absence of an Applied Electric Field, *J. Colloid Interface Sci., 23*:367 (1967).
42. K. T. Whitby and A. R. McFarland, Electrical Measurement of the Mass Concentration of a Self-Preserving Aerosol Size Distribution, *J. Air Pollution Control Assoc., 18*:760 (1968).
43. B. Vonnegut and R. Neubauer, Detection and Measurement of Aerosol Particles by the Use of an Electrically Heated Filament, *Anal. Chem., 24*:1000 (1952).
44. V. W. Goldschmidt, Measurement of Aerosol Concentrations with a Hot Wire Anemometer, *J. Colloid Sci., 20*: 617 (1965).

45. P. C. Reist and W. A. Burgess, A Comparative Evaluation of Three Aerosol Sensing Methods, *Am. Ind. Hyg. Assoc. J., 29*:123 (1968).
46. G. Langer, *Status of Acoustic Particle Counter Research,* International Conference on Powder Technology, Illinois Institute of Technology Research Institute, Chicago, Ill., May 20-23, 1968.
47. A. Lieberman, Fine Particle Technology in the Chemical Process Industries, *Chem. Eng., 74,* March 27, 1967, p. 97; April 10, 1967, p. 209; April 24, 1967, p. 163.

6

Particle Sizing Interpretation

6-1 Introduction

When dealing with a monodispersed system of spherical particles, a single size parameter (diameter) describes the general size characteristics of all particles in sufficient detail. However, when particles of many sizes occur together, a single parameter is inadequate to describe the sizes of all the particles that are present. Measurements must be made over the full range of sizes present, and size interpretation and representation become a matter of applied mathematical statistics. This subject is treated briefly in this chapter and more extensively in standard texts.[1,2]

6-2 Graphical Representation of Particle Size

After particles have been measured, the data are classified into the number of particles in successive size intervals (e.g., 1 to 1.99 μm, 2 to 2.99 μm, etc.) and can be represented graphically by a bar diagram, or histogram. The widths of the rectangles represent the size interval, and the heights represent the frequency of particle occurrence in each size interval. An idealized distribution is shown in Fig. 6.1. A smooth curve drawn through the midpoints of the tops of the bars in Fig. 6.1 resembles a bell-shaped (i.e., normal) probability curve. The abscissa at the center of the symmetrical curve divides the area into two equal parts. This represents the *mean* value of particle size, the sum of the values divided by the number of observations.

However, the mean value by itself does not completely define a particle size distribution, since it is possible to have many curves with the same midpoint, or mean, value; an example of this is shown in Fig. 6.1 by the dashed curve. Therefore, a term is required to describe the scatter of observations about the mean value. This is the standard deviation, the root-mean-square deviation of observations about the mean. One standard deviation either side of the mean of a bell-shaped curve includes ±34.13% of the observations, and for a symmetrical distribution curve, the ratio of the 50% size divided by the 15.87% size is equal to the ratio of the 84.13% size divided by the 50% size. The spread of the curve on either side of the 50% line is defined by the magnitude of this ratio, which is the standard deviation.

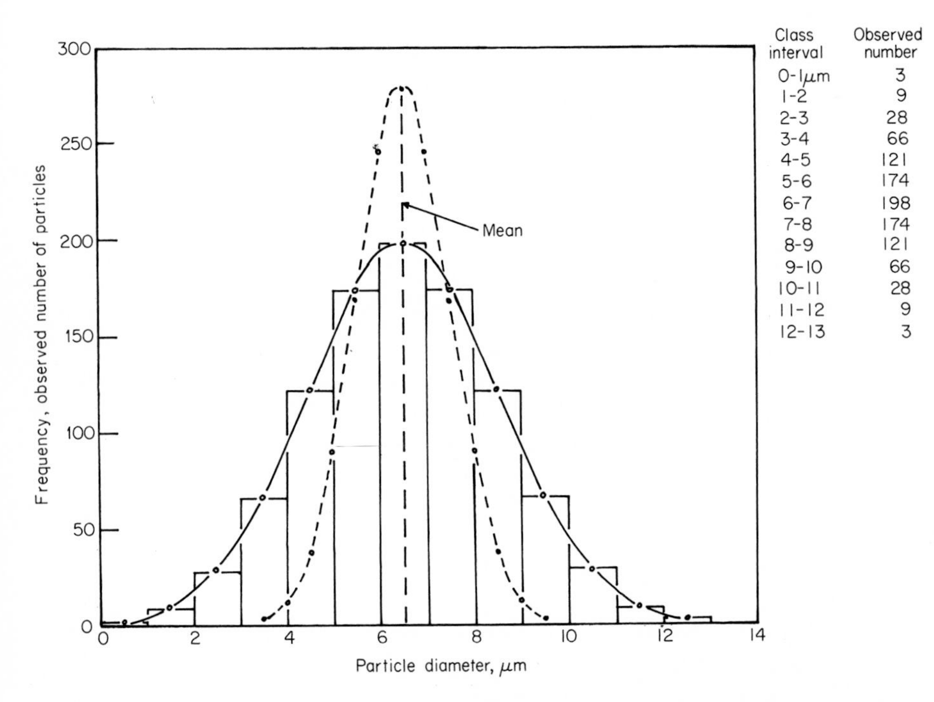

FIG. 6.1. Histogram of a normal probability size distribution.

Real particle size distributions seldom give symmetrical curves. Instead, the size frequency histogram of a typical dust more nearly resembles the one shown in Fig. 6.2, in which the observed frequencies are expressed as percentages rather than as the actual observed numbers. When a smooth line is drawn through the centers of the tops of the bars, the resulting curve is asymmetrical, or skewed. Although asymmetrical curves may be represented graphically with reasonable ease, mathematical analysis leads to considerable complexity. When the particle size distribution is skewed as shown in this figure, the mean value is unduly affected by very large or very small outlying observations. When this is true, the median is a more useful form of average. The median is the center value of a series of observations when the observations are arranged in order from lowest to highest, otherwise as the interpolated center value. For classified data the median is defined as the abscissa that divides the area of the histogram into two equal parts. For the symmetrical distribution shown in Fig. 6.1, mean and median coincide, whereas for the skewed distribution shown in Fig. 6.2, they do not.

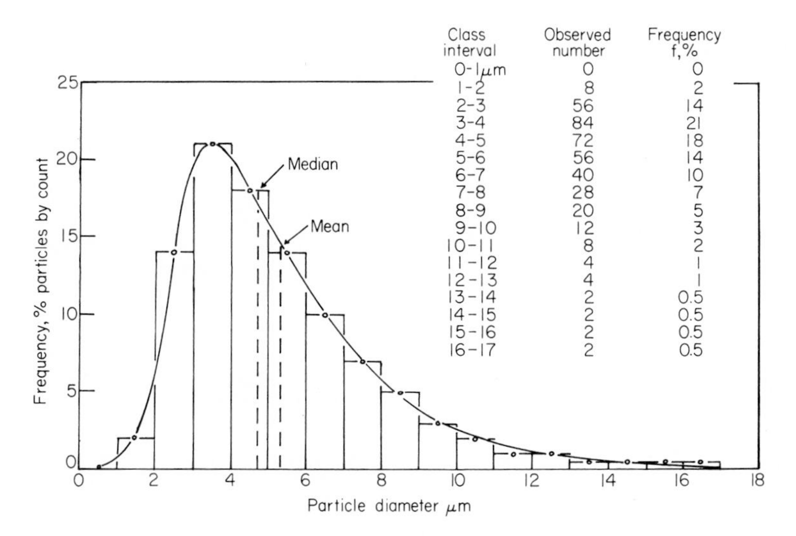

FIG. 6.2. Histogram of a skewed particle size distribution.

Drinker[3] has shown that most asymmetrical particle size frequency curves can be converted into symmetrical curves which resemble a normal probability curve when the logarithm of size is substituted for size and when size frequency data are grouped by uniform logarithmic increment rather than arithmetic increment.[4] Then, the same size distribution shown in Fig. 6.2 appears in Fig. 6.3 as a symmetrical bell-shaped curve, called a logarithmic-normal distribution curve. There is no fundamental reason for all particle sizing data to approximate a logarithmic-normal frequency distribution, but it is commonly observed in practice, especially for dusts produced by comminution.

The bell-shaped distribution curve in Fig. 6.3 can now be analyzed mathematically with relative ease. Plotting log size instead of size alone is

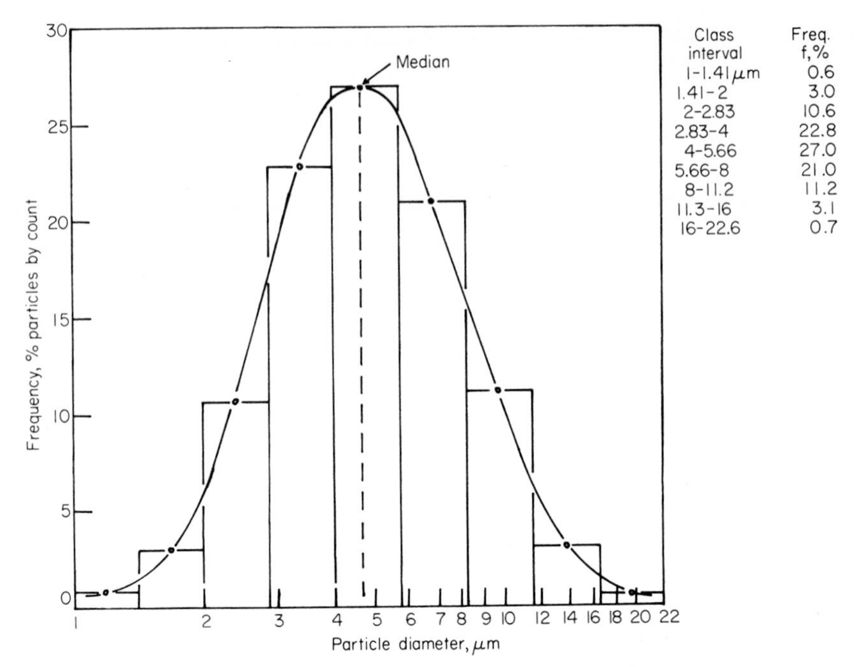

FIG. 6.3. Histogram of a log-normal size distribution.

consistent with the practice of sizing according to a size scale which is in a geometric series, and is preferable because it gives equal prominence to data in all parts of the size range. As shown, the interval from 1 to 2 μm has the same space on the log scale as the interval from 10 to 20 μm, which seems proper from experience with sizing of dusts. In contrast, on an arithmetic scale the intervals 0 to 10 μm and 10 to 20 μm, respectively, would be given equal space.

The cumulative plot is the most useful type of graph for particle size distribution analysis. The ordinate, cumulative percent of dust which is finer (or larger) than a given size, is plotted against the upper limit of the class interval. A semilog cumulative plot is shown in Fig. 6.4 for the size data of Fig. 6.3. The percent of the particles between any two sizes is represented by the difference between the ordinates of the two sizes. The work of determining the two parameters which define a distribution, the median size and the standard deviation, which are the geometric mean and geometric standard deviation when employing logarithms of numbers, can be considerably reduced by utilizing logarithmic-probability paper. This contains a scale derived from the probability integral on one axis and a logarithmic scale on the other to allow a graphical solution of the problem. On logarithmic-probability (log-prob) paper the summation curve obtained by the integration of a logarithmic size frequency distribution plots as a straight line, as shown in Fig. 6.5 using the same data of Figs. 6.3 and 6.4. The median or geometric mean size* is the 50% size, since the median by definition bisects the symmetrical frequency curve. The spread of the distribution, or geometric standard deviation (σ_g), is defined by the slope of the line as indicated by the ratio

$$\frac{50\% \text{ size}}{15.87\% \text{ size}} \quad \text{or} \quad \frac{84.13\% \text{ size}}{50\% \text{ size}}$$

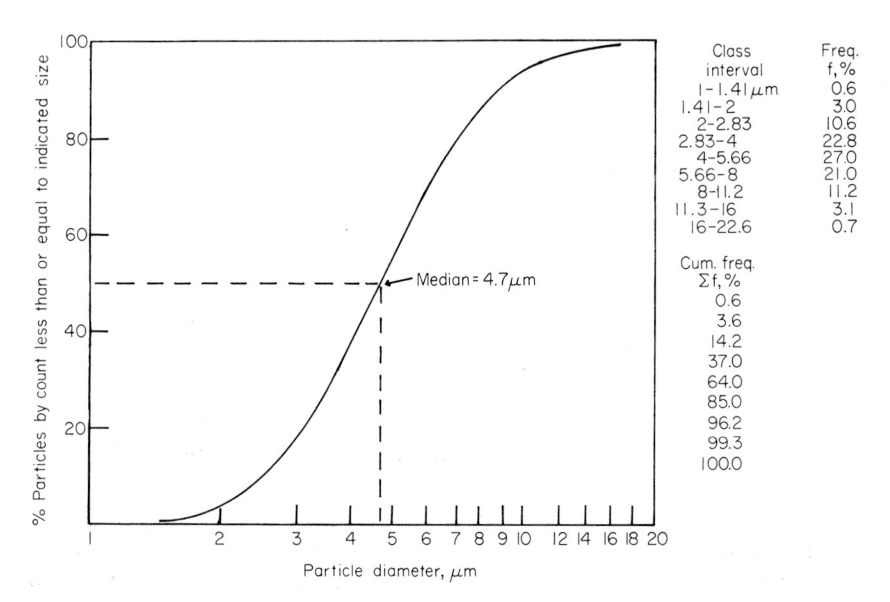

FIG. 6.4. Cumulative log-normal size distribution.

6-3 Mathematical Representation of Particle Size

It is necessary to construct a mathematical model for the relationship between particle size and frequency of occurrence if analysis by other than a simple histogram is desired. A number of curve-fitting equations have been proposed; some have been based on empirical observations of the closeness of fit with observed particle size frequency distributions; others have been

*The geometric mean is defined as the nth root of the the product of n terms: e.g., the square root of the product of two numbers. The logarithm of the geometric mean is equal to the arithmetic mean of the logarithms.

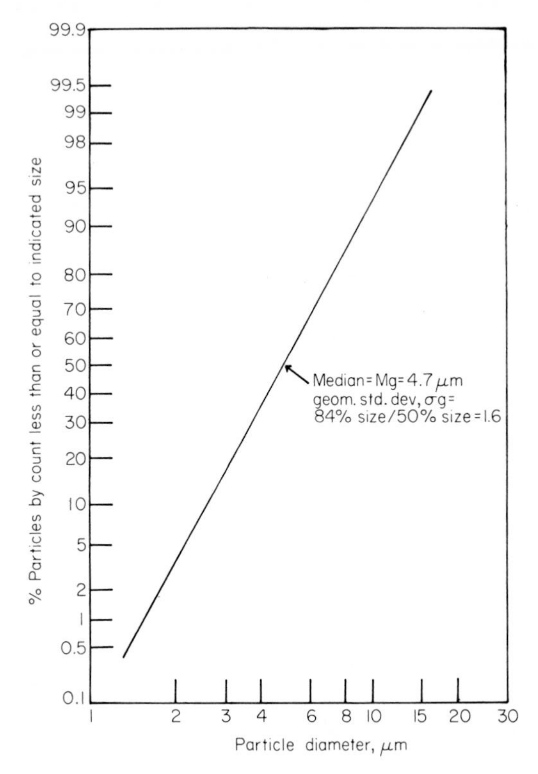

FIG. 6.5. Cumulative particle size distribution plotted on logarithmic probability graph paper.

based on distribution functions of a general statistical nature. Among the former are the distribution laws proposed by Roller[5] and Rosin and Rammler,[6] given in the Appendix to this chapter (p. 257), which were developed to represent the distributions found in mine coal and pulverized coal, respectively. Among the generalized distribution laws that have been applied to fine particulate matter are the binomial and the Gaussian distribution laws. The criterion for selecting one relation in preference to another is, of course, the closeness of the fit of experimental measurements to the distribution theory selected. One law commonly applied to particle size data that usually gives satisfactory results is the log-normal distribution law, proposed by Hatch and Choate[7] for particle distributions resulting from crushing and grinding of mineral substances and later modified by Kottler[8,9] and others, for application to special particle size distributions such as the precipitation of silver halide granules in photographic film emulsions.

The distribution function is characterized in terms of two arbitrary constants which define some average diameter as a measure of central tendency and a measure of the variation of diameters about this average. As a general expression for class intervals of equal size, the fraction of the total number of particles present P in any size category between D and $D + dD$ may be expressed as a function of diameter D, defined such that

$$P = f(D)\, dD \tag{6.1}$$

When a particle size frequency curve exhibits a normal probability distribution,

$$P = \frac{1}{(2\pi\sigma)^{1/2}} \exp\left[-\frac{(D - \bar{D})^2}{2\sigma^2}\right] dD \tag{6.2}$$

where $\bar{D}$ = the characteristic mean diameter, and σ = the Gaussian standard deviation. The cumulative fraction F of particles from size zero to size D becomes

$$F = \int_0^D f(D)\, dD \tag{6.3}$$

Applying normal probability relations, Hatch and Choate[7] substituted the logarithms of D, $\bar{D}$, and σ in equation 6.2 to transform the cumulative size distribution (equation 6.3) into the following form:

$$F = \int_\infty^{\ln D} \frac{1}{(2\pi)^{1/2} \ln \sigma_g} \exp\left[-\frac{(\ln D - \ln D_{\mathrm{nmd}})^2}{2 \ln^2 \sigma_g}\right] d(\ln D) \tag{6.4}$$

where F = the cumulative number of particles with logarithms of diameters less than $\ln D$.

σ_g = the geometric standard deviation; equal to 84.13% size/50% size or the 50% size/15.87% size.

D_{nmd} = number median diameter (i.e., 50% size by count).

Size by count is important in the problem of silicosis, where dustiness in air has traditionally been determined on a count basis in the United States. On a count basis, 50% of the total number of particles are greater than the count median diameter, and 50% are less. For toxic materials that are rated on a total mass basis (e.g., lead dust), distribution by weight may be of more significance. On a weight basis, 50% of the total mass of the material is represented by particles whose diameters are greater than the mass median diameter, and 50% by particles whose diameters are less than the median. This latter constitutes a definition of the mass median diameter, which is one of the parameters frequently used to describe the statistical size of a dust or powder.[10] Another possibility is to express the distribution in terms of specific surface so that half the total surface is represented by particles greater than a stated size, and half by particles smaller than this size. There has been some evidence that toxicity may be a function of total particle surface, in which case this parameter is of considerable importance.

If $(\overline{D^m})^{1/m}$ represents any *mean* diameter, where m has a value of 1, 2, or 3 for number, surface, or volume average diameter, respectively, then

$$\overline{D^m} = \int_{-\infty}^{+\infty} \frac{D^m}{(2\pi)^{1/2} \ln \sigma_g} \exp\left[-\frac{(\ln D - \ln D_{\text{nmd}})^2}{2 \ln^2 \sigma_g}\right] d(\ln D) \quad (6.5)$$

When the following substitutions are made:

$$\begin{aligned} x &= \ln D \\ \exp mx &= D^m \\ a &= \ln \sigma_g \\ b &= \ln D_{\text{nmd}} \end{aligned}$$

then

$$\overline{D^m} = \frac{1}{(2\pi)^{1/2} a} \exp\left(bm + \frac{a^2m^2}{2}\right) \int_{-\infty}^{+\infty} \exp\left\{-\frac{[x-(b+a^2m)]^2}{2a^2}\right\} dx \quad (6.6)$$

The integral portion of equation 6.6 reduces to $(2\pi)^{1/2}(a)$, so that

$$\overline{D^m} = \exp\left(bm + \frac{a^2m^2}{2}\right) \quad (6.7a)$$

and

$$\overline{D^m} = D^m{}_{\text{nmd}} \exp\left(\frac{m^2}{2} \ln^2 \sigma_g\right) \quad (6.7b)$$

Upon converting equation 6.7b to common logarithms for computational purposes, one obtains

$$\log\left(\overline{D^m}\right) = m \log D_{\text{nmd}} + 2.303 \left(\frac{m^2}{2} \log^2 \sigma_g\right) \quad (6.8)$$

When $m = 2$, the surface average particle diameter becomes

$$D_s = (\overline{D^2})^{1/2}$$

and

$$\log D_s = \log D_{\text{nmd}} + 2.303 \log^2 \sigma_g \quad (6.9)$$

Similarly, when $m = 3$, the volume average particle diameter becomes

$$D_v = \left(\overline{D^3}\right)^{1/3}$$

and

$$\log D_v = \log D_{\text{nmd}} + 3.45 \log^2 \sigma_g \tag{6.10}$$

Equations 6.9 and 6.10 enable one to compute surface, D_s, and volume, D_v, diameters from size-count parameters D_{nmd} and σ_g which may be obtained from a cumulative distribution of size frequency by count or the reverse.

The cumulative distributions for particle surface or mass are derived from the following general relationships using equation 6.5:

$$F(D^m) = \frac{1}{(\overline{D_m})} \int_{-\infty}^{+\infty} D^m f(D)\, dD$$

$$= \frac{1}{(\overline{D_m})} \int_{-\infty}^{+\infty} \frac{1}{(2\pi)^{1/2} a} \exp\left[mx - \frac{(x-b)^2}{2a^2}\right] dx \tag{6.11a}$$

$$F(D^m) = \int_{-\infty}^{\ln D_{\text{med}}} \frac{1}{(2\pi)^{1/2} a} \exp\left\{-\frac{[x-(b+a^2m)]^2}{2a^2}\right\} dx = \frac{1}{2} \tag{6.11b}$$

where D_{med} = any median diameter.

The cumulative distribution for particle number (i.e., $m = 1$) may be expressed as

$$F(D) = \int_{-\infty}^{\ln D_{\text{nmd}}} \frac{1}{(2\pi)^{1/2} a} \exp\left[-\frac{(x-b)^2}{2a^2}\right] dx = \frac{1}{2} \tag{6.12}$$

In order that equations 6.11 and 6.12 may remain identities, the terms $(x-b+a^2m)$ and $(x-b)$ must equal zero when $\ln D_{\text{med}}$ and $\ln D_{\text{nmd}}$ are substituted for x. Accordingly

$$D_{\text{med}} = D_{\text{nmd}} \exp\left(m \ln^2 \sigma_g\right) \tag{6.13a}$$

$$\log D_{\text{med}} = \log D_{\text{nmd}} + 2.303\, m \log^2 \sigma_g \tag{6.13b}$$

Defining surface and mass median diameters as D_{smd} and D_{mmd}, when $m = 2$ and $m = 3$, respectively, one obtains

$$\log D_{\text{smd}} = \log D_{\text{nmd}} + 4.60 \log^2 \sigma_g \tag{6.14}$$

and

$$\log D_{\text{mmd}} = \log D_{\text{nmd}} + 6.90 \log^2 \sigma_g \tag{6.15}$$

D_{nmd} and D_{mmd} are commonly designated as M_g and M_g', respectively, and the term Δ is frequently used to represent D_{smd}.

TABLE 6.1
MATHEMATICAL DEFINITIONS OF THE AVERAGE DIAMETERS OF PARTICULATE SUBSTANCES[a]

Average diameter (1)	Symbol (2)	Mathematical definition (3)	Equivalent logarithmic value in terms of statistical parameters of distributions curves[b]	
			By count (M_g and σ_g) (4)	By weight (screen analysis) (M_g' and σ_g') (5)
Geometric mean	M_g	$\text{antilog}\left(\frac{\Sigma n \log d}{\Sigma n}\right)$	$\log M_g$	$\log M_g' - 6.9078 \log^2 \sigma_g'$
Arithmetic mean	σ	$\left(\frac{\Sigma nd}{\Sigma n}\right)$	$\log M_g + 1.1513 \log^2 \sigma_g$	$\log M_g' - 5.7565 \log^2 \sigma_g'$
Specific surface	d_s	$\left(\frac{\Sigma nd^{-1}}{\Sigma n}\right)^{-1}$	$\log M_g - 1.1513 \log^2 \sigma_g$	$\log M_g' - 8.0591 \log^2 \sigma_g'$
Surface area	Δ	$\left(\frac{\Sigma nd^2}{\Sigma n}\right)^{1/2}$	$\log M_g + 2.3026 \log^2 \sigma_g$	$\log M_g' - 4.6052 \log^2 \sigma_g'$
Volume	D_v	$\left(\frac{\Sigma nd^3}{\Sigma n}\right)^{1/3}$	$\log M_g + 3.4539 \log^2 \sigma_g$	$\log M_g' - 3.4539 \log^2 \sigma_g'$
Surface area per unit volume[c]	D_v^3/Δ^2	$\left(\frac{\Sigma nd^3}{\Sigma nd^2}\right)$	$\log M_g + 5.7565 \log^2 \sigma_g$	$\log M_g' - 1.1513 \log^2 \sigma_g'$

[a]From Drinker and Hatch, Ref. 10.

[b]Logarithms to the base 10.

[c]This diameter gives the specific surface for the sample as a whole; it should not be confused with d_s, the diameter of the hypothetical particle having *average* specific surface.

Table 6.1 summarizes mathematical definitions of the average diameters of nonuniform particulate substances in terms of the parameters of the size distribution curves by count and by weight. The derivations of several of the equations in this table have been presented above.

Typical examples of the construction of size frequency histograms and cumulative distributions for normal and log-normal representations of particle size analysis data are discussed by several authors.[1,2,4,11,12]

6-4 Interpretation of Particle Size by Count Data

In this section, a number of methods of data evaluation are discussed. In most cases, a logarithmic-probability distribution has been used for illustrative purposes, as it has wide applicability to the work of environmental health scientists, but the methods that are reviewed here may be applied to other size distributions as well.

The cumulative size distribution curves shown on the logarithmic-probability graph in Fig. 6.6 were plotted from the microscopic sizing data[13] shown in Table 6.2. An electron microscope count (column 6) reveals that a large number of particles are below the limit of resolution of the light microscope (column 4).

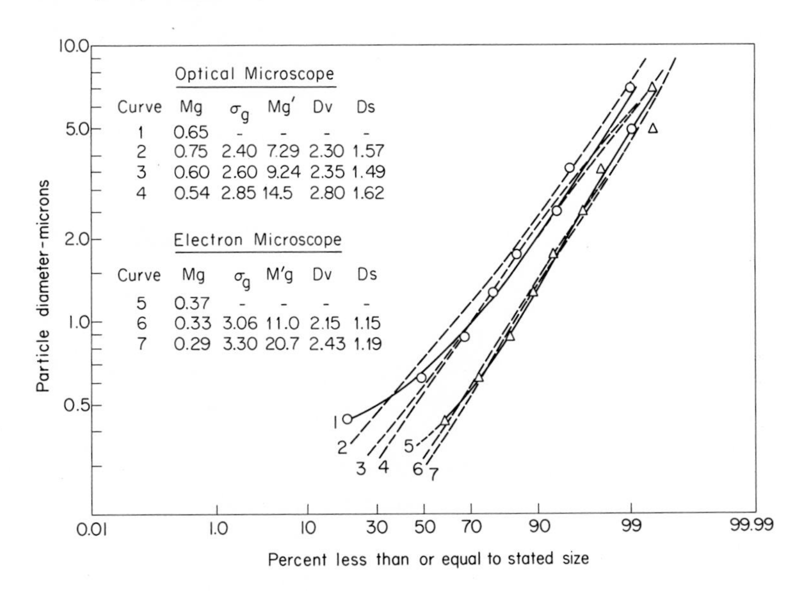

FIG. 6.6. Particle size distributions from data in Table 6-2.

TABLE 6.2
TABULATED SIZE DATA FROM OPTICAL AND ELECTRON MICROSCOPE MEASUREMENTS OF COLLECTED AIRBORNE DUST

		Optical microscope		Electron microscope	
(1)	(2)	(3)	(4)	(5)	(6)
Diameter μm	(2) Number of particles in interval	Cumulative number equal to or less than max. size in Col. 1	Cumulative percent equal to or less than max. size in Col. 1	Cumulative number equal to or less than max. size in Col. 1	Cumulative percent equal to or less than max. size in Col. 1
≤ 0.44	57	57	19	119	59.5
0.45–0.62	87	144	48	148	74.0
0.63–0.88	54	198	66	166	83.0
0.89–1.25	36	234	78	178	89.0
1.26–1.76	21	255	85	185	92.5
1.77–2.50	24	279	93	193	96.5
2.51–3.53	6	285	95	195	97.5
3.54–5.00	12	297	99	199	99.5
5.01–7.07	0	297	99	199	99.5
7.08–10.0	3	300	100	200	100

Curve 1, Fig. 6.6, is an estimated line of best fit for the optical microscope sizing data. It indicates a count median diameter M_g of 0.65 μm. Since curve 1 is not linear, the equations given in Table 6.1 (which presuppose straight lines on logarithmic-probability plotting paper) cannot be used for calculating size parameters—i.e., σ_g, the geometric standard deviation; M_g', the mass median diameter; D_v, the mean volume diameter; and D_s, the mean surface diameter. Straight lines 2, 3, and 4 in Fig. 6.6 have been drawn through the data points to represent approximate normalized (i.e., straight-line) curves from which size parameters may be calculated by exact mathematical functions. The characteristic values derived from each, tabulated in Fig. 6.6, show considerable variation. For example, calculated values of D_v range from 2.3 to 2.8 μm. If these figures were used to compute the total number of particles in a known weight of dust, the highest would exceed the lowest by 180%.

The diffculty associated with calculating size parameters from distributions that do not exactly follow the logarithmic-probability distribution was emphasized during an interlaboratory comparison of methods used by different investigators to estimate the parameters of a uranium oxide particle size distribution.[14] Mercer[15] has interpreted the results of these interlaboratory analyses to mean that all conversions of count frequency distributions to mass or surface area distributions must be accepted with reservations.

Great discrepancies occur between the three straight lines of estimated best fit, numbers 2, 3, and 4 in Fig. 6.6, when extrapolations are made well beyond the observed points. For example, curve 4 indicates that 99.53% of the total particles are less than 10 μm in diameter and 99.99% less than 35 μm in diameter. A mass distribution derived from curve 4 with the formulas summarized in Table 6.1 shows that 50% of the weight of all particles present is represented by diameters greater than 14.5 μm M_g', 63% by diameters greater than 10 μm, and 22% by diameters greater than 35 μm. However, the actual airborne dust contained particles no greater than 10 μm in diameter. Therefore, the cumulative size curve (number 1 of Fig. 6.6) should bend toward the horizontal in the 98% region and become asymptotic to the 10-μm line beyond the 99.99% point. Since the mass median diameter must, under these circumstances, be no greater than 10 μm (the size of the largest particles present), the size-by-count parameters derived only from the straight central portion of the curve must not be used to calculate the size-by-mass or size-by-surface area distributions.

Extreme care must be exercised at all times when deriving size-by-weight data from a size-by-count curve to be certain that statistically significant numbers of size measurements have been made in the upper size range which makes up a substantial fraction of the weight in the calculated size-by-weight distribution. Statistically significant numbers are shown in Table 6.3. Use of the equations in Table 6.1 requires careful adherence to the guidelines contained in Table 6.3. Therefore, size analyses by microscopic methods are not favored by dust collector manufacturers, whose performance guarantees

TABLE 6.3
NUMBER OF PARTICLES TO BE COUNTED TO ACHIEVE ANY GIVEN ACCURACY

Weight percentage of particles in any size range:	2	5	10	15	20
Expected accuracy %	Number of particles to be counted				
2	(3)[a]	(8)	25	56	100
1	(6)	25	100	225	400
0.5	16	100	400	900	1600
0.2	100	625	2500	5600	10000
0.1	400	2500	10000	22500	40000

[a]The values in parentheses have been increased above those calculated from the formula given by Fairs[19] so as to maintain the same overall standard of accuracy as in the remainder of the table.

are given as weight efficiency for specific particle size ranges. Instead, a number of the leading collector manufacturers in the United States have agreed to use the Bahco centrifugal air elutriation instrument, which gives size fraction data directly in percent by weight.

Curve 1, Fig. 6.6, shows marked curvature at the low range. This deviation from a normalized linear form is characteristic of many size distributions and may be attributable to either of two causes: the dust may not be distributed according to the log-normal relation, or the lower limit of size resolution attainable with the selected sizing method may fail to detect a large fraction of the particles that are present, assuming that one satisfies the criteria for statistical reliability and numbers of particles in a given size interval as set out in Table 6.3.

When the dust sample represented by curve 1, Fig. 6.6, was sized by electron microscopy (particle resolution down to 0.005 μm), additional particles in the size range below 0.44 μm were found. The cumulative number of particles in each size range and cumulative percent by electron microscope count are shown in columns 5 and 6 of Table 6.2. Curve 5 of Fig. 6.6 shows two important changes from curve 1 derived from optical microscopic sizing: M_g is lower, and the experimental points more nearly approach a straight line. Curves 6 and 7 represent normalized straight lines for estimating the size parameters tabulated on Fig. 6.6. A method that reveals smaller particles results in an increase in both the total number of particles and the total surface area for the same weight of dust. From this example (i.e., comparison of curves 1 and 5, Fig. 6.6) it may be concluded that underestimation of particle numbers in the low end of the size-by-count curve shifts the size distribution curves to the left on the log-probability plot, and this automatically increases the apparent fraction of particles in the larger size ranges. In the above case, an underestimate of small particle numbers would result in an unsafe approach from the hygienic viewpoint.

6-4.1 Stratified Sampling

It is reasonable to expect that sizing larger numbers of particles in each size group will increase the statistical reliability of the analysis. However, experience indicates that most of the particles will be concentrated in a relatively few size intervals. The data shown in Table 6.2 are typical, since only 5% of all the particles counted were observed in the three largest size groupings (i.e., 3.5 to 10 μm). The numbers of observations in this size region are statistically inadequate. To find approximately ten particles in the two largest size ranges shown in Table 6.2 would require a total count in excess of 1000 particles, greatly increasing the time required to make a single size analysis.

To improve data reliability and yet minimize the number of measurements, a statistically unbiased procedure may be used to provide increased accuracy in the higher percentiles (where numbers are small) with relatively few measurements. The statistical technique that is used is called stratified sampling.[1] When applied to particle sizing it has been called "truncated multiple traversing" by Sichel.[16] Whitby[17] has described a similar procedure.

Sichel's basic criterion is that at least ten particles must be observed in every size range which has a significant influence on the size curve. The system is best explained by referring to the data tabulated in Table 6.4. In the horizontal row designated traverse 1, the data from Table 6.2, column 2, have been entered. Based on the criterion of ten or more particles per size category, sufficient measurements are shown for sizes through 2.5 μm. Therefore, a second traverse is made which corresponds to the method used for the first 300 measurements, with one very important exception. Another area of the same size as the first is searched, but only those particles greater than 2.5 μm are measured. As a result, only 21 particles are measured in the second traverse. Note that a total of 18 particles now appears in the size ranges 2.50 to 3.53 μm and 3.53 to 5.00 μm. During a third traverse, which covers a field of equal area, sizes greater than 5.00 μm are enumerated. After this process has been repeated three more times, the total number of particles counted in each size category will be as shown in the horizontal row at the bottom of Table 6.4 labeled Total. In the next line below, each value has been reduced to the total number of particles per single traverse by dividing each item in the total count by the number of traverses required to achieve the total. In the last line, the cumulative percentages equal to or less than each size group are shown. In this fashion, a more correct relationship results from larger numbers in the upper end of the size spectrum; yet only 33 particles have been measured in the last five of the six searches that make up the total enumeration. If size-selective or stratified sampling had not been employed, it would have been necessary to size a total of $300 \times 6 = 1800$ particles while making six complete size enumerations of all particles in each field.

According to a tentative recommended practice of the American Society for Testing and Materials,[18] even for a distribution containing a wide range of particle sizes, the number of particles measured in the modal class should be not less than 25, and in an average distribution the modal class should contain at least 100 particles.

Fairs[19] states: "The errors which may occur in making a size analysis fall into two classes. The first group includes all errors of observation and

TABLE 6.4
DATA TABULATION FOR PARTICLE SIZING BY TRUNCATED MULTIPLE TRAVERSING SYSTEM[a]

Traverse number	Size range μm										Total number
	≤ 0.44	0.45–0.62	0.63–0.88	0.89–1.25	1.26–1.76	1.77–2.50	2.51–3.53	3.54–5.00	5.01–7.07	7.08–10.0	
1	57	87	54	36	21	24	6	12	0	3	300
2							12	6	3	0	21
3									3	0	3
4									2	1	3
5									2	1	3
6									1	2	3
Total	57	87	54	36	21	24	18	18	11	7	333
Number per traverse	57	87	54	36	21	24	9	9	1.8	1.2	300
Cumulative percent	19	48	66	78	84	92	95	98	99.4	100	

[a]From Sichel, Ref. 16.

manipulation which are personal to the observer, and which can, with experience, be reduced to negligible proportions for the size of count normally carried out. The second group is statistical and entirely independent of the observer; it comprises the inevitable counting error and the accuracy factor." The counting error (referring solely to the reproducibility of the count) can be estimated and controlled by use of statistical relationships between percent of particles in any size range, expected accuracy, and number of particles to be counted, as shown in Table 6.3, taken from Fairs.[19] In order that the analysis may reflect the same degree of accuracy over the entire size range (i.e., the accuracy factor), it is necessary that the inevitable counting error remain approximately constant for each size range.

6-4.2 Confidence Limits

Having constructed a particle size distribution curve on log-probability coordinates, it is necessary to know how closely the true size characteristics of the dust have been approximated. It may be expected that the distribution for the number of particles counted in any stated size category will follow a binomial distribution with repeated samplings and approach a normal probability distribution.[1] This is in direct contrast to the geometric relation which exists among the different size groups. A commonly used criterion is the 95% confidence limit; i.e., the true population mean value can be expected to lie within certain calculated upper and lower limits (m_l and m_u) centered on each single estimate of the mean 95% of the time on successive analyses of the same slide. Or in repeated samples, 95% of such intervals will contain the true population mean.[1] For a 95% confidence limit, there is one chance in twenty that the true mean lies outside the range of values bounded by the confidence limits.

When the size distribution curve can be closely approximated by a straight line on log-probability paper, statistical methods may be used to determine the reliability of the mean or any other percentile value in terms of established confidence limits. The size data in columns 1 and 2 in Table 6.2 have been replotted in Fig. 6.7 as a fractional distribution curve, with the ordinate as the fraction of particles in each size group and the abscissa as the logarithm of particle diameter, so that the size-frequency curve appears as a segment of a symmetrical Gaussian (normal probability) curve. (This plotting method is valid only when the logarithmic size interval between size limits is constant, as shown.)

It makes no difference whether the logarithm of particle diameter is positive or negative, since numerical values are added algebraically. The logarithm of $M_g = 0.60$ (equal to 0.222) and the logarithm of $\sigma_g = 2.60$

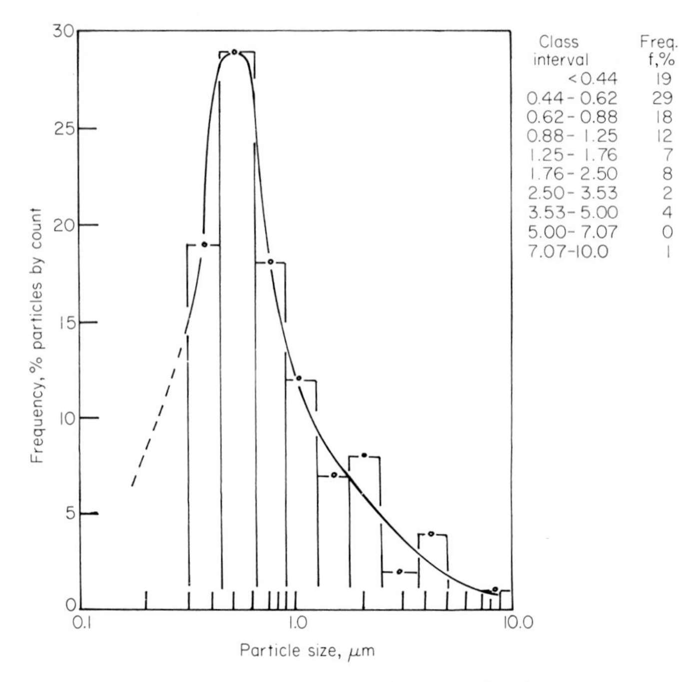

FIG. 6.7. Fractional size curve for data of Table 6.2.

(equal to 0.415) may be treated as the mean (x) and the standard deviation (σ) of a normal probability distribution, and the standard deviation of the mean ($\sigma_{\bar{x}}$) can now be calculated from the following relation:[1]

$$\sigma_{\bar{x}} = \sigma(N)^{-1/2} \tag{6.16}$$

where N represents the total number of particle measurements. Applying equation 6.16 to the total number of particles examined in Table 6.4 without regard to selective sizing, one obtains $\sigma_{\bar{x}} = 0.415/333 = 0.0228$. Since 95% confidence limits have been chosen, the range of expected mean values for x is represented by $x \pm 2\sigma_{\bar{x}}$, or -0.1764 to -0.7676. When these logarithms are converted to their corresponding particle diameters, the 95% confidence limits for (M_g) are found to range from 0.54 to 0.67 μm.

The confidence limits of the standard deviation for a normal frequency distribution σ_σ may be determined from the relation

$$\sigma_\sigma = \sigma(2N)^{-1/2} \tag{6.17}$$

For 333 particle measurements, it is

$$0.415(666)^{-1/2}$$

The 95% confidence limits for σ would then be $\sigma \pm 2\sigma_\sigma = 0.415 \pm 0.032$, or from 0.383 to 0.447, and it is now possible to calculate the size range to be expected for any percentile based on 333 nonselective particle measurements by expressing the logarithm of the cumulative size equal to or less than the fraction under investigation in terms of standard deviation, for which the 95% confidence limits are 2.4 to 2.8.

6-4.3 Graphical Methods for Nonlinear Size Distributions

It has been shown in Section 6-4 that particle size parameters calculated from relationships based on the equations shown in Table 6.1 (which presuppose a straight line on log-probability plotting paper) will be inaccurate when a line connecting points of measured observations shows significant deviations from linearity. Histograms and graphical integrations have proved to be reliable methods for analyzing size data which are not amenable to curve fitting.[20] As noted in Section 6-4, it is especially important to establish with certainty the path of the high end of the size-by-count curve before transformations from count to weight or volume distributions are made. The largest particles present can be estimated rapidly and reliably by sieving or by low-power microscopic examination. From this, the upper limiting size to be used in graphical integrations is established, and errors such as calculating a mass median diameter that is larger than the biggest particle present can be avoided. This occurs because particle surface and volume are proportional to the square and cube of the diameter, and small discrepancies in the estimation of the diameter become greatly magnified. Drinker and Hatch[10] made this clear in the following manner:

"When $\sigma_g = 3$, $M_g' = 40 \times M_g$ and 95 percent of the weight in the dust is contributed by no more than 5 percent of the particles in the largest sizes. Thus, if $M_g = 1$ μm, 95% of the particles by weight are larger than 6 μm and 95% by number are smaller than this size."

When data are plotted on logarithmic-probability paper, the 99.99% point may be accepted as the limiting size. Experience indicates that neglecting the few particles greater than this limiting size introduces no serious errors into the calculations of mass or surface area distributions.

A simple semigraphical procedure for analyzing size distributions that do not give a straight line on log-probability plotting paper, described below, utilizes the best known and most frequently used equations for size analysis combined with a graphical means for establishing straight lines of best fit for curved upper and lower portions of a size-by-count curve.[21] The method is illustrated by the following example, taken from data previously presented.

Other illustrations are contained in reference 21. In Fig. 6.8, two solid straight lines, labeled count (lower) and count (upper), were drawn to represent a curve of best fit for the sizing data shown in Table 6.4. The slope of the lower count branch follows the experimental data quite closely. The slope of the upper count branch was selected on the assumption that the approximate intercept size for the 99.99 percentile should be 9 μm, as no particles greater than 10 μm in diameter were found. Size parameters were assumed from linear extrapolations of each branch, and each branch was analyzed as an independent size distribution. This assumption does not mean that the actual observed distribution would have resulted from a mixture of any two such independent distributions. The assumption is for analytical purposes.

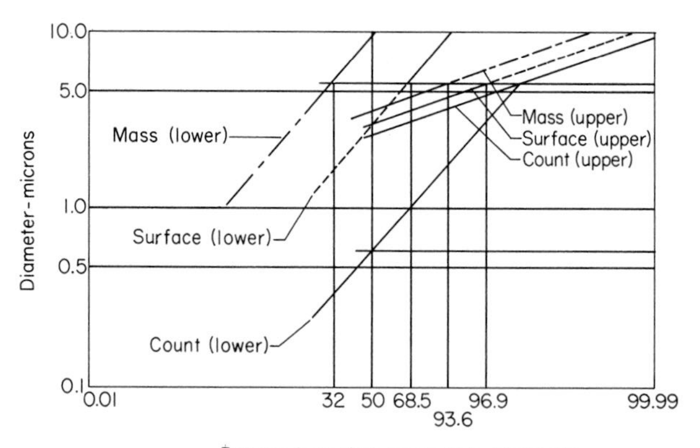

FIG. 6.8. Graphical estimation of average surface and volume diameters (D_s and D_v) when two straight lines are used to represent a nonlinear size-by-count curve on logarithmic-probability paper.

Size Parameters Determined by Hatch-Choate Equations

Lower branch	Upper branch
$M_g = 0.60$ μm	$M_g = 2.52$ μm
$\sigma_g = 2.6$	$\sigma_g = 1.40$
$M_g' = 9.24$ μm	$M_g' = 3.53$ μm
$D_s = 1.49$ μm	$D_s = 2.82$ μm
$D_v = 2.35$ μm	$D_v = 2.98$ μm
$D_{smd} = 3.72$ μm	$D_{smd} = 3.16$ μm
$\Sigma V_l = (D_v)^3 = 13.0$	$\Sigma V_u = (D_v)^3 = 26.5$
$\Sigma S_l = (D_s)^2 = 2.22$	$\Sigma S_u = (D_s)^2 = 7.95$
$\Sigma V_l < 5.9$ μm $= 0.32(13.0) = 4.16$	$\Sigma V_u > 5.9$ μm $= 0.064(26.5) = 1.70$
$\Sigma S_l < 5.9$ μm $= 0.685(2.22) = 1.52$	$\Sigma S_u > 5.9$ μm $= 0.031(7.95) = 0.246$
$\Sigma V_{l+u} = 4.16 + 1.70 = 5.86$	$(D_v)_{l+u} = (5.86)^{1/3} = 1.80$ μm
$\Sigma S_{l+u} = 1.52 + 0.25 = 1.77$	$(D_s)_{l+u} = (1.77)^{1/2} = 1.33$ μm

Calculation of the characteristic parameters (M_g', D_s, D_v, and D_{smd}) was facilitated by use of a special graph (Fig. 6.9) in which the ratio of the appropriate mean or median diameter to the count median diameter is plotted against the geometric standard deviation. These curves were derived from the equations shown in Table 6.1. Numerical values for the upper and lower branches of the size curve shown in Fig. 6.8 are tabulated below the figure.

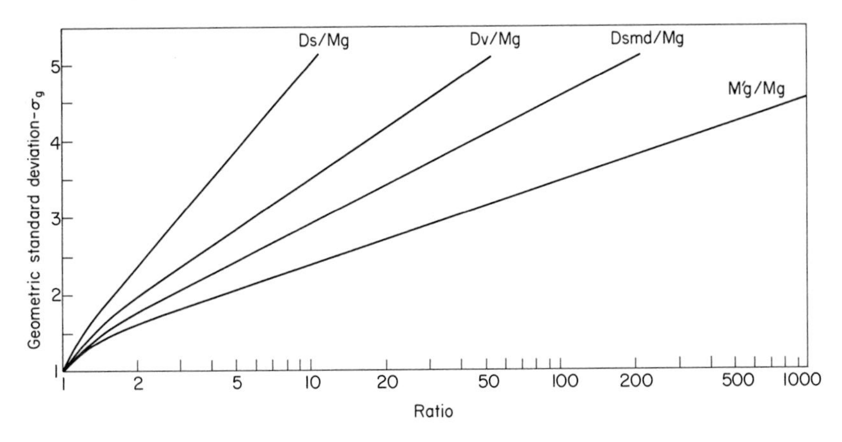

FIG. 6.9. Graphical derivation of characteristic particle diameters (D_s, D_v, D_{smd} and M_g') from count median diameter (M_g) and geometric standard deviation.

The relative mass (or volume) represented by the size distribution depicted by either branch of the count curve is defined by D_v^3 when the product of total number of particles and the shape factor is assigned an arbitrary value of unity. Since mass is directly proportional to volume when specific gravity is constant, cumulative mass and volume distributions are identical. Similarly, the relative surface associated with either branch of the count curve is defined by D_s^2 when the product of total number of particles and shape factor is unity. Numerical values of total relative surface and volume—i.e., $\sum S$ and $\sum V$—are tabulated in Fig. 6.8 with subscripts l and u, designating lower and upper branches, respectively.

Fig. 6.8 shows that the intersection of lower and upper count curves corresponds to a particle diameter of 5.9 μm. If we now construct the surface area and mass (volume) distribution curves corresponding to the lower count curve (by locating the respective values of D_{smd} and M_g' on the 50% line and drawing a straight line parallel to the lower branch), the intersection of the 5.9-μm line with each curve shows the fraction of total surface (i.e., 68.5%) and volume (i.e., 32%) represented by particles smaller than 5.9 μm. The calculated total relative surface (ΣS_l) and volume (ΣV_l) are 1.52 and 4.16.

Treating the upper count branch in a similar fashion, size-by-surface and size-by-volume curves were added to Fig. 6.8, again observing the intersection at the 5.9-μm line. The fractions of total surface and volume related to sizes greater than 5.9 μm are 31% and 64%, respectively. Calculated values for the total relative surface (ΣS_u) and volume (ΣV_u), represented by diameters greater than 5.9 μm, are 0.246 and 1.70.

Since the fraction of particles possessing total surface and volume less than and greater than 5.9 μm must, by definition, equal 100%, the summation of relative surfaces and volumes associated with each fraction is the total relative surface or volume depicted by a size distribution described by two straight lines.

$$\Sigma S_{l+u} = \Sigma(S_l < 5.9\ \mu\text{m}) + \Sigma(S_u > 5.9\ \mu\text{m}) = 1.77$$

$$\Sigma V_{l+u} = \Sigma(V_l < 5.9\ \mu\text{m}) + \Sigma(V_u > 5.9\ \mu\text{m}) = 5.86$$

The mean diameter with respect to surface, D_s, then is simply $(\Sigma S_{l+u})^{1/2}$, or 1.33 μm. Similarly, the mean diameter with respect to volume (or mass), M_g', is $(\Sigma V_{l+u})^{1/3}$, or 1.80 μm.

6-5 Interpretation of Particle Sizing by Weight Data

Many sizing instruments and analytical techniques provide graded fractions of particles that are analyzed by weighing. Sieve analyses, elutriation methods (e.g., Roller and Micro-particle Classifier), settling methods (e.g., Andreasen pipet and Cummings technique), and cascade impactor analyses are among the methods that yield data that may be plotted on log-probability paper directly in size-by-weight units. If size-by-count information is desired from analyses determined on a weight basis, the relationships contained in Table 6.1 and Fig. 6.9 may be used in the manner demonstrated in the preceding section. In this case, the statistical validity of the small end of the size-by-weight distribution must receive the same careful consideration as that required by the large end of the size-by-count distribution when converting to a weight relationship. All the mathematical and graphical methods noted in the previous section are recommended for the analysis of size-by-weight data that do not plot as a straight line.

Cumulative distributions by weight are obtainable directly by sedimentation methods of analysis such as the Oden sedimentation balance and the Micromerograph, which record the cumulative weight of particles which settle, with time, from a tranquil settling chamber onto a sensitive balance pan. When the size of the settling chamber is fixed and the true specific gravity of the particles is known, settling times may be converted into characteristic diameters by reference to Stokes' law and the weight-time observations plotted as a cumulative size-by-weight distribution curve. The mathematical relationships for cases of tranquil and stirred settling have been derived by Ranz.[11]

Appendix

The following is a partial list of size distribution functions which make use of two parameters to characterize the range and spread of particle sizes and which have been found useful in describing the particle size of highly disperse systems.

1. Logarithmic probability distribution function [T. Hatch and S. P. Choate, *J. Franklin Inst.*, *207*:369 (1929)].

$$f(D)\,dD = \frac{1}{(2\pi)^{1/2}\ln\sigma_g}\exp\left[-\frac{(\ln D-\ln\bar{D})^2}{2\ln^2\sigma_g}\right]d(\ln D)$$

where $\ln\bar{D} = \overline{\ln D}$ = log-geometric mean diameter.

$$\ln\sigma_g = \left[\overline{(\ln D-\ln D)^2}\right]^{1/2} = \text{log-geometric mean deviation}$$

2. Rosin-Rammler distribution function [P. Rosin and E. Rammler, *J. Inst. Fuel*, *7*:29 (1933)].

$$f(D) = abD^{b-1}\exp(-aD^b)$$

where $a = f(D,b) = \frac{1}{\bar{D}^b}\left(\frac{1}{b}!\right)^b.$

D = number average particle diameter.
b = constant.

3. Roller's distribution function [P. W. Roller, *J. Franklin Inst.*, *223*:609 (1937)].

$$f(D) = a\left(\frac{1}{2D^{7/2}/2}+\frac{b}{D^{9/2}/2}\right)\exp(-b/D)$$

where a and b are arbitrary constants, and b is directly proportional to the diameter where $f(D)$ is a maximum.

4. Nukiyama–Tanasawa distribution function [S. Nukiyama and Y. Tanasawa, *Trans. Soc. Mech. Engrs. (Japan), 5*:63 (1939)].

$$f(D) = aD^2 \exp(-bD^q)$$

where $a = f(D_{vs},q)$.

$b = f(D_{vs},q)$.

D_{vs} = diameter of a particle with the same ratio of volume to surface as that of the total sum of all the particles.

q = constant.

[For a more general discussion of size distribution functions, see C. E. Lapple, *Heating, Piping, Air Conditioning, 18*:108 (1946); and J. Dalle Valle, *Micromeritics,* 2nd ed., Pittman, New York, 1948.]

Since methods of making experimental size determination are generally inaccurate, and since it is difficult to obtain a check by two different methods, a discussion of the relative correctness of any one of the functional forms for $f(D)$ listed above is of little profit. The form to use is the one which appears to fit the known data reasonably well. This can be established from experimental values without too much trouble, and can be used to determine other information desired without excessive mathematical computations.

References

1. P. G. Hoel, *Introduction to Mathematical Statistics,* 3rd ed., John Wiley & Sons, Inc., New York, 1962.
2. G. Herdan, *Small Particle Statistics,* 2nd ed., Butterworths, London, 1960.
3. P. Drinker, The Size Frequency and Identification of Certain Phagocytosed Dusts, *J. Ind. Hyg., 7*:305 (1925).
4. J. E. Smith and M. L. Jordan, Mathematical and Graphical Interpretation of the Log-Normal Law for Particle Size Distribution Analysis, *J. Colloid Sci., 19*:549 (1964).
5. P. S. Roller, Law of Size Distribution and Statistical Description of Particulate Materials, *J. Franklin Inst., 223*:609 (1937).
6. P. Rosin and E. Rammler, The Laws Governing the Fineness of Powdered Coal, *J. Inst. Fuel, 7*:29 (1933).
7. T. Hatch and S. P. Choate, Statistical Description of the Size Properties of Non-Uniform Particulate Substances, *J. Franklin Inst., 207*:369 (1929).
8. F. J. Kottler, The Distribution of Particle Sizes, *J. Franklin Inst., 250*:339 (1950).
9. F. J. Kottler, The Goodness of Fit and the Distribution of Particle Size, *J. Franklin Inst., 251*:499 (1951).
10. P. Drinker and T. Hatch, *Industrial Dust,* 2nd ed., McGraw-Hill Book Company, Inc., New York, 1954.

11. W. E. Ranz, *I, Calculations of the Size Distribution of Aerosol Particles from Tranquil Settling Data; II, Calculation of the Mass Median Diameter of an Aerosol for Stirred Settling Data,* USAEC Report SO-1000, Technical Report No. 1, Engineering Experiment Station, University of Illinois, August 1950.
12. J. M. DalleValle, *Micromeritics,* 2nd ed., Pitman Publishing Corporation, New York, 1948.
13. R. Dennis, *Notes on the Analyses of Particle Size Distributions,* Harvard School of Public Health, Boston, Mass., May 1963.
14. H. J. Ettinger and S. Posner, Evaluation of Particle Sizing and Aerosol Sampling Techniques, *Am. Ind. Hyg. Assoc. J., 26*:17 (1965).
15. T. T. Mercer, Aerosol Production and Characterization: Some Considerations for Improving Correlation of Field and Laboratory Derived Data, *Health Phys., 10*:873 (1964).
16. H. S. Sichel, On the Size Distribution of Airborne Mine Dust, *J. S. African Inst. Mining Met., 58*:171 (1957).
17. K. T. Whitby, *Determination of Particle Size Distribution—Apparatus and Techniques for Flour Mill Dust,* Bulletin No. 38, Engineering Experiment Station, University of Minnesota, January 1950.
18. *Tentative Recommended Practice for Analysis by Microscopical Methods,* Number E20-58T, 1958 Book of ASTM Standards, Part 8, American Society for Testing and Materials, Philadelphia, Pa., 1958.
19. G. L. Fairs, Developments in the Technique of Particle Size Analysis by Microscopic Examinations, *J. Roy. Microscop. Soc., 71*:209 (1951).
20. M. Corn, Statistical Reliability of Particle Size Distributions Determined by Microscopic Techniques, *Am. Ind. Hyg. Assoc. J., 26*:8 (1965).
21. R. Dennis, G. A. Johnston, M. W. First, and L. Silverman, *Performance of Commercial Dust Collectors — Report of Field Tests,* USAEC Report NYO-1488, Harvard University, November, 1953.

7

Applications of Particle Size Analysis

7-1 Introduction

Suspended particles are encountered in association with many processes and locations. These include process vessels and vent lines; exhaust ducts and stacks following control equipment; the interior of plants and laboratory workrooms; stopes and drifts of mines; and the ambient air after effluent aerosols have undergone dispersal in the atmosphere.

Particle size is of special concern in industrial hygiene and toxicology laboratories for studying the physical and physiological behavior of fine particles; in chemical and biological laboratories using or creating solid and liquid particle dispersions or radioactively tagged microbial aerosols; in "ultra-clean" or "white room" environments during manufacture or assembly of precision instruments or components in a dust-free atmosphere; in evaluating the behavior of particulate fallout from nuclear weapons tests; in studies of atmospheric precipitation processes, weather modification, and meteorological dispersion and circulation; and in the study of stratospheric particle transport. A description of particle size is important in the study of effects from the accidental release of materials from nuclear reactors involving fuel element failure, fire, or explosion. These studies are generally performed with analytical models or as a part of simulated accident experiments. Other areas of interest involving particle size include the behavior of exhaust products from certain solid-fueled rocket designs; burn-up on re-entry of orbiting nuclear reactors; fate of disseminated toxic chemicals such as pesticides, lead, and so on.

Particle sizing in the environmental health sciences can be divided into broad areas of technical interest, as shown in Table 7.1, where certain of the important criteria in each category are noted.

The following sections contain a general discussion of typical applications of particle size analysis to industrial hygiene, radiation protection, air pollution control, industrial toxicology, and related areas. These studies are cited to illustrate the application of principles and practices described in previous chapters. It is not possible to provide a detailed review of all techniques used in these studies, but sufficient references have been included to assist the investigator with a specific problem in application.

7-2 Applications of Particle Sizing to Industrial Hygiene and Radiation Protection

7-2.1 Deposition of Inhaled Particles

The principal route of entry into the body of airborne particulate matter is by inhalation during normal respiration. Many of the following discussions describe applications of particle sizing techniques used to determine the quantity of respirable particles in a sample. The sizes of inhaled particles which can affect health are established by the aerodynamic behavior of the particles in relation to the characteristics of the respiratory system and its geometry. Descriptions of the pulmonary system and the significant factors in deposition of respirable particles have been presented by Hatch and Gross.[1]

The predominant sizes of insoluble inhaled particles which affect health are generally less than 10 μm Stokes' or aerodynamic diameter (i.e., related to unit density spheres). The health hazard of an environment can be evaluated properly only after noting the relative numbers of small and large

TABLE 7.1

APPLICATIONS OF PARTICLE SIZE ANALYSES FOR ENVIRONMENTAL SCIENCE AND ENGINEERING STUDIES

Application or location	Types of sample collector most useful	Notes on air pump and flowmeter	Special considerations and remarks
	A. Industrial and Radiological Hygiene		
Sampling from air at the work place	Membrane filter, also can precede with size selective collector and examine respirable fraction separately.	Any suitable flowmeter and pump. Can use personal air sampling unit worn by the worker.	Collect samples at times and in places which adequately represent the conditions to be studied, breathing zone usually, while operation takes place. Include background samples.
Sampling in stacks and ducts	Use in-stack collector whenever possible.	Make pressure and temperature corrections to indicated volume.	Similar to discussion below under air pollution. Concentrations may be considerably lower in radiation hygiene work.
Fallout and stack dispersion studies	Adhesive paper, open-top containers, or air samples with filters.	Usually high volumes required.	Background of extraneous material may be quite high, so discrimination in analysis required. Portability and remote locations also desired.
High-altitude sampling	Filters, impactors, electrostatic precipitators, freezeout collectors, and condensation nuclei counters.	Usually high volume necessary because particle concentration is low. Temperature, pressure (altitude), and aircraft speed needed to calculate air volume sampled, corrected to STP.	Usually special devices required.
Respirator testing	Membrane filter.	See references in text.	Testing should be performed while respirator is in use on a subject doing work to evaluate overall efficiency of device and effects of face-fit, etc.

Measurement of radioactivity of individual particles	Membrane filter.	Any suitable.	Light deposit required, analysis by contact autoradiography using nuclear track emulsion.
		B. Air Pollution	
Stacks and process effluents	Cyclones and filters located inside the stack. Electrical precipitators, thermal precipitator, impinger, cyclone, and filter used outside.	Rugged but portable pump. Flowmeter corrections required for temperature and pressure.	Sample isokinetically at equal area points in duct where velocity has been previously determined by pitot traverse, or use an isokinetic sampling probe. Use of Stairmand disk (annular orifice) provides uniform velocity and concentration profile 4–6 diameters downstream. Stay far away from bends and fittings. Avoid deposition and condensation in sampling lines.
Area or community air sampling, meteorological tracer studies, prompt fallout, and local deposition	Membrane or absolute filter. For larger particles, open-top container or adhesive surface.	Battery power useful to follow plume or fallout, mobile equipment in truck, helicopter or airplane required.	For a community air quality monitoring network many samplers required, and they may need to run continuously.
Mists and fogs	Direct airborne analysis with automatic counting and sizing instruments, collection with cascade impactor, deposition on coated slides or treated tape in electrostatic precipitator.	Any suitable.	Photograph for airborne or deposited droplets, later counting and sizing, use a size-selector collector, and weigh, count, or otherwise measure incremental deposits.

TABLE 7.1 (CONTINUED)

Application or location	Types of sample collector most useful	Notes on air pump and flowmeter	Special considerations and remarks
Condensation nuclei	Overpressure required to produce condensation in an adaptation of a Wilson cloud chamber is a function of nuclei size.	Any suitable.	For particles less than 1 μm, very strongly affected by pressure or ionized air molecules which promote condensation.
Combustion sources and products	May need high-temperature all-glass filter paper, although membranes may be suitable for quick exposure. Cyclones and glass bag filters used for flyash.	Any suitable.	High temperature usually a problem as well as condensation of water from combustion. For open burning, products of combustion and furnance effluents released to community, see area sampling.
		C. Toxicological Studies	
Measurement of generated aerosols	For direct collection on electron microscope grid, use electrostatic precipitator or thermal precipitator. For larger particles use cascade impactor, membrane filter, or precipitator.	Any suitable.	Cascade impactor can be used as routine analysis tool if first calibrated by optical microscopy and effect of particle deposition on walls assessed. For monodisperse aerosols use light-scattering methods.
Measurement of size and surface in animal tissue	Fix tissue, stain, embed, and section; use light or electron microscope as required. Scanning electron microscope *in situ*.	Any suitable.	Similar to methods for cytology and subject to same limitations due to alteration of tissue during preparation. Digest tissue and examine particle residues.

Application or location	Types of sample collector most useful	Notes on air pump and flowmeter	Special considerations and remarks
Measurement of respired particles	Same as for generated aerosols, above.	Any suitable.	Principal problems arise in dividing airflow to determine exhaled particles.
Microbiological aerosol sizing	Collect directly on agar surfaces with cascade impactor; collect on membrane filter for *in situ* sizing of spores and pollens.	Any suitable.	For vegetative forms, desirable to keep moisture content above 70% as prolonged desiccation on filter is destructive. Protect specimens from thermal or ultraviolet radiation.

insoluble particles present. Large insoluble particles will be trapped in the nasal passages and upper airways and be rapidly eliminated by coughing, sneezing, or swallowing. They are of concern in the upper-respiratory tract if they produce local irritation (e.g., chromic acid mist droplets) or irradiation of adjacent contacted tissues. However, if large particles are soluble, they can be absorbed and transported to a critical or concentrating organ after solution in tissue fluids at any deposition site. Small insoluble particles may deposit in the deeper spaces of the lung. They can be removed by protective physiological processes (clearance) or may be retained in the body for long periods. The combination of these several processes governs the potential hazard. Particle deposition in the connected compartments of the respiratory system, (i.e., the nasopharyngeal space, the tracheobronchial spaces, and the final pulmonary gas-exchange volume), is shown in Fig. 7.1. These curves represent the results of computations of particle deposition from analytical models of the deposition process combined with the aerodynamic and geometric parameters of a model respiratory system.[2] As indicated, there is surprisingly little variability in the deposition fraction at a given mass median aerodynamic diameter as a consequence of large variations in the distribution parameter for log-normal particle size distributions. The effects of specific chemicals or radioactive species deposited in the respiratory system depend on the interaction of the material with lung tissue, the properties of the

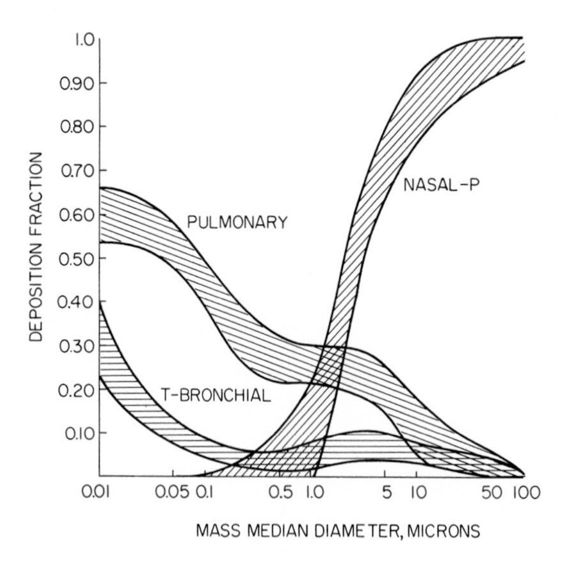

FIG. 7.1. Calculated deposition of aerosol particles in the human respiratory system. (From ICRP Task Group on Lung Dynamics, Ref. 2.)

material, the biological defense mechanisms, and the transport or storage of the material within the body. The relative rapidity of clearance of inhaled radioactive dust, or the biological retention period in relation to size, has also been modeled and analyzed.[2-4] Much information on the effects of specific nonradioactive particulate materials (e.g., silica, asbestos, and coal dust) in relation to particle size may be found in the current literature of industrial hygiene and toxicology. This information has been reviewed in two symposia.[5] The respiratory deposition curves illustrated in Fig. 7.2 present the effect of particle size on total, upper-respiratory, and deep lung deposition for a number of breathing frequencies as studied in healthy young male subjects under sedentary conditions. Hatch[6] described methods of computing particle deposition in the pulmonary spaces and summarized the data presented in the curves of Fig. 7.2 as follows:

"(1) The highest probability for deposition of inhaled particles in the respiratory spaces of the lungs occurs within the size range of 1 to 2 μm (gravity settlement) and in the submicroscopic region below 0.2 μm (precipitation by diffusion).

"(2) Above 1 to 2 μm, penetration to and deposition in the lobules falls off with increasing size simply because fewer particles escape upper-

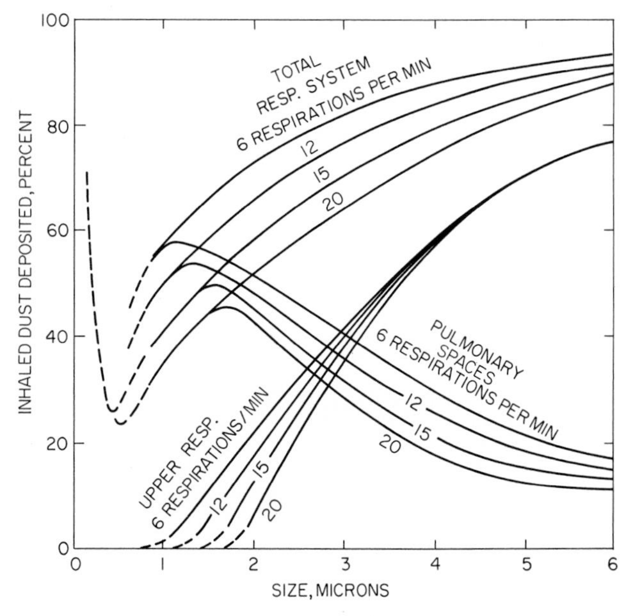

Fig. 7.2. Regional and total deposition of inhaled aerosol particles in relation to particle size and breathing frequency. (From Hatch and Gross, Ref. 1.)

respiratory trapping. Above 10 μm, the probability for penetration to the lobules is essentially zero.

"(3) Below 1 to 2 μm, lobular deposition falls off because the efficiency of removal by gravity settlement within the lobules themselves decreases.

"(4) The lowest probability for deposition of inhaled particles in the respiratory system occurs at ¼ to ½ μm, where the combined forces of precipitation by gravity and diffusion are at a minimum.

"(5) The probability of lobular deposition increases below ¼ μm because deposition by diffusion increases as particle size goes down (in contrast to gravitational settlement, which decreases with size)."

These conclusions have been used to assess the effects of respirable particles in many of the studies cited below.

7-2.2 Size Measurement of Airborne Particles

Sampling and analysis for investigating the size and size distribution of airborne particulate materials are often employed in mining, milling, and manufacturing operations to assess the health hazard to workers exposed to toxic dusts, fumes, smokes, and mists. The principles which have been developed for these situations are applicable to air sampling for radioactive and biological materials as well. In the nuclear energy industry, particle size analyses are required for the metal dusts and fumes customarily investigated in ordinary industrial hygiene exposures (such as lead and cadmium used in shielding fabrication) and for the most energetic radioactive materials which become a significant hazard after the inhalation of only a few particles. This last point is illustrated by the tolerance levels for various metals presented in Table 7.2 in terms of the daily maximum number of 1-μm particles.[7] From this table, it can be seen that inhalation of a single 1-μm ^{210}Po particle represents a health exposure ten times the daily tolerance. It is evident that control procedures required for many radioactive particles will be more restrictive than for more common industrial toxic dusts. The proper management of facilities using radioactive materials requires careful consideration of particle size. Size analyses provide data for the rational design of new facilities, processes, and operations; for the evaluation of existing control measures; and for the estimation of the degree of control required when a process or operation is changed or modified.

Requirements for radiation protection that apply to holders of United States Atomic Energy Commission licenses for the use of radioactive materials are given in the United States Federal Register,[8] which indicates the need to consider particle size in sampling and exposure as follows:

TABLE 7.2
DAILY TOLERANCES FOR VARIOUS INSOLUBLE MATERIALS[a]

Isotope	Form	Number of 1-μm particles for daily tolerance
Natural uranium	UO_2	3.49×10^8
^{235}U	UO_2	9.35×10^7
^{233}U	UO_2	2.10×10^4
^{239}Pu	PuO_2	1.19×10^3
^{210}Po	Po	0.09
^{24}Na	NaCl	2.53
^{90}Sr	SrO	10.0
^{226}Ra	$RaBr_2$	173.0
Beryllium	BeO	3.54×10^7
Lead	PbO	4.35×10^8

[a]From Hyatt, Ref. 7. Based on the International Commission on Radiological Protection values in *Health Physics, 3*:41 (1956). Calculation is based on daily air intake of 10 m^3 and volume of 1 particle equal to 5.22×10^{-13} cm^3.

"The commission may authorize a licensee to expose an individual . . . to air-borne concentrations in excess of the limits specified . . . upon . . . demonstrating that the concentration is composed . . . of particles of such size that . . . they are not respirable . . . ." Thus there is an explicit economic incentive to obtain and utilize particle size data for comprehensive health hazard control. It should be kept in mind that application of this principle does not yet encompass the American Conference of Governmental Industrial Hygienists' Threshold Limit Values adopted as of 1970.

Appraisal of dust exposures in industrial hygiene surveys is discussed by Drinker and Hatch[9] for nonradioactive materials. For the more severe health requirements imposed by the use of certain radioactive materials or modern chemical and biological agents, for which only a few particles represent a daily tolerance dose, careful evaluation of the maximum exposure, even though brief, is of extreme importance. This requirement has led to an increasing use of personal air samplers (i.e., small battery-powered air sampling units which can be worn by employees while working). These sample continuously and provide an estimate of the exposure experienced in the breathing zone over the entire work period. When dealing with highly radioactive materials, it is absolutely essential that a sample collector be selected that will remove all, or almost all, of the small particles involved. For radioactive materials of lower total activity, such as occur during uranium ore milling exposures, a collector with somewhat lower efficiency for fine particles may be acceptable if other advantages, such as reduced size, larger sample, etc., are obtained. When less-efficient samplers are used, knowledge of the particle size collection efficiency characteristics of the device is needed so that appropriate corrections can be applied to the analytical results.

Analysis for particle size may be performed by the methods and techniques described in previous chapters. When size is correlated with information on the toxicity of airborne materials, this helps to define requirements for process changes related to operation or control and provides an important guide for the design or selection of appropriate control equipment.

From the foregoing, it will be evident that many studies require determinations of the particle size of airborne materials in terms of aerodynamic behavior to make it possible to estimate the likelihood of particulate deposition in the lung. For irregular particles, the kind most commonly encountered, air sampling with an aerodynamic size classifier is widely used to estimate the fraction of airborne particles that will reach the lung. Size classifiers have been developed which have captured efficiency curves closely related to the nasopharyngeal curve presented in Fig. 7.1; these are customarily followed by a second collector to retain the fraction penetrating the simulated upper-respiratory tract. These devices have been described in Chapter 4. Each size-selective sampling device has unique size retention characteristics and must be calibrated carefully for best results.[10–18]

The Mine Safety Act of 1969 (PL-91-173) recognizes the importance of size distribution of airborne dust in black lung disease causation by specifying selective air sampling for measuring coal mine airborne dust exposures. During its 1968 annual meeting the American Conference of Governmental Industrial Hygienists accepted a report of its Committee on Threshold Limit Values for quartz which specified that concentration and percent quartz be determined from the fraction passing a size selector with the following characteristics:

Aerodynamic diameter, μm (unit density sphere)	% passing selector
$\lesssim 2$	90
2.5	75
3.5	50
5.0	25
10	0

The use of size-selective sampling devices to fractionate respirable airborne materials must be interpreted carefully, because actual deposition is a function of an individual worker's pulsating pulmonary air flow patterns, and these are related to the amount and type of work, state of health, etc.

7-2.3 Measurement of Particle Size in Tissues and Biological Fluids

Preparations for measuring the particle size of material deposited in the respiratory system and other tissues include chemical digestion of the tissue and microscopic examination of the residue, and preparation of standard histological sections with examination of the thin tissue slice under the microscope. The first method may change the chemical and physical nature of the deposited particles (especially size), whereas the second method is tedious. The use of radioactive particles in the preparation of test aerosols makes the location and identification of particles in tissue sections much simpler, since autoradiography can be used. However, many of the analytical techniques that are applied to nonradioactive particles collected in air samples or fallout samples can be utilized successfully for particle size measurements in tissues.

Techniques utilized for particle size analysis in tissues by microscopy are illustrated by the following examples. Cartwright and Nagelschmidt[19] prepared lung residues of coal miners with pneumonoconiosis for size analysis by refluxing 5 g of dried ground lung with 100 ml of 50% hydrochloric acid for 24 hr, followed by repeated washing with alcohol.

"Part of the sample was dried and mulled in nitrocellulose for electron microscope counting. The remainder was dispersed in glycerol with a mechanically driven stirrer, designed to provide large viscous drag forces but to avoid the risk of fracture of particles by the application of pressure to them; the optical counts were made in glycerol as mounting medium. Counts were made only when the adequacy of dispersal had been confirmed microscopically. The lung dusts, as counted, were probably more highly dispersed than when they were originally inhaled but the original state of aggregation can neither be known nor reproduced.

"The counts were made by comparing the projected surface of the dust particles with that of standard circles on a May graticule in sixteen size ranges down to 0.08 μm with overlap between optical and electron microscopy in three size ranges. Special efforts were made to obtain reliable counts for the larger particles by examining large areas. Fig. 7.3 shows typical pictures of the material at three different magnifications."

Arnold[20] has developed an autoradiographic procedure for radioactive particles in tissues which "eliminates all emulsion-tissue separation and allows the preparation of emulsions less than 0.5 μm in thickness." The tissues are fixed in acetone to minimize migration of particles and chemical fogging of the emulsion. They are then embedded in paraffin or some other suitable medium, and cut into sections 2 to 5 μm thick. These are cemented to glass slides with egg albumin and immersed in a solvent such as xylol to dissolve the embedding medium. After drying, Eastman Kodak Co. NTB (beta-sensitive) emulsion or NTA (alpha-sensitive) emulsion, liquefied at

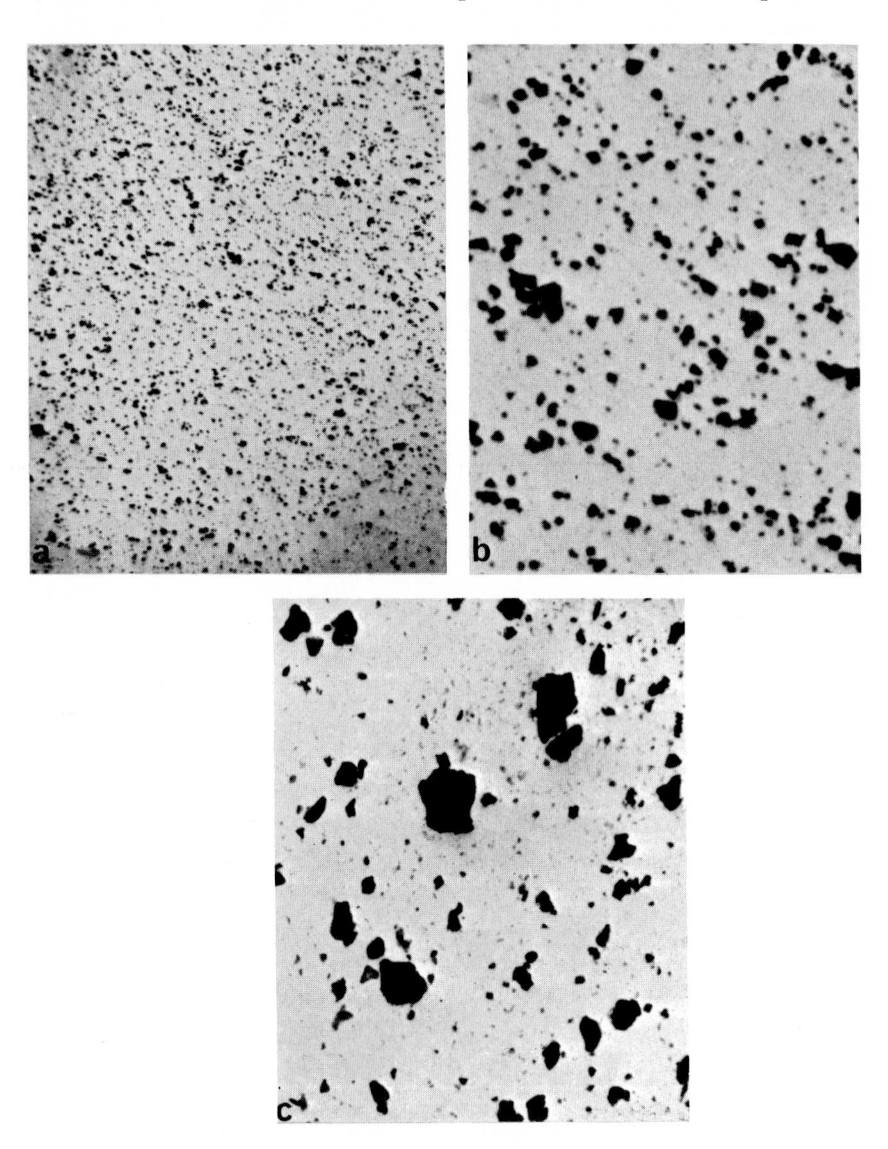

FIG. 7.3. Dispersed lung dust of coal miner. (a) Optical micrograph ×250; (b) optical micrograph ×1000; (c) electron micrograph ×4000. (From Cartwright and Nagelschmidt, Ref. 19.)

a temperature of 38°C, is applied to the surface of the tissue with a brush. After the emulsion has been hardened in a gentle current of cool air, the slides are stored in a sealed box containing an open vessel of saturated calcium chloride to maintain a relative humidity of 40% and placed in a refrigerator at 0° to 5°C for the desired exposure period, the length of which depends on the activity of the radioisotopes and the image intensity desired.

The exposed slides are developed for 1 min in D-19 developer (Eastman Kodak), fixed and hardened for 2 min, and placed in two washes of 4% formalin for 4 hr to harden the thin emulsion. After this treatment, the tissue section may be stained by conventional techniques and the preparation examined under high magnification for microscopic counting and sizing of the particles revealed by the alpha and beta tracks.

Evans[21] described a method of mounting thin tissue sections containing radioactive particles directly on photographic emulsion in a darkroom by floating paraffin ribbons on water and slipping an emulsion-coated lantern slide underneath. "After suitable exposure, the plate is developed, and the tissue stained, etc. The preparation may be studied microscopically as the autographic image is in place just below the tissue."

A number of methods that are suitable for concentrating particulate materials in biological fluids and preparing them for light or electron microscopic examination on membrane filters have been summarized by the Millipore Filter Corporation.[22] The specific techniques described there were developed for the isolation and examination of body cells and bacteria, but they are adaptable, with only minor changes, to the examination of all manner of particulate material for size analysis. In practice, small volumes (a few milliliters) of relatively clear fluids may be filtered directly through a membrane filter without preliminary treatment and washed with dust-free water, saline, or any other appropriate fluid. The all-glass vacuum filtering apparatus, shown in Fig. 7.4, is suitable for this purpose. Samples which are heavily contaminated with tissue materials can be cleared, in many cases, by digestion of excess mucoprotein with 5% Prolase-300* in physiological saline prior to filtration through a membrane.

7-2.4 Respirator Testing

The testing of dust respirators for materials not significantly more toxic than lead is covered by approval schedules of the United States Bureau of Mines.[23] For more dangerous substances, when inhalation of only a few particles represents a substantial part of the daily tolerance limit, more

---

*Wallerstein Laboratories, New York, N.Y.

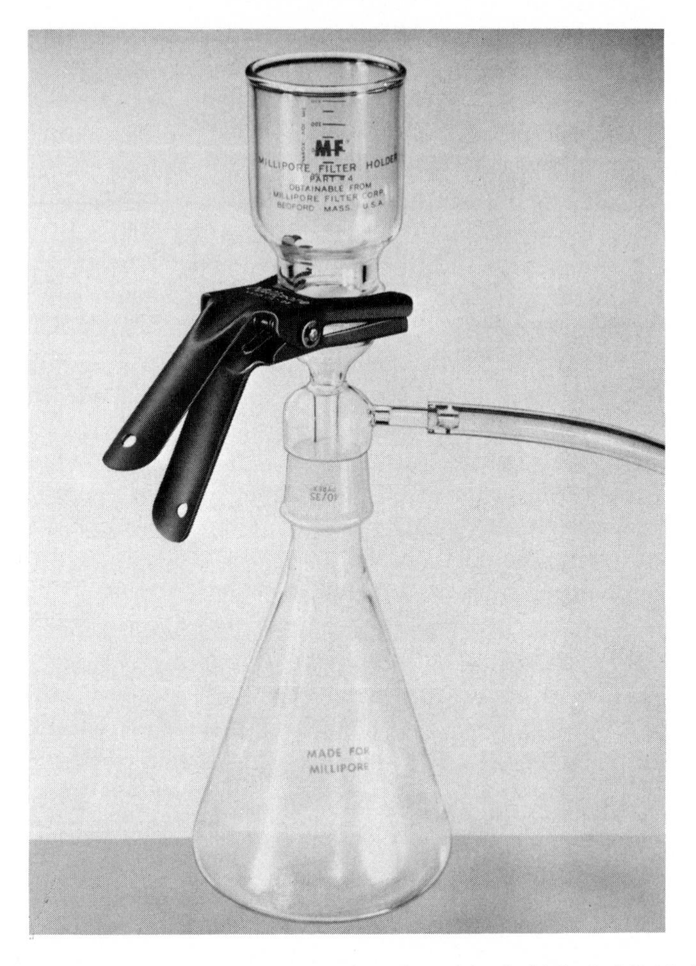

FIG. 7.4. Filtering apparatus for preparation of particles in biological fluids for size analysis. (From Millipore Filter Co., Ref. 22.)

severe criteria and more sensitive testing methods have been developed. Special tests are described in revised Schedule 21B[24] for respirators designed to protect against dusts, fumes, and mists having a threshold limit value less than 0.1 mg per cubic meter or against radionuclides.

Davies[25] has reviewed the development of industrial and military respiratory protective devices to about 1948, and Hyatt[7] extended the summary of respirators for industrial applications to 1962. Recent British experience in this field has been covered by Hermiston *et al.*[26] Testing methods and application of respirators for protection against organophosphorous pesticides have been described by the United States Department of Agriculture.[27] Other information on the development and testing of respirators for protection against military agents is contained in two reviews.[28, 29]

A dust respirator test method using calcium carbonate dust particles to assess the performance of respirators recommended for nuisance type dusts, described in Silverman,[30] is illustrative of the manner in which particle sizing techniques may be used. These respirators, constructed of form-fitting filter materials such as nonwoven rayon cotton fibers, surgical gauze, foamed rubber, and plastic, were tested on human subjects with limestone dust having a mean size by count and weight, when aerosolized, of 1.1 μm and 2.3 μm, respectively, and a standard deviation of 1.6. A small filter holder unit of the type shown in Fig. 7.5 with disks of Type AA Millipore filter material* was placed inside the mask, making it possible to take samples from the interior during normal breathing of the subjects. A simultaneous sample was taken of the air outside the mask, and the two dust samples

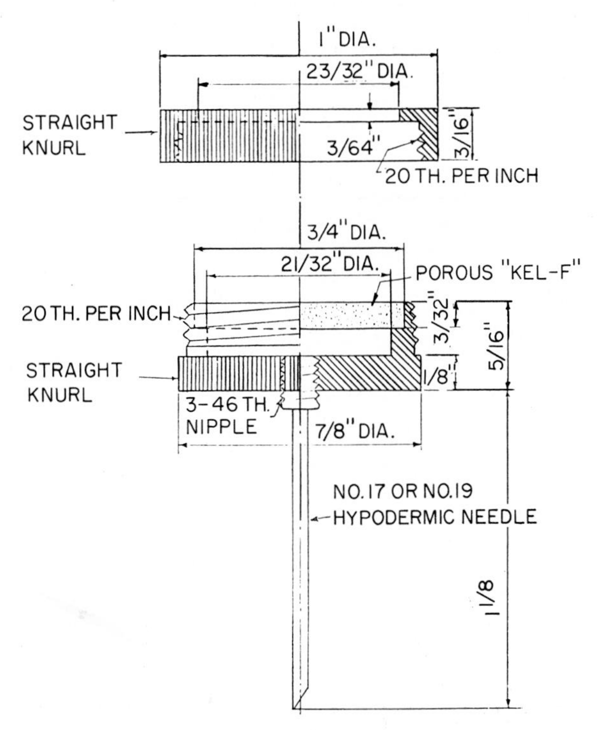

FIG. 7.5. Respirator sampler filter holder. (From Silverman and Burgess, Ref. 32.)

*Millipore Filter Corporation, Bedford, Massachusetts.

were counted and sized to determine the particle size efficiency of the dust mask. A typical result is shown in Fig. 7.6. The efficiency of the mask, whose inside and outside dust size distributions are shown in Fig. 7.6, was 72.8% on a weight basis. It may be seen from the size distribution curves that virtually no particles greater than 2 μm (which represented 60% of the dust cloud) penetrated the mask and that some of the particles less than 2 μm were filtered out as well.

The particle size efficiency characteristics of highly efficient valved dust respirators, which meet United States Bureau of Mines approval tests, may be determined by similar methods, using the electron microscope to reveal particles below 0.3 μm. The Respiratory Protective Devices Manual[31] lists monodisperse particles (e.g., dioctyl phthalate, *Bacillius globigii,* spores, pollens) and polydisperse particles (e.g., methylene blue, sodium chloride, uranine, copper sulfate) that have been used at one time or another for respirator testing and gives information on mean size and size distribution.

Silverman and Burgess[32] designed a respirator "sampling director" which utilizes a sensitive pressure switch connected to the interior of the mask to regulate samplings through the in-mask device shown in Fig. 7.5 during inhalation periods only. By timing the sampling periods and measuring

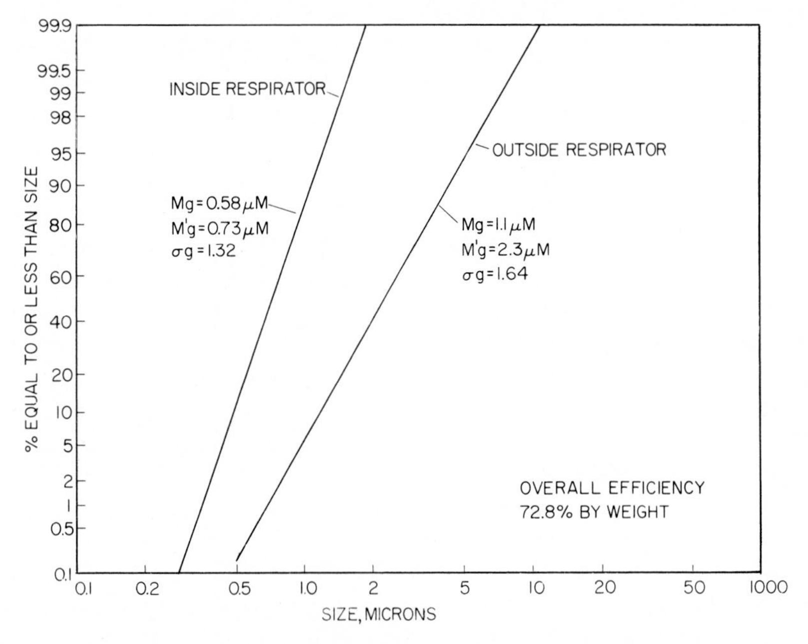

FIG. 7.6. Particle size efficiency study of nuisance dust respirator. (From Silverman, Ref. 30.)

inspiratory air flow simultaneously, the respiratory minute volume can be calculated. When this is multiplied by the test dust concentration inside the mask, the total mass of dust penetrating the respirator in a given time period is known.

7-2.5 Identification and Measurement of Individual Radioactive Particles

Determination of which particles in an air sample are radioactive, and how they are distributed in the observed size distribution, is often necessary in order to evaluate a health exposure to a dust of mixed composition. In one method, samples are collected on membrane filters and placed upside down in a modified sample holder on the microscope stage. A shielded fine slit connected to a thin end-window counter for gamma emitters is attached to the microscope, and, as particles are observed and sized, their activity can be assessed simultaneously.

For low-energy beta-emitters and for alpha-emitters, techniques of autoradiography described by Leary[33] can be used. The collected sample on its membrane filter is placed in contact with a nuclear emulsion* and then exposed for several hours or days. Particle size is inferred from the number of radial tracks per particle when counted under the microscope, by using a known specific activity. A limitation of this contact method is the inability to determine simultaneously radioactive and nonradioactive particles or agglomerates of mixtures which may be present. Moss *et al.*[34] have presented a modified method in which the photographic emulsion is first softened in water and then pressed tightly against the membrane filter. Over 90% of the particles are transferred to the emulsion when it gets tacky during drying. Photographic development and sizing are carried out in the manner described above, but both radioactive and inert particles can be observed simultaneously under the microscope. Although the particle transfer by this technique is not 100% effective, particles remaining on the membrane filter can be estimated by examining it under the microscope and applying suitable correction factors to the results obtained from the photographic film.

In a similar technique, developed by George[35] for electron microscopy, particles are collected on membrane filters and transferred to a carbon-coated electron microscope grid by slow dissolution of the filter in acetone vapor. Next, a drop of warmed NTA emulsion (45°C), diluted 1:1 with distilled water plus 0.5% surfactant, is applied to the specimen grid by means of a transfer loop. After suitable exposure (3 days) the grids are developed

---

*Kodak NTA or Type A; "Tech-bits" No. 2, Rochester, New York, 1963.

and examined directly in the electron microscope. A typical microphotograph of an autoradiograph prepared by this method is shown in Fig. 7.7. This technique is likely to be equally effective for particles collected on carbon-coated grids in a point-to-plane electrostatic precipitator, described previously. Techniques suitable for autoradiography of particles in biological specimens are described in Section 7.3, above.

FIG. 7.7. Radioactive and non radioactive particles on membrane filter using autoradiography. [From L. A. George, II; Science *133*:1423 (1961).]

Transfer techniques for isolation for analysis of individual "hot" particles from fallout filter samples are described by Brar *et al.*[36] They also discuss techniques used by previous investigators.

7-3 Applications of Particle Sizing to Air Pollution Control and Fallout Studies

7-3.1 Sampling of Stacks and Process Effluents

It is frequently necessary in air pollution control activities to sample suspended solid and liquid particulate materials flowing in stacks, ducts, flues, pipes, etc. The basic principles of stack sampling for size analysis have been reviewed in Chapter 2. For radioactive materials, an additional consideration arises because some radioactive gases decay to particulate daughters within a short period of time. These daughter products are of molecular size when formed, but they may become attached to naturally occurring dust particles. The speed with which this transformation occurs may be an important consideration when selecting sampling locations.

Particle size information from process vents and off-gas stacks and ducts is needed for selecting proper gas-cleaning equipment. The most suitable stack sampling procedures and equipment vary for different situations. For high loadings of inert material, customarily found in process and furnace effluents, it is convenient to select a sampler with sufficient storage capacity to permit at least one complete traverse of the total duct cross section. If this is not possible, it is good practice to take individual isokinetic samples at the centers of equal duct areas in the same manner as a Pitot-static tube traverse.

Condensation, losses in the probe, and other complications are best avoided by immersing the collector in the stream to be sampled. To meet this requirement, the usual stack gas collectors are filter papers (including those composed of all-glass fibers); filter thimbles of cellulose, glass, or ceramic; glass fabric filter bags; and packed beds of fibers or granules. In instances where large and small particles occur together, as is the case with the effluent from power boilers burning pulverized coal, a cyclone collector may be placed ahead of a filter to separate the mass of particles greater than 10 μm and reduce the filter deposit.

Out-of-stack particle collectors must be used when conditions of temperature or gas composition make in-stack collectors impractical. Commonly used out-of-stack collectors are filters, electrostatic precipitators, impingers, scrubbers, condensation traps, and combinations of these. Deposition of particles and condensation of moisture in a sampling probe leading to an externally located collector must be noted in all cases in order to recover losses of sample and prevent plugging of the probe. External heating of the probe and intermittent vibration or rapping are some of the procedures that have been used to minimize these difficulties. Additional comments on duct and stack sampling are included in Chapter 2.

It is not feasible to summarize here the features of the many devices and techniques that have been developed for specific situations. Details of practice are contained in publications of industrial associations and technical societies,[37–40] in literature of manufacturers of dust collecting equipment,[41,42] and in texts on air pollution control.[43] In all cases where the particle size spectrum is unknown, sampling should be conducted isokinetically, as explained in Chapter 2.

In the laboratory, or in certain repetitive field sampling situations when velocity and dust loading profiles are known, it is possible to withdraw a sample from a single point in the duct and apply a suitable correction factor when calculating average results. For other situations, uniform duct velocity and dust loading profiles may be secured by use of an annular orifice 4 to 6 diameters upstream of the sampling point in a straight duct. The annular orifice is formed by a thin disk having a diameter equal to the pipe diameter divided by the square root of 2 ($D/\sqrt{2}$) and placed perpendicular to flow.[44] The resistance to flow of this device (2 to 3 pipe velocity heads) and its susceptibility to abrasive damage make it impractical for use in high-volume systems with low static exhausters.

7-3.2 Community Air Sampling

Sampling in areas outside plant boundaries is needed for air quality studies of deposition and dissemination of radioactive and inert particles. These particles may be toxic to crops, animals, or man; they may create a nuisance, or damage property by reason of dust accumulations, soot-fall, or odors.

The kinds of information that are needed will vary for each situation; hence, the choice of locations for samples, sampling times, and types of collectors depends on specific requirements. When measuring concentration and particle size downwind of a stack emission source, the techniques developed for large-area surveys are applicable. A case in point is environmental evaluations for proposed nuclear power plants. Here, local micrometeorology is of great importance in interpretation, and it has become the practice to carry out experimental release studies with inert particles and gases in conjunction with meteorological measurements. Finely divided tracer materials of proper size that have been used in the past include fluorescent zinc–cadmium sulfide, collected on membrane filters and analyzed under ultraviolet light; oil smoke, collected on membrane filters, washed off with a solvent, such as CCl_4, and analyzed in a fluorimeter; uranine or rhodamine B, collected on membrane filters and analyzed by fluorescence; and antimony oxide, analyzed by neutron activation analysis for ^{122}Sb after sample collection. These and other tracer techniques have been reviewed by Bloomfield in reference 43, Vol. 2, p. 532. Methods for size-selective sampling, cited in Section 7-2.2, may be applied to atmospheric aerosol spectral analysis.

7-3.3 Particle Sizing of Mists, Fogs, and Other Liquid Droplets

Particle sizes of liquid droplets can be obtained by many of the methods listed in Chapter 3. Houghton[45] determined the droplet sizes of natural clouds and mists by photographing a narrow illuminated section of the suspension. This technique is applicable to many situations encountered in industrial hygiene, such as spray tower or scrubber evaluations, or raindrop washout studies, when the droplets are greater than about 30 μm. Smaller droplets, such as occur in sea fogs (3 to 10 μm), and certain natural precipitation processes require more intense illumination. The use of laser holographic microscopy[46] provides sufficient energy to photograph suspended particulate material as small as ten times the wavelength of the illuminating beam (particle size about 3 μm is a practical lower limit). Commercial instruments have been produced for this application.[47] Laser microscopic holography is unique, as it permits the observation of a great depth of field on a single hologram.

Radar and laser echoes (return of coherent radiation from a cloud) may be used for estimating average drop sizes in a cloud, as the energy of the scattered radiation varies with the particle size. Radar is most effective for estimating the size of larger droplets (50 μm). Lidar (laser radar) offers a general technique for measuring remotely the transmittance and reflectance of nonprecipitating fine particle clouds for the purpose of determining their nature, extent, and internal structure.[48–51] Since lidar is a single-ended system relatively unaffected by solar radiation, it can be used both night and day. Applications of lidar to the measurement of atmospheric dispersions of solid particles from stack plumes are discussed below.

Liquid droplet size distributions can be measured with a vertical cascade impactor having round jets and shallow cups for collection of the impacted drops. Devices of this type have been described by Ranz and Wong[52] and by Mitchell and Pilcher[53] as modifications of May's original design.[54] Many of these instruments are commercially available. The deposited liquid can be analyzed chemically, colorimetrically, or gravimetrically, as appropriate. Water dropets from scrubber effluents can be size-segregated in a series of cyclones of varying diameters, followed by a standard impinger (dry).

Other techniques used to determine sizes of liquid droplets include the deposition of dyed liquids on glossy white paper (with subsequent measurement of the diameters of stains or spots) and the deposition of water droplets on dyed or gelatin-coated paper. It is possible to provide a hydrophobic surface treatment for glass slides which prevents deposited droplets from spreading and allows direct microscopic observation in a constant-humidity cell. Other investigators have deposited water droplets in oil-filled cells to permit direct observation. Maybank and Fenrick[55] described a technique for the collection of droplets in a dish of liquid nitrogen. The frozen droplets may be examined microscopically in a cold box or analyzed by screening at low temperature. In addition, it may be possible to utilize this technique with the Coulter counter (Chapter 5) for automatic sizing of frozen droplets.

It is possible to determine the size distribution of liquid aerosol particles by passing them through the automatic optical particle size counters described in Chapter 5, but provision must be made for continuous maintenance of the appropriate partial pressure of the vapor in order to minimize evaporation during transit of the optical cell. Usually this is not a difficult problem. The recirculating flow system described by O'Konski (Chapter 5) provides this feature, as discussed in his original paper, and it can be incorporated in most commercial instruments.

The size distribution of submicroscopic liquid droplets (i.e., those less than 0.3 μm) can be determined by growing the droplets in a saturated vapor, as described by LaMer[56] in his pioneering experimental verification of the Kelvin equation. Small liquid aerosol particles grow when exposed to vapor of a solution of the nonvolatile substance in a solvent. Growth is completed when the partial pressure of each substance in the droplet is equal to that in the solution, and initial particle size is calculated by means of the Kelvin equation. LaMer discussed particle growth in dioctyl phthalate–toluene and sulfuric acid–water systems, but the principle may be applied to submicroscopic solid particles, to condensation nuclei, and to individual particles in a vapor condenser type of aerosol generator.

7-3.4 Stack Dispersion Studies

The dispersal of fine particulate effluents from a stack may be considered to occur in two consecutive steps: first, an initial plume rise as a consequence of the momentum and buoyancy of the gas at the stack exit, and second, convective transport of the material downwind, combined with the simultaneous diffusion of the constituents into the surrounding atmosphere by the process of atmospheric turbulence. Downwind concentrations may be calculated from atmospheric diffusion equations for a release from an elevated point source in which assumptions are made for continuity of mass of the species and for a Gaussian distribution of concentration across the vertical and lateral axes of the plume. Meteorological parameters such as wind speed, stability, and standard deviations of the azimuth and elevation angles of the wind direction are used to define the rate of plume expansion with distance. Parameters used in practical diffusion model equations have been developed from extensive measurements of plume expansion from fairly low-level releases (a few hundred meters) which have been carried out close to the ground over nearly uniform terrain and over relatively short distances (a few miles). The increase in size of modern stationary power generation sources has resulted in the use of 1000-ft stacks and effective plume rises above the physical stack height of 2000 ft or more. Investigations of the effectiveness of very tall stacks are being conducted by measurement of plume geometry and the distribution of finer particulate material within the plume over large downwind distances, and lidar probing techniques are being applied successfully to these problems.[57–62]

The laser return signal (Mie backscatter) is a function of particle size, particle concentration, and incident wavelength. Therefore, the signal produced by a single-wavelength laser is not a unique indication of average particle size in the cloud. Although this disadvantage is common to all optical aerosol measuring instruments, rapidity and economy in plume tracking make the laser a useful tool for stack dispersion studies. Information about the aerosol concentration distribution in the plume can be obtained from the fine structure of the signal return. By making measurements with several laser wavelengths, the particle size in the plume can be estimated.

Particulate material is removed from the plume by gravitational sedimentation for sizes larger than about 100 μm. Smaller particles coagulate to form larger particles that are susceptible to sedimentation or washout by rain. Particles are removed from the atmosphere naturally by impaction on surfaces or as nucleation centers for ice or rain. The size of the larger particles can be determined from downwind samples collected with adhesive-coated plates or fallout jars. Size analysis of the collected material may be performed by techniques indicated in previous chapters, such as microscopy or the Coulter counter. Finer particles have been collected with field sampling devices such as membrane filters or Roto-rod samplers (a small H-shaped rod coated with adhesive and rotated at high speed for the collection of particles greater than 2 μm by impaction). Analysis of the finer particles has generally been performed by optical microscopy.

High-volume air samplers with all-glass absolute filter papers have been employed extensively for the collection of total suspended particulate matter in the atmosphere, but the collected dust is not ideal for particle size determination because of the difficulty of removing a sample from the filter paper and the highly compacted and agglomerated state of the deposit. Size fractionating collectors have been suggested for routine atmospheric sampling by using cyclones, parallel-plate settling chambers, or graded-efficiency cascaded filters as the initial collection stages. These techniques must be calibrated on known aerosol particles generated in the laboratory, and for this reason they have not as yet found wide acceptance. Many stack dispersion studies made to identify specific sources and to determine local meteorological behavior have been performed utilizing identifiable particulate tracer materials such as zinc–cadmium sulfide fluorescent particles, aerosolized dye solutions, fine glass beads, etc. In these instances the particle size of the dispersed material is established as a function of the operation of the dissemination device, and particles with a narrow size range are usually selected.

7-3.5 Condensation Nuclei

If a small volume of air containing submicron-sized particles and air ions is saturated with moisture and adiabatically expanded, the resultant change in internal energy of the gas will cause cooling below the saturation temperature and moisture will condense on the small particles. The kinetics of the condensation process make it possible to predict (as an idealized first approximation) that all surfaces will accept moisture at the same rate. Therefore, particles tend to grow to a uniform size because smaller particles grow in diameter faster than larger ones. After the particles have all grown to the same size (about 5 μm), the attenuation of a beam of light may be used to assess their numbers. From a knowledge of the total number in the aerosol and a supplementary gravimetric analysis of the particulate phase it is possible to calculate the average diameter of the nuclei. Instruments for measuring atmospheric condensation nuclei concentrations have been described.[63, 64] These devices have been employed to evaluate the performance of particle collectors[65] and for the in-place testing of high-efficiency filter installations.[66]

It is possible to obtain some information on the size distribution of nuclei by varying the adiabatic saturation pressure or by employing a preliminary size-selective collector, such as a diffusion battery. It has been suggested, however, that not all particles act as centers for condensation (presumably owing to some surface energy requirement for the molecular accommodation coefficient), and there may be an unknown error in counts and average sizes obtained by this means. Alternatively, condensation nuclei in the atmosphere may be collected on membrane filters or in electrostatic precipitators and measured by electron microscopy using the techniques described earlier. For nonvolatile droplets, a gelatin replication technique has been developed which is suitable for electron microscopy.[67]

The role of condensation nuclei in precipitation processes has received considerable attention during recent years. For example, it has been demonstrated that these fine particles carry condensed radionuclides, such as ^{131}I; they may contribute to formation of photochemical smog; and industrial, automotive, and aircraft effluents in the atmosphere have been implicated in the inadvertent modification of weather.

7-3.6 Fallout Studies

The terrestrial deposition of airborne radioactive particulate material from detonation of nuclear weapons or from accidental releases from nuclear reactors or processing facilities (fallout) is of continuing concern in human ecology. In addition, the distribution of radionuclide energy in this debris is an index of the type of nuclear event. For these reasons, fallout particulate matter has been studied in detail throughout the world during recent years with the support of federal agencies (for example, in the United States by the Division of Biology and Medicine of the Atomic Energy Commission and the radiation protection program of the Environmental Protection Agency) in collaboration with industry and universities. Initially, products from a large release, as a weapon test, may rise 30 to 40 km into the stratosphere because of the thermal buoyancy of the gases. Prompt fallout of larger particles, down to about 50 μm, occurs near the site and may spread several miles downwind. Particle sizes of these larger products can be determined from sedimentation samplers, such as adhesive paper or coated plates, using optical microscopy combined with radiation detectors, or through use of nuclear track emulsions, as described previously. Prediction of the deposition of the larger particles in the size distribution can be made from a ballistic calculation as a function of particle size and the local wind structure. Intermediate-sized particles may travel far. For example, Mamuro *et al.*[68] have reported that the size of prompt fallout particles in Japan from Chinese and Russian nuclear explosions included "hot" particles of 15 to 20 μm. Brar[36] reported individual "hot" particles of 2 to 10 μm from samples obtained in late 1967.

Submicroscopic material will be carried much further and, with stratospheric injection, will become widely dispersed in the stratosphere. Material re-entering the troposphere will eventually be deposited on the earth's surface by sedimentation after particle growth by agglomeration with other atmospheric particles, through precipitation processes, or by diffusion and turbulent impaction on objects. The distribution of radionuclides and their concentrations and sizes have been studied throughout the world as a function of time, meteorology, climatology, etc., in order to determine their behavior in terms of global atmospheric circulation and ecological effects. The size of fine radioactive fallout particles is difficult to obtain because of attachment to the inert particles normally present in all samples taken near the surface. In addition, particles collected near the earth's surface contain relatively larger amounts of naturally radioactive products originating from the decay of gaseous radon and thoron that continually diffuse out of the soil minerals of the earth and become attached to the ambient aerosol. Methods that have been developed to determine the relationship between natural and fission product radioactivity and particle size of ambient tropospheric aerosol particles involve the separation of sizes during sampling and activity analysis of serial fractions of the total deposit. The size distribution of the radioactive species is determined from the size-separating characteristics of the sampling device, as obtained by calibration on generated aerosols. Devices that have been used include electrostatic precipitators,[69] cascade impactors,[70, 71] parallel-plate sedimentation batteries followed by a final filter, standard impingers operated at differing velocities followed by filters, and a series of aerosol filters of successively finer fiber size.[11–13]

7-3.7 High-Altitude Sampling

Problems associated with high-altitude sampling for particle sizing relate to design of devices to permit control of sampling time and location, instrument size and weight limitations, and the extreme scarcity of particles in the upper air.

Junge *et al.*[72] have shown that usually there are very few particles per milliliter at high altitudes (compared to several hundred per milliliter at ground level and up to 10^6 nuclei >0.002 μm in New York City) and that average particle size decreases with altitude. Particles that remain in the stratosphere have been reported by Junge[73] to average approximately 0.3 μm. This conclusion is based on model analysis of supporting forces plus results obtained from examining high-altitude samples. Junge also discusses the stratospheric sulfate aerosol layer, probably formed from photo-oxidation of terrestrial H_2S and SO_2 with particles in the range of 0.2 to 4 μm, as well as extraterrestrial particle flux.

Sample collectors that have been used include modified designs of filters, electrical precipitators, impactors, thermal precipitators, and condensation nuclei counters, as well as optical particle size analyzers.[72–83] Very high sampling rates are required to collect sufficient material for analysis. Special flowmeters are also necessary; these are described in the references cited above. Sampling platforms include high-altitude aircraft, balloons, and sounding rocket probes.

Although the subject of high-altitude sampling for particulate material has been of primary concern to geophysicists, it must be considered by environmental health scientists also because of the long-term effects of particles on human ecology and their transport to and from the biosphere.

7-4 Applications of Particle Sizing to Toxicological Studies

7-4.1 Measurement of Generated Aerosols

Study of the hazards of inhaled aerosols and their effects on man and experimental animals requires prolonged generation of aerosols with controlled properties. Many studies have indicated the role of size in lung deposition,[1, 84] and others [85] have shown that lung clearance is both size- and concentration-dependent. For example, particles of 0.09 μm clear faster than those of 0.6 μm.

Although investigators frequently generate polydisperse solid particles and relate the observed biological effects to the count or mass median diameter of the cloud, only the use of controlled monodisperse spherical particles in toxicological studies produces particle size–toxicity results of an unequivocal nature. Nevertheless, clouds with monodisperse characteristics are difficult to generate, and it is acknowledged that almost always polydisperse clouds occur in practice.

Methods for generating polydisperse and monodisperse aerosols suitable for toxicological and other studies have been summarized by several investigators.[86–91] References cited describe in detail many devices which can be used to generate or disseminate solids and liquid solutions and suspensions. Monodisperse polystyrene latex spehres* have recently become available in large quantities for use in aerosol and hydrosol research, and they are useful for toxicological studies when they can be tagged. Sizes available range from 0.12 to 3.0 μm, with a standard deviation between 1 and 10% of the mean size. When dispersed from suspension by atomization, they usually acquire electrostatic charges which require discharge by humidification or air ions if neutral particles are needed. These spheres have been mentioned in Chapter 5 in connection with the calibration of automatic optical particle size analyzers, and they are also used for size standards in electron microscopy.

Measurement of aerosol particle size after generation of a test cloud (suitably diluted to keep interparticle agglomeration to a minimum) is performed by the methods described in earlier chapters. It should be

*Dow Chemical Company, Midland, Michigan.

emphasized that when conducting toxicological and pharmaceutical studies, it is the aerosol itself that must be measured and not the parent material from which it is generated, because it is difficult to disperse very fine powders and hydrosols by air dispersers so as to produce single particles. Therefore, when evaluating the effects of particle size on deposition in the human and animal respiratory system, frequent aerosol samples are needed to provide direct checks on the performance of the aerosol generating equipment, since valid conclusions can be arrived at only when dealing with unitary particles.

7-4.2 Measurement of Respired Particles

The deposition and behavior of inspired particles are important factors in the interpretation of maximum allowable concentrations and threshold limit values for toxic and radioactive materials. In the case of quartz dust, the threshold limit concentration of free silica was established many years ago on the basis of microscopic counts of particles between 1 and 10 μm. The original limits were established from extended observations of workers in mines and quarries, combined with limited animal experiments. Because higher toxicity materials are in use today, particle size has become the most important factor in determining effects—i.e., whether particles can be inhaled and upon retention yield a toxic effect. For certain materials, such as plutonium, it has not yet been possible to assign a lower limit of particle size where toxicity is negligible. Current aerosol inhalation research is directed almost exclusively to studies of submicron particles in recognition of their toxicity and of the uncertainty of our knowledge regarding the physiological response to this size range.

Many studies of particle deposition in the respiratory system were reviewed by Hatch and Gross.[1] The following methods are indicative of those used.

Brown[92] measured respiratory deposition in man by having his subjects inhale dust-laden air from a chamber containing a cloud of the desired concentration and size and exhale through an electrostatic precipitator. Particle size and concentration in the inspired air were determined by taking samples directly from the dust chamber through a second electrostatic precipitator tube. At the conclusion of the exposure period, precipitated dust samples representing inspired and expired air were counted and sized, and particle size retention curves were prepared which indicated size retention for subjects at rest and under measured work loads. The experiments showed that percent retention was directly proportional to dust size over the particle range investigated (about 1 to 6 μm).

Altshuler *et al.*[93] investigated the influence of particle size on regional deposition within the respiratory system by continuous direct optical size analyses of exhaled clouds. Instantaneous air flow measurements were made by pneumotachography, and the expired dust was measured by a photoelectric analyzer. When these measurements were correlated with the CO_2 level in the expired air, it was possible to determine in which region of the respiratory system each increment of expired air originated and the average particle size and numbers associated with it. In this manner it was possible to determine deposition by particle size at different depths in the respiratory system. Brown *et al.*[94] employed similar widely used methodologies for their retention studies.

7-4.3 Aerobiology

Size has an important influence on the transmission of airborne pathogenic organisms. Bacteria and virus-containing droplets of saliva discharged during coughing, sneezing, and speaking range in size from approximately 10 μm to greater than 100 μm.[95] Droplets of this size settle relatively rapidly under the influence of gravity. For example, as shown in Table 7.3, 12-μm water droplets settle at the rate of 0.9 ft/min, and 100-μm droplets at 59 ft/min. Therefore, if expelled droplets remained at their original size, disease transmission would not occur by this mechanism over distances much greater than a few feet. Wells[95] has shown, however, that small droplets of saliva evaporate almost instantaneously to form small dried residues of the order of a few microns which he terms "droplet nuclei." These settle at a greatly reduced rate. The relationship between rate of evaporation and settling distance for droplets of various sizes has been calculated by Wells and is shown in Table 7.3. The figures in this table were derived for drops of water, but the drying of saliva is inhibited by the presence of bacteria,

TABLE 7.3
Evaporation Time of Droplets and Falling Distance before Evaporation[a,b]

Diameter of droplet (μm)	Falling speed (ft/min)	Evaporation time (sec)	Distance in feet droplets will fall before evaporation (at RH 50%)
200		5.2	21.7
100	59	1.3	1.4
50		0.31	0.085
25		0.08	0.0053
12	(1)	0.02	0.00028

[a]From Wells, Ref. 95.
[b]Water droplets in unsaturated, still air at 22°C.

viruses, and constituents of the mucus, so that actual drying rates are somewhat slower. Nevertheless, drying of small droplets occurs relatively rapidly, and the small residues consist of bacteria or virus particles encased in semi-dried mucus, which tends to preserve viability. These droplet nuclei have a mean size of 5.5 μm[96] and remain suspended in air indefinitely because normal room convection currents (about 25 to 50 ft/min) greatly exceed settling rates. Particles of this size are inhaled readily, and the smallest are capable of penetrating to all parts of the respiratory tract.

Although the maximum period of viability of free-floating pathogens in droplet nuclei is not known with certainty, there is evidence suggesting that some may survive in indoor atmospheres for several days and drift hundreds of feet from the point of emission. The tendency of droplet nuclei, because of their small size, to follow even modest air currents makes high rates of room ventilation (20 to 40 air changes per hour) an excellent method for reducing the infectious potential of indoor air.

Particle sizing of individual organisms from a cultured growth is not indicative of the actual size of the bacterium which is the agent for airborne infection. Techniques of particle sizing for airborne organisms must take into account ambient effects on the droplet nuclei when sampled. General

FIG. 7.8. Andersen sampler.

methods for collection of airborne microorganisms have been described in detail by Wolf *et al.*[97] These authors note that, although particles aerosolized from clean suspensions of microorganisms are usually spherical, most bacteria found in the atmosphere are associated with foreign material and the combination may have a very irregular shape.

Many of the methods used for the size analyses of bacterial particles drawn from the air are identical with those used for nonviable particles (e.g., direct microscopic measurement of organisms collected on membrane filters or deposited on glass slides by thermal or electrical precipitation). Identification of microorganisms among other air contaminants, such as

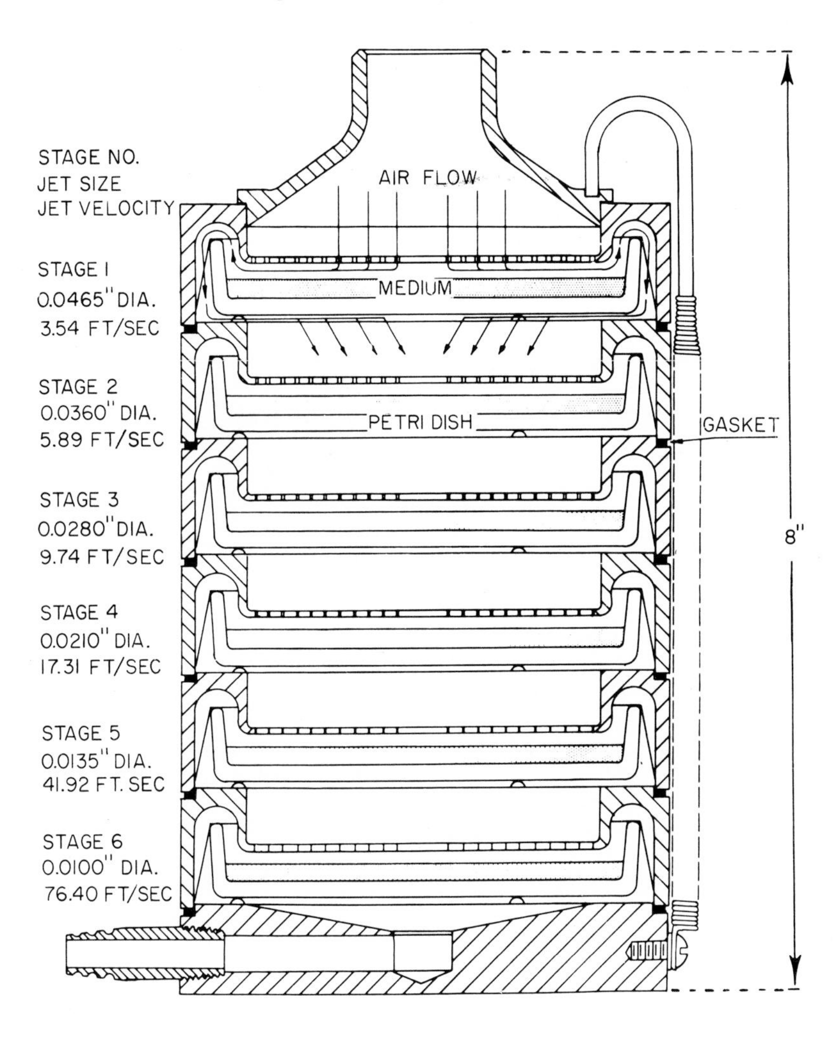

soot and lint, which are normally much more numerous, is aided by differential staining techniques but is at best difficult.

The most efficient way of identifying viable airborne microorganisms is to culture the sample and observe the colonies, but this method destroys all evidences of the original size of the airborne particle. Size discrimination, as well as organism culturing for identification, is possible by the use of devices such as the Conifuge, the TDL cascaded slit,[97] the Casella Slit Sampler and the Raynier Sampler. The Andersen*[98] cascade impactor (Fig. 7.8) separates viable particles into six size fractions and collects them by direct impaction on culture media. The manufacturer supplies calibration data with the instrument for a sampling flow rate of 1 cfm. May's cascade impactor and all similar instruments can be used to make a size separation of airborne particles during sampling. Biological assay of the sized fractions can be carried out by washing the collecting surfaces with a sterile liquid and culturing the washings on a solid medium suitable for the microorganisms under study. Survival of vegetative bacterial cells is likely to be poor when they are subjected to dehydration associated with prolonged high air flow rates, as on a dry impactor slide. For this reason ordinary impactors, membrane filters, and similar dry dust collecting instruments may be unsuitable for determining total airborne counts of viable organisms.

Other discussions of the methods of particle size analysis used in current research in aerobiology are contained in recent texts,[99-102] and symposia,[103-106] and in a sample of research reports dealing with particle size in relation to hospital contamination; hospital environmental control for patients with low resistance; sewage treatment; air pollutants and air cleaning; aerosol generation, sampling, and analysis; and ambient environmental effects on viability and infectivity of organisms.[107-123]

References

1. T. Hatch and P. Gross, *Pulmonary Deposition and Retention of Inhaled Aerosols,* Academic Press, New York, 1964.
2. International Commission on Radiological Protection, Task Group on Lung Dynamics, Deposition and Retention Models for Internal Dosimetry of the Human Respiratory Tract, *Health Phys., 12*:173 (1966); Errata, *Health Phys., 13*:1251 (1967).
3. T. Hatch, Respiratory Dust Retention and Elimination, in *Proceedings of Pneumoconiosis Conference, Johannesburg, 1959,* A. J. Orenstein (Ed.), J. & A. Churchill, Ltd., London, 1960.

---

*2000 Inc., Salt Lake City, Utah.

4. National Academy of Sciences — National Research Council, Subcommittee on Inhalation Hazards, *Effects of Inhaled Radioactive Particles,* Publication 848, NAS-NRC, Washington, D.C., 1961.
5. C. N. Davies (Ed.), *Inhaled Particles and Vapours,* Pergamon Press, Oxford, 1961; *Inhaled Particles and Vapours,* II, Pergamon Press, Oxford, 1967.
6. T. Hatch, Distribution and Deposition of Inhaled Particles in the Respiratory Tract, *Bacteriol. Rev., 25*:237 (1961).
7. E. C. Hyatt, Air Purifying Respirators for Protection Against Airborne Radioactive Contaminants, *Health Phys., 9*:425 (1963).
8. United States Federal Register, 10-CFR-20, USAEC Regulations (1963), Paragraph 20, 103, Section G, subparagraph (2).
9. P. Drinker and T. Hatch, *Industrial Dust,* 2nd ed., Chapter 7, McGraw-Hill Book Company, Inc., New York, 1954.
10. J. M. Beekmans, Correction Factor for Size-Selective Sampling Results, Based on a New Computed Alveolar Deposition Curve, *Ann. Occupat. Hyg., 8*:221 (1965).
11. L. B. Lockhart, Jr., R. L. Patterson, Jr., and A. W. Saunders, Jr., The Size Distribution of Radioactive Atmospheric Aerosols, *J. Geophys. Res., 70*:6033 (1965).
12. L. B. Lockhart, Jr., R. L. Patterson, Jr., and A. W. Saunders, Jr., Distribution of Airborne Radioactivity with Particle Size, in *Radioactive Fallout from Nuclear Weapons Tests,* USAEC Report CONF-765, Clearinghouse for Federal Scientific and Technical Information, Springfield, Va., 1965.
13. B. Shleien, T. P. Glavin, and A. G. Friend, Particle Size Fractionation of Airborne Gamma Emitting Radionuclides by Graded Filters, *Science, 147*:290 (1965).
14. J. F. Roesler, Application of Polyurethane Foam Filters for Respirable Dust Separation, *J. Air Pollution Control Assoc., 16*:32 (1966).
15. R. Reiter and K. Pötzl, Bestimmung der effectiven Gefährlichkeit von Aerosolen und Stauben mit Hilfes eines Atemtraktmodells, *Naturwiss., 52*:107 (1965).
16. R. Reiter, Erste Entwicklungsstuffe eins physikalischen Atemtraktmodells als Retentionssimulator, *Z. Biol. Aerosol-Forsch., 13*:133 (1966).
17. B. Shleien, J. A. Cochran, L. Benander, L. L. Bernard, J. D. Lutz, and A. G. Friend, Determination of the Respiratory Tract Deposition and Particle Size Deposition of Airborne Materials by Graded Filtration, in *Atmospheric Radioactivity Analysis and Instrumentation, Status Report III,* Northeast Radiological Health Laboratory, Report No. NERHL-67-2, January 1967.
18. B. Shleien and A. G. Friend, A Method for the Estimation of the Respiratory Deposition of Airborne Materials, *Health Phys., 13*:513 (1967).
19. J. Cartwright and G. Nagelschmidt, The Size and Shape of Dust from Human Lungs and Its Relation to Relative Sampling, in C. N. Davies (Ed.), *Inhaled Particles and Vapours,* Pergamon Press, Oxford, 1961.
20. J. S. Arnold, An Improved Technique for Liquid Emulsion Autoradiography, *Proc. Soc. Exptl. Biol. Med., 85*:113 (1954).
21. T. C. Evans, Radioautograph in Which the Tissue is Mounted Directly on The Photographic Plate, *Proc. Soc. Exptl. Biol. Med., 64*:313 (1947).
22. *Techniques for Microbiological Analysis,* Bull. ADM-40, Millipore Filter Corporation, Bedford, Mass., 1965.

23. *Procedure for Testing Filter-type Dust, Fume, and Mist Respirators for Permissibility,* United States Bureau of Mines, Schedule 21, Aug. 20, 1934.
24. *Respiratory-Protective Apparatus; Tests for Permissibility; Fees; Part 14 — Filter Type Dust, Fume, and Mist Respirators,* United States Bureau of Mines, Schedule 21B, Subchapter B, Jan. 19, 1965.
25. C. N. Davies, Fibrous Filters for Dust and Smoke, in *Ninth International Congress of Industrial Medicine, London, Sept. 13-17, 1948,* John Wright and Sons, Ltd., London, 1949.
26. S. T. Hermiston, R. F. Hounam, and R. P. Rowlands, The Use of Respirators for Protection Against Airborne Radioactivity, in C. N. Davies (Ed.), *Design and Use of Respirators,* p. 160, Pergamon Press, Oxford, 1962.
27. A. H. Yeomans, R. A. Fulton, F. F. Smith, and R. L. Busbey, *Respiratory Devices for Protection Against Certain Pesticides,* United States Department of Agriculture, Agricultural Research Service Publication ARS 33-76-2, Beltsville, Md., 1966.
28. G. J. Fleming, A Comparison of Performance Standards and Tests Specified by Bureau of Mines Schedule 14F and by the Chemical Corps for the M9A1 Mask, *Am. Ind. Hyg. Assoc. J., 19*:130 (1958).
29. C. J. Shoemaker, Recent Trends in Army Research on Respirators, American Industrial Hygiene Conference, St. Louis, Mo., May 13-17, 1968.
30. L. Silverman, Investigation of the Performance of Nuisance Dust Filter Masks, Unpublished Report, 1961.
31. American Industrial Hygiene Assocation and American Conference of Governmental Industrial Hygienists, *Respiratory Protective Devices Manual,* Braun and Brimfield, Inc., Ann Arbor, Mich., 1963.
32. L. Silverman and W. A. Burgess, *Respiratory Protective Equipment, Final Report,* USAEC Report NYO-9323, Harvard School of Public Health, Boston, Mass., June 30, 1964.
33. J. A. Leary, Particle Size Determination in Radioactive Aerosols by Radioautograph, *Anal. Chem., 23*:850 (1951).
34. W. D. Moss, E. C. Hyatt, and H. F. Schulte, Particle Size Studies on Plutonium Aerosols, *Health Phys., 5*:212 (1961).
35. L. A. George, II, Electron Microscopy and Autoradiography, *Science, 133*:1423 (1961).
36. S. S. Brar, G. Svihla, D. M. Nelson, and P. F. Gustafson, Isolation of "Hot" Radioactive Particles from Particulate Matter Collected at Argonne, Illinois and Fort Worth, Texas, in *USAEC Health and Safety Laboratory Fallout Program Quarterly Report,* USAEC Report HASL-193, April, 1968.
37. P. G. W. Hawksley, S. Badzioch, and T. H. Blackett, *Measurement of Solids in Flue Gases,* British Coal Utilization Research Association, Leatherhead, Surrey, England, 1961.
38. *Determining the Properties of Fine Particulate Matter,* ASME Power Test Codes, PTC 28-1965; *Determining Dust Concentration in a Gas Stream,* ASME Power Test Codes, PTC 27-1957; American Society of Mechanical Engineers, New York.
39. American Industrial Hygiene Association, *Air Pollution Manual, Part I, Evaluation,* Detroit, 1960.
40. Manufacturing Chemists' Association, *Air Pollution Abatement Manual,* Chapter 6, Sampling Procedures and Measuring Equipment, Washington, D.C., 1952.

41. *Methods for Determination of Velocity, Volume, Dust and Mist Content of Gases,* Bull. WP-50, 7th ed., Western Precipitation Corporation, Los Angeles, Calif., 1968.
42. *Test Methods,* Research-Cottrell, Inc., Bound Brook, N.J. (Revised Aug. 7, 1957.)
43. A. C. Stern (Ed.), *Air Pollution,* Vols. 1 and 2, 2nd ed., Chapters 3, 4, 16, 19-22, 25, 26, 28, Academic Press, New York, 1968.
44. Anonymous (C. J. Stairmand), Sampling Gas-Borne Particles, *Engineering (London),* Aug. 22, 1941, pp. 141-143; Sept. 5, 1941, pp. 181-183.
45. H. G. Houghton, The Size and Size Distribution of Fog Particles, *Physics, 2*:467 (1932).
46. G. W. Stroke, *An Introduction To Coherent Optics and Holography,* Academic Press, New York, 1966.
47. B. A. Silverman, B. J. Thompson, and J. A. Ward, A Laser Fog Disdrometer, *J. Appl. Meteorol., 3*:792 (1964).
48. G. G. Goyer and R. Watson, The Laser and Its Application to Meteorology, *Bull. Am. Meteorol. Soc., 44*:564 (1963).
49. G. Fiocce and G. Graus, Observations of the Aerosol Layer at 20 Km by Optical Radar, *J. Atmos. Sci., 21*:323 (1964).
50. J. E. Masterson, J. L. Karney, and W. E. Hoehene, The Laser as an Operational Meteorological Tool, *Bull. Am. Meteorol. Soc., 47*:695 (1966).
51. C. A. Northend, R. C. Honey, and W. E. Evans, Laser Radar (Lidar) for Meterological Observations, *Rev. Sci. Instr., 37*:393 (1966).
52. W. E. Ranz and J. B. Wong, *Jet Impactors for Determining the Particle Size Distributions of Aerosols,* Technical Report No. 4, USAEC Report SO-1005, University of Illinois, 1951.
53. R. I. Mitchell and J. M. Pilcher, Design and Calibration of an Improved Cascade Impactor for Size Analysis of Aerosols, in *Fifth Atomic Energy Commission Air Cleaning Conference,* USAEC Report TID-7551, 1958.
54. K. R. May, The Cascade Impactor, *J. Sci. Instr., 22*:187 (1945).
55. J. Maybank and W. J. Fenrick, *A New Technique for the Collection of Liquid Droplets and the Subsequent Measurement of Their Size,* Suffield Technical Paper No. 186, Defense Research Board, Department of National Defense, Suffield Experimental Station, Ralston, Alberta, Canada, 1960.
56. V. K. LaMer, *Studies on Filtration of Monodisperse Aerosols,* USAEC Report NYO-512, Columbia University, 1951; V. K. LaMer, E. C. Y. Inn, and I. Wilson, *J. Colloid Sci., 5*:471 (1950); V. K. LaMer and R. Gruen, *Trans. Faraday Soc., 48*:410 (1952).
57. D. H. Lucas, K. W. James, and I. Davies, The Measurement of Plume Rise and Dispersion at Tilbury Power Station, *Atmos. Environ., 1*:353 (1967) et seq.
58. P. M. Hamilton, K. W. James, and D. J. Moore, Observations of Power Station Plumes Using a Pulsed Ruby Laser Rangefinder, *Nature, 210*:723 (May 14, 1966).
59. W. D. Conner and J. R. Hodkinson, *Optical Properties and Visual Effects of Smoke-stack Plumes,* United States Department of Health, Education, and Welfare, National Center for Air Pollution Control, Cincinnati, Ohio, 1967.
60. E. W. Barrett and O. Ben-Dov, Application of Lidar to Air Pollution Measurements, *J. Appl. Meteorol., 6*:500 (1967).

61. P. M. Hamilton, The Use of Lidar in Air Pollution Studies, *Inter. J. Air Water Pollution, 10*:427 (1966).
62. D. H. Slade (Ed.), *Meteorology and Atomic Energy,* United States Atomic Energy Commission, Division of Technical Information, July 1968.
63. G. F. Skala, A New Instrument for the Continuous Measurement of Condensation Nuclei, *Anal. Chem., 35*:702 (1963).
64. P. J. Nolan and L. W. Pollak, The Calibration of a Photo-electric Nucleus Counter, *Proc. Roy. Irish Acad., 51* (Section A): 9 (1946).
65. G. McGreevy, The Evaluation of the Performance of an Electrostatic Precipitator Using a Pollak-Nolan Nucleus Counter, *Atmos. Environ., 1*:87 (1967).
66. L. Silverman and G. McGreevey, Application of the Pollak-Nolan Nucleus Counter to the Routine Testing of Air Filters, *Atmos. Environ., 1*:1 (1967).
67. M. T. Harris, Method for Size-distribution Determinations of Non-volatile Droplets by Electron Microscopy, *Brit. J. Appl. Phys., 10*:139 (1959).
68. T. Mamuro, K. Yoshikawa, T. Matsunami, A. Fujita, and T. Azuma, Electron Microscopic Examination of Highly Radioactive Fallout Particles, *Nature, 197*:478 (1963); Fractionation Phenomena in Highly Radioactive Fallout Particles, *Ibid., 197*:964 (1963).
69. M. H. Wilkening, Natural Radioactivity as a Tracer in the Sorting of Aerosols According to Mobility, *Rev. Sci. Instr., 23*:13 (1952).
70. M. I. Kalkstein, P. J. Drevinsky, E. A. Martell, C. W. Chagnon, J. E. Manson, and C. E. Junge, *Natural Aerosols and Nuclear Debris Studies, Progress Report II,* Research Notes No. 24, Air Force Cambridge Research Laboratories, Geophysics Research Laboratories, Bedford, Mass., 1959.
71. J. Rosinski and J. Stockham, *Studies Related to Radioactive Fallout,* USAEC Report TID-12333, 1961.
72. C. E. Junge, C. W. Chagnon, and J. E. Manson, Stratospheric Aerosols, *J. Meterol., 18*:81 (1961).
73. C. E. Junge, *Atmospheric Chemistry and Radioactivity,* p. 193, Academic Press, New York, 1963.
74. L. R. Solon, P. Lilienfeld, and H. J. DiGiovanni, A System for Large Volume Aerosol Sampling in the Stratosphere Using Electrostatic Precipitation, *Arch. Meteorol., Geophys., Bioklimatol., A17*:23 (1968).
75. C. W. Chagnon, *Balloon-borne Air Sampling Apparatus,* Instrumentation for Geophysical Research, Bull. No. 6, Air Force Cambridge Research Center Report No. AFCRC-TR-57-215, Bedford, Mass., 1957.
76. P. W. Hodge and T. S. Rinehart, High Altitude Collection of Extraterrestrial Particulate Materials, *Astron. J., 63*:306 (1958).
77. H. W. Zeller, A. I. Schekman, and S. C. Stern, *Balloon-borne Particulate Fractionator,* Final Report, Contract No. AF19(604)-4943, General Mills, Inc., Minneapolis, Minn., 1959.
78. H. Dolezalek and A. L. Oster, Mobility Measurements in the Upper Atmosphere, *J. Res. Atmos.,* Vol. II (April-September, 1966).
79. J. M. Rosen, The Vertical Distribution of Dust to 30 Kilometers, *J. Geophys. Res., 69*:4673 (1964).
80. J. M. Rosen, Stratospheric Dust, *Atmospheric Physics Anual Report,* University of Minnesota, 1967.
81. A. W. Klement, Jr., and J. Z. Holland, The AEC Program of Atmospheric Radioactivity and Fallout Research, *Health Phys., 10*:629 (1964).

82. A. W. Klement (Ed.), *Radioactive Fallout from Nuclear Weapons Tests,* USAEC Report CONF-765, Clearinghouse for Federal Scientific and Technical Information, Springfield, Va., 1965.
83. P. W. Krey, High Altitude Balloon Sampling Program, in *USAEC Health and Safety Laboratory Fallout Program Quarterly Summary Report,* USAEC Report HASL-173, pp. 11-21, October 1966.
84. L. Dautreband, *Microaerosols,* Academic Press, New York, 1962.
85. W. J. Bair and D. H. Willard, Plutonium Inhalation Studies, III, *Health Phys., 9*:253 (1963).
86. L. Silverman, Experimental Test Methods, in P. L. Magill, F. R. Holden, and C. Ackley, (Eds.), *Air Pollution Handbook,* Sec. 12, McGraw-Hill Book Company, Inc., New York, 1956.
87. L. Silverman and C. E. Billings, Methods of Generating Solid Aerosols, *J. Air Pollution Control Assoc., 6*:76 (1956).
88. K. T. Whitby, D. A. Lundgren, and C. W. Peterson, Homogeneous Aerosol Generators, *Intern. J. Air Water Pollution, 9*:263 (1965).
89. N. A. Fuchs and A. G. Sutugin, Generation and Use of Monodisperse Aerosols, in C. N. Davies (Ed.), *Aerosol Science,* Chapter 1, Academic Press, New York, 1966.
90. B.Y.H. Liu, Methods of Generating Monodisperse Aerosols, in *8th Conference on Methods in Air Pollution and Industrial Hygiene Studies, Oakland, Calif., Feb. 1967,* and Publication No. 104, Particle Technology Laboratory, Department of Mechanical Engineering, University of Minnesota.
91. M. Corn, Aerosol Generation, in R. Dennis (Ed.), *Handbook on Aerosols,* United States Atomic Energy Commission, 1971 (in preparation).
92. C. E. Brown, Quantitative Measurements of the Inhalation, Retention, and Exhalation of Dusts and Fumes by Man, *J. Ind. Hyg., 13*:285, 293 (1931).
93. B. Altshuler, L. Yarmus, E. D. Palmes, and N. Nelson, Aerosol Deposition in the Human Respiratory Tract, *A.M.A. Arch. Ind. Health, 15*:293 (1957).
94. J. H. Brown, K. M. Cook, F. G. Ney, and T. Hatch, Influence of Particle Size upon the Retention of Particulate Matter in the Human Lung, *Am. J. Public Health, 40*:450 (1950).
95. W. F. Wells, *Airborne Contagion and Air Hygiene,* Harvard University Press, Cambridge, Mass., 1955.
96. J. P. Duguid, The Size and Duration of Air-carriage of Respiratory Droplets and Droplet Nuclei, *J. Hyg., 44*:471 (1946).
97. H. W. Wolf, P. Skaliy, L. B. Hall, M. M. Harris, H. M. Decker, L. M. Buchanan, and C. M. Dahlgren, *Sampling Microbiological Aerosols,* Public Health Monograph No. 60, United States Government Printing Office, Washington, D. C., 1959.
98. A. A. Andersen, New Sampler for the Collection, Sizing, and Enumeration of Viable Airborne Particles, *J. Bacteriol., 76*:471 (1958).
99. T. Rosebury, *Experimental Airborne Infection,* The Williams and Wilkins Company, Baltimore, Md., 1947.
100. R. L. Riley and F. O'Grady, *Airborne Infection,* The Macmillan Company, New York, 1961.
101. P. H. Gregory, *The Microbiology of the Atmosphere,* Leonard Hill, London, 1961.

102. A. R. Jacobson, Viable Particles in Air, in A. C. Stern (Ed.), *Air Pollution,* 2nd ed., Vol. 1, Chapter 4, Academic Press, New York, 1968.
103. W. McDermott (Ed.), Conference on Airborne Infection Held in Miami Beach, Florida, 7-10, Dec. 1960, *Bacteriol. Rev., 25*:173 (1961).
104. R. L. Dimmick (Ed.), *First International Symposium on Aerobiology, University of California, Berkeley, Calif., Oct. 2-5, 1963,* Naval Biological Laboratory, Naval Supply Center, Oakland, Calif., 1963.
105. H. M. Tsuchiya and A. H. Brown (Co-Chairmen), *Proceedings of the Atmospheric Biology Conference, 13-15 April, 1964, University of Minnesota.*
106. M. H. Lepper and E. K. Wolfe (Eds.), Second International Conference on Aerobiology (Airborne Infection), Chicago, Ill., 29-31 March, 1966, *Bacteriol. Rev., 30*:485 (1966).
107. J. M. Beebe, Stability of Disseminated Aerosols of *Pasteurella tularensis* Subjected to Simulated Solar Radiation at Various Humidities, *J. Bacteriol., 78*:18 (1959).
108. C. E. Brown, Human Retention from Single Inhalations of *Bacillus subtilis* Spore Aerosols, in C. N. Davies (Ed.), *Inhaled Particles and Vapours,* p. 122, Pergamon Press, Oxford, 1961.
109. L. Zimmerman, Survival of *Serratia marcescens* after Freeze-Drying or Aerosolization at Unfavorable Humidity, *J. Bacteriol., 84*:1297 (1962).
110. H. M. Decker, L. M. Buchanan, L. B. Hall, and K. R. Goddard, *Air Filtration of Microbial Particles,* PHS Publication No. 953, United States Government Printing Office, Washington, D.C., 1962.
111. G. Phillips, G. J. Harris, and M. W. Jones, Effects of Air Ions on Bacterial Aerosols, *Intern. J. Biometerol., 8*:27 (1964).
112. R. G. Bond, M. M. Halbert, H. D. Putnam, O. R. Ruschmeyer, and D. Vesley, *Survey of Microbial Contamination in the Surgical Suites of 23 Hospitals,* University of Minnesota School of Public Health, March 1964.
113. R. G. Bond and G. S. Michaelsen, *Bacterial Contamination from Hospital Solid Wastes, Final Report,* University of Minnesota School of Public Health, August 1964.
114. J. E. Malligo and L. S. Idoine, Single-Stage Impaction Device for Particle Sizing Biological Aerosols, *App. Microbiol., 12*:32 (1964).
115. K. R. May, Calibration of a Modified Andersen Bacterial Aerosol Sampler, *Appl. Microbiol., 12*:37 (1964).
116. G. E. Hess, Effects of Oxygen on Aerosolized *Serratia marcescens, Appl. Microbiol., 13*:781 (1965).
117. J. B. Harstad, Sampling Submicron T1 Bacteriophage Aerosols, *Appl. Microbiol., 13*:899 (1965).
118. G. S. Michaelsen, O. R. Ruschmeyer, and D. Vesley, *The Bacteriology of "Clean Rooms", Final Report,* University of Minnesota School of Public Health, July 1966.
119. G. S. Michaelsen and D. Vesley, Dissemination of Airborne Microorganisms in an Institutional Environment, in B. R. Fish (Ed.), *Surface Contamination,* p. 285, Pergamon Press, Oxford, 1966.
120. G. S. Michaelsen, D. Vesley, and M. M. Halbert, The Laminar Flow Room Concept for the Care of Low Resistance Hospital Patients, annual meeting of American Public Health Association, San Francisco, Calif., Nov. 1, 1966.

121. J. W. Happ, J. B. Harstad, and L. M. Buchanan, Effect of Air Ions on Submicron T1 Bacteriophage Aerosols, *Appl. Microbiol.*, *14*:888 (1966).
122. C. W. Randall and J. O. Ledbetter, Bacterial Air Pollution from Activated Sludge Units, *Am. Ind. Hyg. Assoc. J.*, *27*:506 (1966).
123. M. Corn and M. Y. Flynn, Dispersion of Aerosols in Hospitals Using Aerosol Inhalation Therapy, *Am. Ind. Hyg. Assoc. J.*, *28*:386 (1967).

GLOSSARY

Aerodynamic Diameter, Kinetic Diameter. The diameter of a hypothetical sphere of unit density having the same terminal settling velocity as a particle in question, regardless of its geometric size, shape, and true density. For certain particles of unknown shape, size, and density, aerodynamic diameter may be obtained from measures of settling velocity, impaction behavior, etc.

Aerosol. A stable suspension of solid or liquid particles in a gas.

Agglomerate (n.), *Aggregate* (n.). A particle consisting of two or more unitary structures (e.g., a doublet, triplet, etc.). See Fig. 1.1.

Agglomeration (n.) The process of interparticle contact and cohesion.

*Count Median Size.** A measurement of particle size for samples of particulate matter, consisting of that diameter of particle such that one half of the number of particles is larger and half is smaller.

*Dust.** A loose term applied to solid particles predominantly larger than colloidal and capable of temporary suspension in air or other gases. Dusts do not tend to flocculate except under electrostatic forces; they do not diffuse but settle under the influence of gravity. Derivation from larger masses through the application of physical force is usually implied.

Feret's Diameter. The normal distance between two parallel tangents to the extreme points on the particle measured in a consistent manner. See Fig. 3.7b.

*Flocculation.** Synonymous with agglomeration.

*Fog.** A loose term applied to visible aerosols in which the dispersed phase is liquid. Formation by condensation is usually implied. In meterology, a dispersion of water or ice.

*Fume.** Properly, the solid particles generated by condensation from the gaseous state, generally after volatilization from melted substances, and often accompanied by a chemical reaction such as oxidation. Fumes flocculate and sometimes coalesce. Popularly, the term is used in reference to any or all types of contaminant, and in many laws or regulations with the added qualification that the contaminant have some unwanted action.

---

*From ASTM Designation: D1356-67a.

Gas.[*] One of the three states of aggregation of matter, having neither independent shape nor volume and tending to expand indefinitely.

Green's Diameter. The average diameter of a hypothetical particle that is in some way representative of the particles in the sample.

Impaction.[*] A forcible contact of particles of matter, a term often used synonymously with impingement.

Isokinetic.[*] A term describing a condition of sampling, in which the flow of gas into the sampling device (at the opening or face of the inlet) has the same flow rate and direction as the ambient atmosphere being sampled.

Martin's Diameter. The distance between opposite sides of the particle, measured in a consistent direction, such that the diameter bisects the projected area. See Fig. 3.7d.

Mass Median Size.[*] A measurement of particle size for samples of particulate matter, consisting of that diameter such that the mass of all larger particles is equal to the mass of all smaller particles.

Mist.[*] Liquid, usually water in the form of particles suspended in the atmosphere at or near the surface of the earth; small water droplets floating or falling, approaching the form of rain, and sometimes distinguished from fog as being more transparent or as having particles perceptibly moving downward.

Particle. An aerosol particle may consist of a single continuous unit of solid or liquid containing many molecules held together by intermolecular forces and primarily larger than molecular dimensions ($>0.001\ \mu m$). A particle may also be considered to consist of two or more such unit structures held together by interparticle adhesive forces such that it behaves as a single unit in suspension or upon deposit. See Fig. 1.1.

Particle Size, Particle Diameter. Some consistent measure of the spatial extent of the matter constituting the particle, such as a measured diameter, area, or volume; or some consistent measure of the behavior of the particle as a function of the spatial extent. If particles are spherical or cubical, particle size corresponds to a single measurable geometric length. If particles are irregularly shaped, diameter is defined by consistent arbitrary measures. Other regular geometric shapes may be defined in terms of dimension ratios.

---

[*]From ASTM Designation: D1356-67a.

Particle Size Distribution. The frequency with which particles of various sizes are distributed in a given sample.

Projected (Area) Diameter. The diameter of a circle of area equal to the profile of the projected area of the particle. See Fig. 3.7a.

*Settling Velocity.** The terminal rate of fall of a particle through a fluid as induced by gravity or other external force; the rate at which frictional drag balances the accelerating force (or the external force).

*Smog.** A term derived from smoke and fog, applied to extensive atmospheric contamination by aerosols, these aerosols arising partly through natural processes and partly from the activities of human subjects. Now sometimes used loosely for any contamination of the air.

*Smoke.** Small gas-borne particles resulting from incomplete combustion, consistently predominantly of carbon and other combustible material, and present in sufficient quantity to be observable independently of the presence of other solids.

*Soot.** Agglomerations of particles of carbon impregnated with "tar," formed in the incomplete combustion of carbonaceous material.

Stokes' Equivalent Diameter. The diameter of a hypothetical sphere having the same terminal settling velocity as the particle in question and having the *same* density as the particle material, whatever its size and shape.

*Vapor.** The gaseous phase of matter which normally exists in a liquid or solid state.

---

*From ASTM Designation: D1356-67a.

Author Index

Numbers in parentheses are reference numbers. Numbers in italics refer to the pages on which the complete references are listed.

Subject Index

O

P

R

S